BIOTHERMAL–FLUID SCIENCES
Principles and Applications

BIOTHERMAL–FLUID SCIENCES
Principles and Applications

Wen-Jei Yang

Department of Mechanical Engineering and Applied Mechanics,
University of Michigan,
Ann Arbor, Michigan

⊙ HEMISPHERE PUBLISHING CORPORATION
A member of the Taylor & Francis Group
New York   Washington   Philadelphia   London

BIOTHERMAL–FLUID SCIENCES: Principles and Applications

1 2 3 4 5 6 7 8 9 0   B C B C   8 9 8 7 6 5 4 3 2 1 0 9

This book was set in Times Roman by General Graphic Services.
The editors were Brenda Brienza and Carolyn Ormes.
BookCrafters, Inc. was printer and binder.
Cover design by Renée E. Winfield.

Library of Congress Cataloging-in-Publication Data

Yang, Wen-Jei, date
   Biothermal-fluid sciences: principles and applications/
Wen-Jei Yang.
      p.   cm.
   Includes indexes.

   1. Rheology (Biology)   2. Thermodynamics.   3. Biochemical
engineering.   4. Body fluid flow.   I. Title.
   QP517.R48Y36   1989
612'.014—dc19   89-1734
                                                                        CIP

ISBN 0-89116-869-9 (cloth)
ISBN 0-89116-974-1 (paper)

CONTENTS

Appendixes

Subject Index   393

Author Index   405

PREFACE

The flows of fluids, heat and materials in the human body are complex processes. Publications on the subject are abundant. However, a composite entity on all transport processes is rare.

This book is an amplification of lecture notes that the author used in teaching a one-semester course in "Thermal-Fluid Sciences in Bioengineering" for students at the senior and first-year graduate level. The material has been organized as a text, but it can also serve as a convenient reference for researchers in industry, biology, and medicine interested in fundamental techniques of analyzing bio-transport problems. It is assumed that the reader has an elementary knowledge of calculus and differential equations.

The purpose of the book is to present the fundamentals and applications of fluid dynamics, thermodynamics, and heat and mass transfer in biomedical systems, at both macro- and micro- (cellular) levels. Since each of the four fields constitutes a broad and independent branch of science, an in-depth coverage of the basic principles is avoided. Instead, references are provided for the reader to consult the standard textbooks for detail derivations which are assigned as exercise problems in the book. The presentation endeavors to convey to the reader an understanding of the transport processes and to provide him with the tools necessary to obtain quantitative information for biomedical problems involving the transfer of momentum, heat and material. Emphasis is placed on prosthetic assisting devices, flow and thermal measurements, flow visualization, mechanisms of defects in the tissues and organs for blood circulation, body temperature regulation, heat and mass exchanges in vivo and in prosthetic devices, and clinical applications of engineering methods such as the modeling of flow and thermal regulations. The uniqueness of the book is that theories and equations formulated, which are based on physical and physiological principles, are highly relevant to the development of clinical procedures and devices, such as for monitoring, diagnosis, and assist. The text examples and problems are couched in terms of actual physiological and physical processes either in vivo or in vitro.

The introductory chapter describes some vital transport processes occurring in the human body to illustrate the importance of thermal sciences in the biomedical field. The vast differences in the nature of transport processes between the engineering and biomedical systems warrant a need for a course on the subject. For convenience and brevity, the expressions for conservation of momentum, heat and material are summarized in Appendix B, and the transport-rate equations and the equations of state are summarized in Tables 1.1 and 1.2.

In the treatment of blood circulation, the uniqueness of pulsatile flow, the complex composition of blood, flow bifurcation, and complexity in vascular walls and beds have been emphasized. The evaluation of flow rate and pressure drop is taken up in separate sections for systemic circulation and microcirculation. Compartmental modeling of flow regulation is introduced together with hemolysis, thrombosis, artificial hearts, and cardiac assist devices.

The chapters on thermodynamics present some concepts, work-heat conversion, the first and second laws and cellular energetics. The practice of the closed and open system analysis has been demonstrated on various organs and tissues.

In the chapter dealing with heat transfer, three basic modes, namely conduction, convection, and radiation, are presented. The roles of insensible heat, metabolic heat generation, and blood cooling on body temperature regulation have been emphasized. The general bio-heat transfer equation has been applied to the macro and micro (cellular) systems, internal and external heat exchanges for homeotherms, hyperthermia, hypothermia, and cryosurgical applications. Clinical heat exchangers and thermal measurements are also presented.

A separate chapter on flow visualization and image processing manifests the importance of the subject in the biomedical field. The use of contrast media is essential to identify the organ or tissue systems of interest. Image enhancement is important in dealing with micro-structures. Methods of flow visualization, thermography, and image processing techniques in biomedical applications are presented.

The last chapter takes on mass transfer processes, which are abundant in the human body. The subject is commonly presented in physiology and bio-chemical engineering and needs less elaboration. Applications have been demonstrated in various organs and tissues with the use of either compartmental model or distributed analysis. Simultaneous heat and mass transfer processes are encountered in body heat regulation through normal skin under various environmental conditions as well as in burn injury. Artificial organs such as the artificial lung (or oxygenators), liver and kidney are studied. Engineering stresses problem solving in natural or applied sciences.

Appendix A contains nomenclature and Appendix C contains conversion factors. Appendix D presents a collection of physical and physiological properties. The property tables provide a convenient source of data for solving the problems at the end of each chapter.

Due to the broadness of thermal-fluid sciences, the selection of the material and the balance in the materials among the four fields has been a most difficult task. Problems that are of a complicated nature or mathematically complex are avoided. Pertinent material has been selected from the literature and put together in a teachable form. Credit should belong to the original sources.

I thank my students for their encouragement and comments on the material.

Wen-Jei Yang

ONE

INTRODUCTION

Metabolism must occur in all living organisms to sustain life. It can be described as a chemical reaction or thermodynamic process in which chemical energy is converted into work and heat. The excess energy in the cells is diffused into the surrounding body fluid and eventually is transferred to the body surface to be dissipated into the environment in order to maintain a constant body temperature. This is a typical heat transfer process. Blood carries nutrients to metabolic sites and transports waste products to the disposal sites. Here, fluid dynamics as well as mass transfer become involved.

We will describe transport phenomena at the macroscopic as well as the microscopic level. Some examples of the macroscopic aspects of transfer processes are blood flow in the arteries and veins and heat and mass transfer in tissues and organs. Microscopic transport phenomena include microcirculation in the capillaries and intracellular and intercellular heat and mass transfer.

1.1 SOME IMPORTANT TRANSPORT PROCESSES IN THE HUMAN BODY

At the outset of the introductory chapter, a number of vital processes in the human body, which involve fluid dynamics, thermodynamics, heat transfer and mass transfer, are presented. There are two reasons for including such a section. First, the reader will realize the important role of biothermal sciences in the biomedical field, as he obtains some familiarity with the actual processes and the prosthetic devices involved. Second, this section provides an introduction to biothermal sciences, including the use of certain terms (which are defined in later chapters), some of the problems for which thermal sciences are relevant, and some accomplishments that have results, at least in part, from the application of thermal sciences. However, one should bear in mind that this section presents only a few examples and, therefore, serves only as a very incomplete introduction.

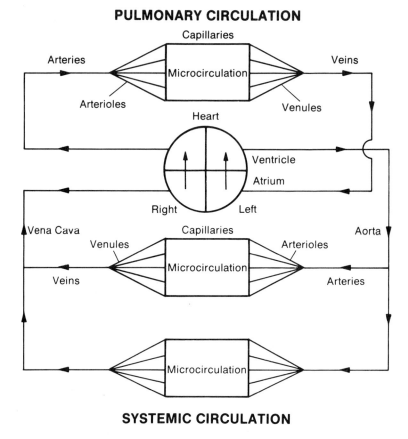

Figure 1.1 Schematic of the cardiovascular system.

1.1.1 Blood Circulation

Blood is a carrier of the nutrients for the cells and metabolic wastes from the cells. Figure 1.1 is a diagram of the circulatory system. The heart consists of two separate pumps arranged side by side. The first pump, the right ventricle, forces blood into the lungs to exchange CO_2 for O_2 and returns it to the left atrium of the heart. This flow passage constitutes the pulmonary circulation. The second pump, the left ventricle, forces the oxygenated blood through the body to exchange O_2 and CO_2 in the cells. From the arteries, the blood flows into the capillaries, then into the veins and finally back to the heart (right atrium). This flow passage constitutes the systemic circulation. Fluid dynamics plays an important role in understanding the mechanisms of flow regulation of blood.

At present, artificial heart valves, blood pumps, left ventricular assist devices, artificial hearts, etc., are either in use or in experimental stages to assist or to replace diseased natural hearts. The design of these prosthetic devices requires the knowledge of flow characteristics and flow control mechanisms in the cardiac system (Ghista et al., 1983).

1.1.2 Cell Energetics

Metabolism is the chemical reaction inside individual cells that produces energy to perform bodily activities and to build new structures. The cell is never in static equilibrium with its environment. It maintains a steady state and is in a continual energy exchange with its environment. Biological reactions in the cell take place in solution and may be considered at constant pressure and volume. Figure 1.2 illustrates a series of biological reactions in which cell work is accomplished at the expense of such bond energy. For example, the combustion of glucose can release 673 kcal/mol of energy at 25°C and 760 mmHg. The first law of thermodynamics states the conservation of energy in the course of the chemical to thermal energy conversion. On the other hand, the second law of thermodynamics provides the information about the direction in which this process can spontaneously occur, that is, the reversibility aspects of the thermodynamic process. The latter also defines the efficiency with which the available energy is converted into useful work in the process.

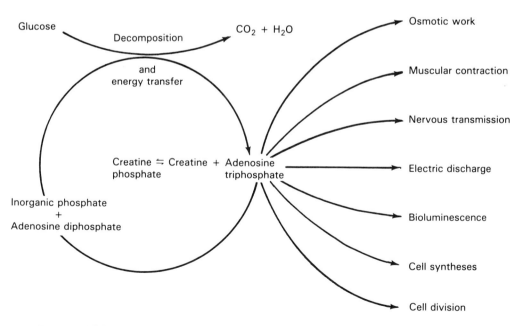

Figure 1.2 High-energy phosphate bonds and cell work.

1.1.3 Temperature Distribution in the Human Body

The intracorporeal transport of heat by circulating blood is an important factor in body temperature regulation. Thermal equilibrium, namely the homeothermic condition, is established when the rate of heat loss, through the body surface and respiratory system, equals the total rate of heat production caused by metabolic reactions. Figure 1.3 depicts Wissler's (1961) thermal model of the six-element person. The model includes the distribution of metabolic heat gen-

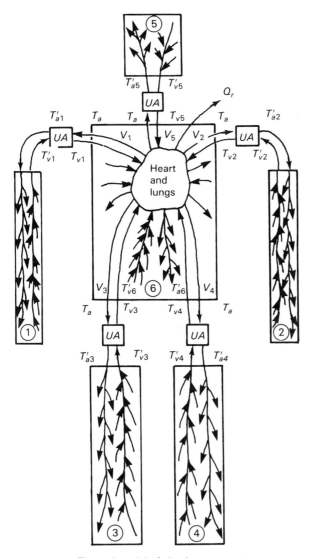

Thermal model of six-element person

Figure 1.3 Thermal model of six-element person.

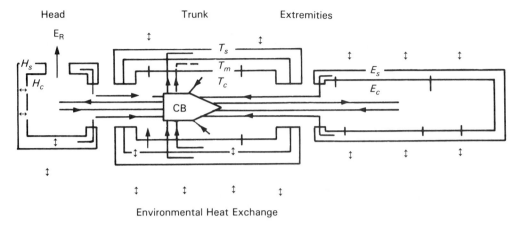

Figure 1.4 Compartmental model of three-element person.

eration, heat conduction in tissue, heat convection by flowing blood, heat losses by radiation, convection, and evaporation at the surface, heat loss through the respiratory tract, and countercurrent heat exchange between large arteries and veins. This model is an example of digital simulation of a biomedical transport process. For unsteady-state thermal regulation, Wissler (1966) developed a 15-element model.

Figure 1.4 shows a compartmental model of the three-element person (Stolwijk and Hardy, 1966), taking into account the physiological thermoregulation mechanism. Each element was divided into two or three concentric layers to characterize the functional differences in their thermoregulation. For example, H_c, H_s, and CB denote the head core, head skin, and central blood compartments, respectively. In contrast to the use of a digital computer for the solutions by Wissler (1961, 1966), this model was solved through an analogic method using the circuit shown in Fig. 1.5.

Closely related to cell energetics in Section 1.1.2, is microheat transfer which refers to the intracellular and intercellular migration of energy. Specifically, it deals with heat production in cells, cell hypothermia, cell hyperthermia in normal and tumor cells, and thermal denaturation of macromolecules or proteins. Heat transfer in cell growth and division and heat transfer resulting from the firing of the action potential are of interest in thermal physiology.

1.1.4 Mass Transfer

In order to sustain life, living organisms must be supplied with nutrients and must remove waste products. It is, then, natural that mass transfer is a vitally important process. The purpose of investigating mass transfer mechanisms is twofold: to understand the factors regulating the operation of living systems

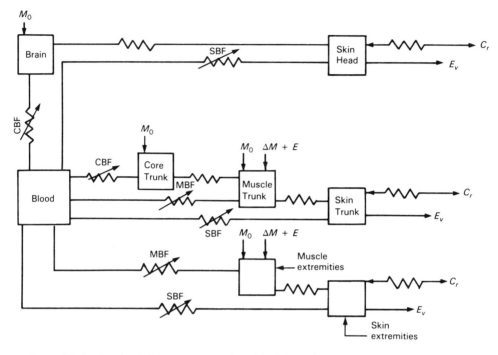

Figure 1.5 Analog circuit for compartmental model of three-element person.

and to design and operate prosthetic devices that are meant to supplement or replace bodily functions. Mass transfer processes are typically observed (1) macroscopically in the respiratory, circulatory, metabolic, gastrointestinal, and excretory systems, and (2) microscopically through membranes in all living cells. The control and regulation of these systems are mostly influenced by the regulation of their mass transfer characteristics through (1) the variation of the permeabilities or diffusivities of membranes, (2) the regulation of species concentrations, (3) the change of chemical reaction rates, or (4) their combinations.

1.1.4.1 Mass transfer in the circulatory system. Blood serves to transport O_2, CO_2, nutrients, metabolic wastes, and control/regulatory species. Therefore, mass transfer in the circulatory system involves exchanging CO_2 for O_2 in the lungs and vice versa at the cell sites, providing nutrients to and removing metabolic wastes from the internal organs and the cells of the body, and carrying control/regulatory substances from their production sites to the control sites. These mass transfer tasks are schematically illustrated in Fig. 1.6.

For the control and regulation of this system, it is useful to develop mathematical models, such as the compartmental model, to determine the amount of various substances presented in the system so that the relative importance of these phenomena and their effects on the overall system can be manifested.

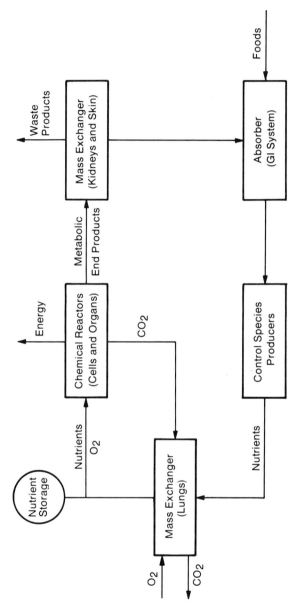

Figure 1.6 Mass transfer processes in the circulatory system.

7

1.1.4.2 Mass transfer in the respiratory system. In the respiratory system, air is moved in and out of the lungs by contraction and relaxation of the respiratory muscles, while blood flows continually through the vessels. Figure 1.7a depicts the anatomy of the respiratory surfaces. Only a thin membrane separates the air from the blood (Fig. 1.7b). This membrane is porous to gases and allows free passage of O_2 into the blood and of CO_2 from the blood into the air. Simple and more elaborate compartmental models have been employed to determine transient changes in the amounts of O_2 and CO_2.

When the lungs as well as the heart are bypassed in cardiac surgery, artificial devices are installed for the extracorporeal oxygenation of blood. Figure 1.8 is a schematic diagram of an artificial heart-lung machine. It consists of a blood pump which substitutes for the heart's function and a gas exchange unit which serves to replace the natural lungs. A blood heat exchanger is used to shorten the time normally required to cool a patient before open-heart surgery and to rewarm the patient after surgery.

The ideal oxygenator should be efficient in gas exchange and gentle to blood. The mass transer rate N can be determined by the equation

$$N = h_D A(P_g - P_b)$$

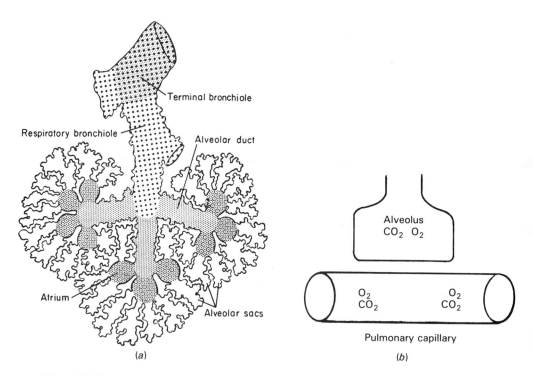

Figure 1.7 Respiratory surfaces and gas-blood mass exchange. (a) Respiratory surfaces. (b) Gas-blood mass exchange.

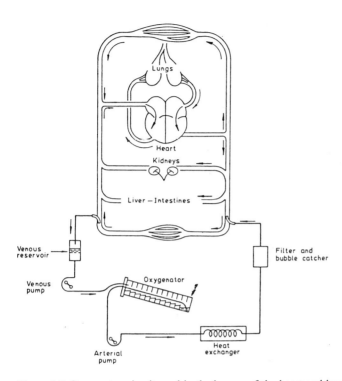

Figure 1.8 Oxygenator circuit used in the bypass of the heart and lungs.

where h_D denotes the mass transfer coefficient; A, the blood-gas contact area; P_g, the partial pressure of oxygen in the gas phase; and P_b, the partial pressure of oxygen in the blood. In order to generate A to enhance N, various methods can be utilized: by dispersing the gas phase in a pool of blood; by dispersing blood into a gas phase via foaming or spraying; by spreading blood in a thin film on a stationary or moving solid plate, screen, etc.; by allowing blood and gas to flow on opposite sides of a large semipermeable membrane; and others. h_D can be augmented through blood-side mixing by mechanical means, turbulent gas flow, or by moving or rotating the solid surfaces with blood films. The amount of oxygen transferred is equal to Nt where t is the blood-gas contact time; this can be prolonged by continually recycling some blood from the outlet back to the inlet of the device. Yang and Yang (1982) surveyed various types of oxygenators that have been in use.

1.1.4.3 Mass transfer in the excretory system. The kidneys constitute an excretory system for ridding the blood of unwanted substances, mostly the end products of metabolic reactions and for regulating the concentrations of most electrolytes in the body fluids. Figure 1.9 shows the kidney and its nephron. Figure 1.9(a) gives the pressures at different points in the vessels and tubules

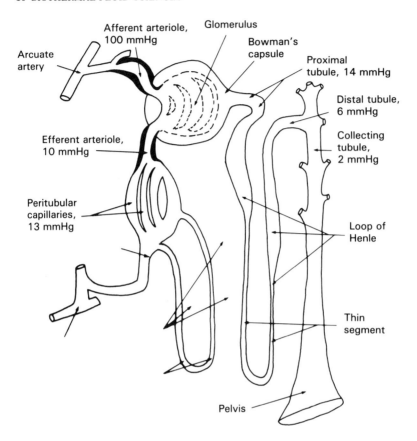

(a)

(b)

Figure 1.9 (a) The kidney and (b) its nephron.

of the functional nephron and in the interstitial fluid. Various modes of mass transfer take place in the kidney, including osmosis, filtration, convection, diffusion, and active transport. Quantitative studies on all these important mass transfer processes have been performed for the understanding of the elements of normal renal function and its associated control and regulation.

When the natural kidney fails, an artificial kidney is employed to normalize blood chemistry and fluid volume by circulating the blood extracorporeally for dialysis. Figure 1.10 shows the basic components and flow paths of the pa-

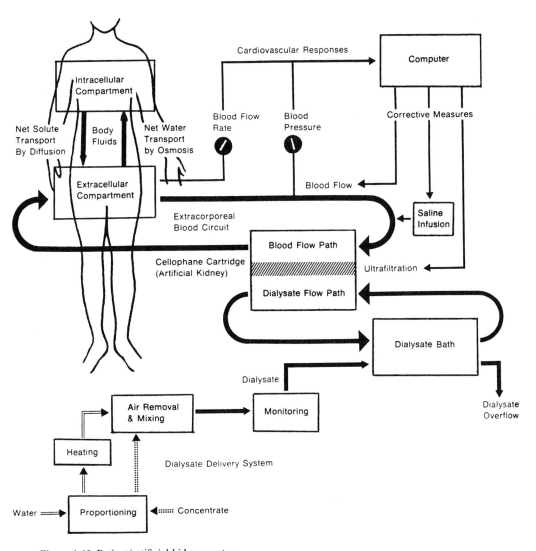

Figure 1.10 Patient/artificial kidney system.

tient/artificial kidney system, which functions in a different fashion from that of the natural kidney. The dialyzer operates on the principle of equilibration. Upon passing through the dialyzer, the blood is processed with a dialysis solution that has the desired composition. The blood is separated from the dialysis fluid by only a thin semipermeable membrane which is permeable to all species except proteins and blood cells. The patient is monitored for cardiovascular performance during dialysis.

1.1.4.4 Mass transfer in the metabolic system. The liver secretes a solution called bile that contains a large quantity of bile salts (special chemicals for changing the absorbed foods into new substances needed by the cells for metabolism). The pancreas secretes digestive juices into the intestines and insulin (to control the transport of glucose through the cellular membrane) directly into the body fluids. Other glands secrete various hormones, some of which are useful in metabolism. The secretion of species is a mass transfer process.

Artificial livers are near the stage of practical usage. Artificial pancreases, which are nothing but insulin injection devices, are in the process of development for automatic and optimal injection of insulin controlled by a microcomputer.

Micromass transfer deals with the intercellular, intracellular, and transmembrane transfer of substances. Figure 1.11 illustrates the migration of oxygen from the tissue capillary to the tissue cell, formation of carbon dioxide inside the cell, and migration of carbon dioxide back to the capillary.

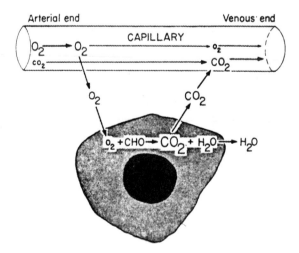

Figure 1.11 Capillary–cell mass transfer and intracellular metabolic reaction.

1.2 MAJOR DIFFERENCES IN TRANSPORT PHENOMENA IN ENGINEERING AND PHYSIOLOGY

In view of the complex organization of the human body, physiological transport phenomena raise problems very different from those encountered in the huge assemblage of engineering transport phenomena. For example, in engineering fluid dynamics, one deals with a Newtonian fluid flowing through a long, straight tube of constant cross section and rigid wall. The fluid velocity may vary with time and location, but is typically in the range of laminar to turbulent flow. The situation is far different in physiological fluid flows. Lighthill (1972) suggests the following five major causes:

1. *Unusual range of Reynolds number.* In the body, only very sporadic or highly localized bursts of turbulence take place. The high Reynolds number end of the laminar flow regime, ranging from a few hundred to several thousands, exists both in the large arteries and in the large airways of the lungs. Thus, an enormously long entrance region is required to establish any simple flow field such as Poiseuille flow in a straight tube, and intense, secondary flow effects are produced due to centrifugal action in curved tubes.
2. *Unusual multiplicity of tube branchings.* The branched networks of tubes from the cardiovascular system and the lungs are extremely intricate and complex. The flow pattern suffers a major distortion at each bifurcation. Before a full recovery is gained, the flow branches once again.
3. *Unusual wall properties of containing vessels.* The networks of vessels containing blood flows and airflows exhibit rather complicated distensibility relations. The wall of the arteries is complicated by viscoelasticity and nonlinear features.
4. *Unusual fluid properties.* Whole blood consists of a suspension of 40–50% by volume of small deformable bodies, mainly red cells in the transparent plasma. While the plasma obeys Newtonian behavior, whole blood is described by effective viscosity which increases substantially with a decreasing rate of strain due to increased formation of various aggregations of red cells at the lower strain rates. The air inhaled into the lungs is also a suspension of dust particles, which deposit at different levels in the bronchial tree depending on particle sizes.
5. *Unusual pulsatility.* Pulmonary inspiration and expiration cause the regular flow reversals of the air in the respiratory system. In the circulatory system, however, arterial blood flow exhibits the pulsatile character. Attenuation of the pulse wave makes such pulsatility far less important in microcirculation and in the veins.

Other differences relate to heat transfer. In engineering, heat dissipation from a solid with internal heat generation to the ambient is by conduction alone.

The situation is quite different in thermal physiology. There are five major causes:

1. *Unusually low conduction heat flux.* In the human body, temperatures vary from the maximum value of 37°C in the core region to the minimum value of about 34°C at extremities. The temperature gradient, $\partial T/\partial n$, is on the order of 0.1°C/cm. Thermal conductivity of tissues k is very low, about that of water at room temperature, 0.6 W/m °C. The conduction heat flux, $k \; \partial T/\partial n$, is on the order of 0.06 W/m², too low to dispose of the basal metabolic rate of 84 W (equivalent to 72 kcal/h) over the body surface area of 1.8 m² (for a 70-kg adult).

2. *Unusual blood cooling.* When blood flows through tissues or organs, it functions not only as a carrier of nutrients and metabolic wastes, but also as a coolant to remove the heat produced by the metabolism of those tissues or organs. Blood gains heat, which is transferred by circulation to the skin, where it is dissipated into the environment. Therefore, blood cooling is vital to homeotherms (to maintain the core temperature within the normal range), because of unusually low conduction heat flux. One of the most popular models of the blood cooling rate per unit volume of the perfused tissue is

$$\dot{q}_b = (\dot{W}\rho C)_b(T_A - T_t) \qquad \text{(W/m}^3 \text{ of tissue)}$$

Here, $\dot{W}_b$ denotes the volumetric rate of blood per unit volume of perfused tissue; ρ_b and C_b, density and specific heat of the blood, respectively; T_A, arterial blood temperature; and T_t, tissue temperature. The unusual feature of blood cooling is its variability. $\dot{q}_b$ changes with $\dot{W}_b$; $\dot{W}_b$ can be modified by the vasomotor activity of the tissue vessels: vasoconstriction decreases $\dot{W}_b$, while vasodilation increases it.

3. *Unusual thermoregulatory mechanisms.* Skin temperature T_s is one of the most important factors in the regulation of body temperature. It varies with a change in the environmental conditions. Figure 1.12 depicts the thermoregulatory mechanisms in the human body. When a resting person is in a thermal steady state (called thermal neutrality when the body temperatures stabilize at about 37°C for the core and about 34°C for skin surface), the basal metabolic heat generation heat q_m balances with the sensible heat loss (by convection and radiation). Heat storage is then at zero. However, when the ambient temperature T_a undergoes a sudden change, for example, an imbalance between thermal input and output, this causes the heat storage to change from zero. As a result, T_s and the core temperature T_c change, which in turn activates the peripheral and central thermoreceptors, respectively. The signals from these thermoreceptors are integrated by the thermoregulatory centers (mainly located in the hypothalamus), which trigger the necessary regulatory mechanisms. The control mechanisms include (a) the vasomotor systems, which induce either vasodilation or vasoconstriction, and (b) active regulations, sweating in the case of $T_a > T_s$, and

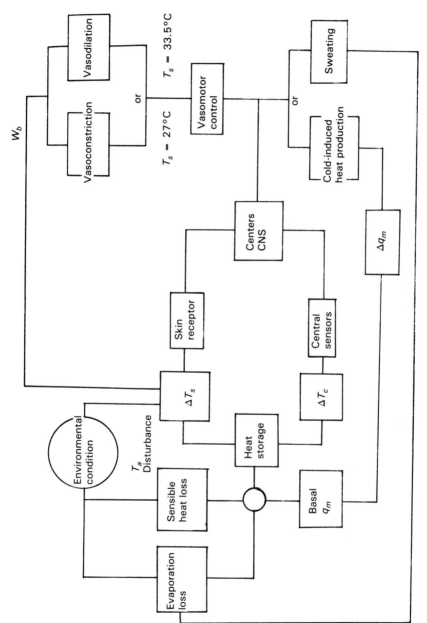

Figure 1.12 Thermoregulatory loop.

15

cold-induced metabolic heat generation (including both shivering and non-shivering) in the case of $T_a < T_s$.

4. *Unusual thermal properties.* Most tissues' materials are so heterogeneous in composition that their thermal properties are not isotropic and age dependent. Noninvasive measurements of thermal properties in vivo are most desirable, but are difficult to accomplish. In vivo test results are often different from those obtained in vitro, which may vary with time after incision.

5. *Unusual range of system size.* Microheat transfer in a system of cell size is unheard of in engineering. The temperature change and gradient inside the cells are not measurable by today's instrumentation.

Instead of being discouraged by complications in physiological transport phenomena, one should take it as a challenge. The basic principles and concepts of thermal sciences are presented in this book. With the understanding of the physiological terms being used, one will be able to solve the problems pertinent to transport processes in biomedical systems.

The reader is referred to some other important literature on thermal sciences and biothermal sciences: Fung (1981, 1984), Ghista et al. (1979), Griffith (1980), Guyton (1971), Houdas and Ring (1982), Kreith (1965), Lih (1975), Newburgh (1968), Seagrave (1971), Streeter and Wylie (1985), Van Wylen and Sonntag (1986), and Yang (1979a,b).

1.3 MOMENTUM, HEAT, AND MASS TRANSFER EQUATIONS WITH SOME APPLICATIONS

All natural laws have been derived empirically. They may be classified into two categories: general and particular. General laws can be applied to any medium under consideration, irrespective of its nature, while the application of particular laws is restricted by the nature of the medium under consideration. Some problems of nature can be formulated completely by employing only general laws, but others must be described by a combination of both laws. Figure 1.13 lists the natural laws pertinent to thermal fluid sciences.

Derivations for momentum, heat, and mass conservation equations are available in all standard textbooks on thermal sciences and thus are treated in brief here. For example, the principle of conservation of mass states

$$
\begin{Bmatrix} \text{Rate of mass} \\ \text{flow across} \\ \text{control surface} \\ \text{into control} \\ \text{volume} \end{Bmatrix} - \begin{Bmatrix} \text{rate of mass} \\ \text{flow across} \\ \text{control surface} \\ \text{out of control} \\ \text{volume} \end{Bmatrix} + \begin{Bmatrix} \text{rate of mass} \\ \text{generation} \\ \text{within control} \\ \text{volume} \end{Bmatrix}
$$

$$
= \begin{Bmatrix} \text{rate of mass} \\ \text{accumulated} \\ \text{within control} \\ \text{volume} \end{Bmatrix} \tag{1.1}
$$

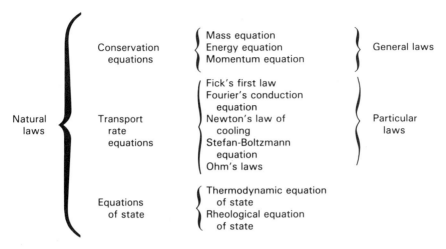

Figure 1.13 Natural laws pertinent to thermal fluid sciences.

For a fluid flowing in and out of a control volume dV with a relative (to control surface) velocity v at an angle α normal to a small element of the control surface dA, the equation can be expressed in mathematical form as (Van Wylen and Sonntag, 1986)

$$- \int_A \int \rho v \cos \alpha \, dA + m'''V = \frac{d}{dt} \int \int_V \int \rho \, dV \qquad (1.2)$$

where m''' denotes the volumetric rate of mass generation within the control volume. When "mass" is replaced by "energy" and "momentum," Eq. (1.1) states the principle of conservation of energy and momentum, respectively. Equation (1.2) is in an integral form applied to the entire system. However, its differential form can be obtained by defining an infinitesimal volume within the system. A few derivations of the conservation equations in differential form are illustrated later for simple transfer problems. The general expressions for momentum, heat, and mass balance are summarized in Appendix B.

Biomedical engineering is the discipline that uses engineering methods to solve problems in medicine and biology. To demonstrate such an approach, the principles of conservation of mass, momentum, and energy are applied to some flow phenomena in blood vessels, for example, steady blood flow in a vessel network. Equation (1.1) states that the total of the currents flowing into a junction must equal the sum of the currents flowing out of the junction. This rule is analogous to Kirchhoff's law of electric circuits:

$$\Sigma \dot{m}_i = \Sigma \dot{m}_e \qquad (1.3)$$

where $\dot{m}$ denotes the mass flow rate, and the subscripts i and e refer to the "into" and "out of" the junction, respectively. If flow is steady and incompressible in a single rigid tube of variable cross section, Eq. (1.1) is reduced to

$$\rho v(x)A(x) = \text{constant} \qquad (1.4)$$

It means that the local velocity $v(x)$ is inversely proportional to the local cross-sectional area $A(x)$, since the fluid density ρ is constant. x denotes the location from a reference position such as an entrance. Equations (1.3) and (1.4) are examples of the conservation of mass.

Newton's second law of motion states that

$$\text{Force} = \text{mass} \times \text{acceleration} \qquad (1.5)$$

If a multiple of forces act on a mass, Newton's law becomes

$$\Sigma \text{ Force} = \text{mass} \times \text{acceleration} \qquad (1.6)$$

There are forces acting on the surface and volume of a fluid mass, called the surface and body forces, respectively. Stress is the force acting on a surface divided by the surface area. It has two components: one is tangent to the surface called shear stress and the other is perpendicular to the surface called normal stress. Pressure is a normal stress. A positive pressure is a negative normal force and a positive normal stress is a tensile stress.

The surface and body forces acting on a fluid mass cause it to move, whereas the stresses induced by fluid viscosity and turbulence oppose flow. Hence, one may write Eq. (1.6) for a fluid mass as

$$\text{Surface force} + \text{body force} - \text{stresses} = \text{mass} \times \text{acceleration} \qquad (1.7)$$

The equation signifies that if the surface and body forces are balanced with shear stresses the flow is steady. Otherwise, the fluid would either accelerate or decelerate. Equation (1.7) can be rewritten in a more elaborate form as

$$- \begin{array}{c} \text{Pressure} \\ \text{gradient} \end{array} + \begin{array}{c} \text{rate of} \\ \text{change of} \\ \text{normal} \\ \text{and shear} \\ \text{stresses} \end{array} + \begin{array}{c} \text{gravita-} \\ \text{tional} \\ \text{force} \end{array} = \text{density} \times \left(\begin{array}{c} \text{local} \\ \text{acceler-} \\ \text{ation} \end{array} + \begin{array}{c} \text{convective} \\ \text{acceleration} \end{array} \right)$$

$$(1.8)$$

Here the pressure gradient is the difference of pressure per unit distance in the flow direction. The gravitational force is the weight per unit volume of the fluid, that is, the product of fluid density and gravitational acceleration. Two kinds of acceleration exist: local and convective. Local or transient acceleration refers to the rate of change of velocity with respect to time, while convective acceleration is the rate of change of velocity from one location to another as a fluid mass moves in a flow field.

In the absence of the gravitational and frictional forces, Eq. (1.8) is reduced to the following expression:

$$\text{density} \times \text{acceleration} = -\text{pressure gradient} \qquad (1.9)$$

Equation (1.9) implies that when the flow is decelerated in its path, a positive pressure gradient is generated. If the situation occurs on one side of a membrane, the pressure on the same side of the membrane becomes higher than that acting on the other side of the membrane. Then the net force acts in the direction of decreasing pressure. This is the principle that operates the mitral and aortic valves as well as the valves of the vein and lymphatics.

Newton's second law of motion, Eq. (1.6), can be expressed in an alternate form as

$$\Sigma F = \frac{m v_2 - m v_1}{t} \tag{1.10}$$

where v is the velocity and t is the time. The expression means that the sum of forces acting on a mass m is equal to the rate of change of momentum of the body, which is the conservation of momentum.

Equation (1.1) for the balance of energy can be expressed in the finite difference form as

$$\Delta KE + \Delta PE + \Delta E = -W + Q \tag{1.11}$$

or in the rate form as

$$\frac{D(KE + PE + E)}{Dt} = -\dot{W} + \dot{Q} \tag{1.12}$$

Here, KE, PE, and E stand for the kinetic, potential, and system energy; W is the sum of the work done by the fluid pressure over the two terminal cross sections of areas A_1, A_2 and the work done by normal and shear stresses over the vessel walls; Q is the heat transfer into or out of the fluid mass; D/Dt is the total (or material) derivative; and $\dot{W}$ and $\dot{Q}$ are time derivatives of W and Q, respectively.

When the heat term is neglected, the balance Eq. (1.11) is reduced to

$$\Delta P + \frac{\rho \Delta v^2}{2} + g \Delta h + \frac{\partial(\rho v^2)}{\partial t} + F = 0 \tag{1.13}$$

F signifies the integrated frictional loss between the two end cross sections. The equation can also be obtained through the integration of Eq. (1.8) along a streamline. For a flow along a streamline, the neglect of the last two terms results in the Bernoulli equation

$$\Delta P + \frac{\rho \Delta v^2}{2} + g \Delta h = 0 \tag{1.14}$$

or

$$P + \frac{\rho v^2}{2} + g h = \text{constant} \tag{1.15}$$

The Bernoulli equation states that for a steady, inviscid flow, the kinetic and potential energy can be converted into pressure. The frictional loss term becomes predominant in microvessels and Eq. (1.13) is reduced to

$$-\Delta P \cong F \qquad (1.16)$$

Similarly, for the systemic circulation from the exit of the aortic valve to the entrance of the right atrium, $\Delta v = 0$ and $\Delta h = 0$ between the two end points. If $\rho v^2 = $ constant, then Eq. (1.13) becomes

$$-\Delta P = F = \Sigma F_i \qquad (1.17)$$

where F_i is the frictional loss in each segment of vessels of the network. In physiology, it is commonly written that

$$-\Delta P = CO \cdot R \qquad (1.18)$$

in which CO denotes the cardiac output (e.g., volumetric flow rate) and R is the total peripheral vascular resistance. $-\Delta P$ is the total pressure drop between the aortic valve and the right atrium including multiple millions of capillary vessels. It is commonly referred to as the systemic arterial pressure.

The cardiac output is calculated from Hagen-Poiseuille law. As is shown later in Chapter 2, it reads

$$CO = \frac{\pi d^4 \, \Delta P}{128 \, \mu l} \qquad (1.19)$$

for steady laminar flow in a long, rigid, circular tube. Here d and l represent the tube diameter and length, respectively. A combination of Eqs. (1.18) and (1.19) yields the laminar flow resistance to be

$$R = \frac{128 \mu l}{\pi d^4} \qquad (1.20)$$

If the tube is branched into n individual ones, then each tube has the resistance of R/n.

Blood flow is also related to thermodynamics. Take blood flow in a heart valve as an example. The left ventricle functions like a pump. The cycle of a pump consists of four processes: suction, compression, discharge, and expansion. Admission and delivery to the cylinder are through valves. The pump performance is expressed by an indicator diagram, for example, a plot of cylinder pressure (head) versus cylinder volume (capacity). Similarly, the cycle of the left ventricle is portrayed in Fig. 1.14, with ventricular filling (diastole, 1-2), isovolumic systole (2-3), ejection (systole, 3-4), and isovolumic relaxation (4-1). Point 3 represents the opening of the aortic valve, whereas the 4 designates the closing of the aortic valve. The pressure at 4 approximates the maximum cyclic (isovolumic) pressure resulting from the maximum isomeric tension of heart valve. The mitral valve opens at 1 and closes at 2. The cardiac muscle is at rest during the process 1-2. The area 1-2-b-a-1 represents the work done by the blood on the ventricular wall. The cycle 1-2-b-a-1 is clockwise, indicating a positive work (work done by blood). The area 3-4-a-b-3 denotes the work done on the blood by the ventricular work. The process 3-4-a-b-3 is counter-

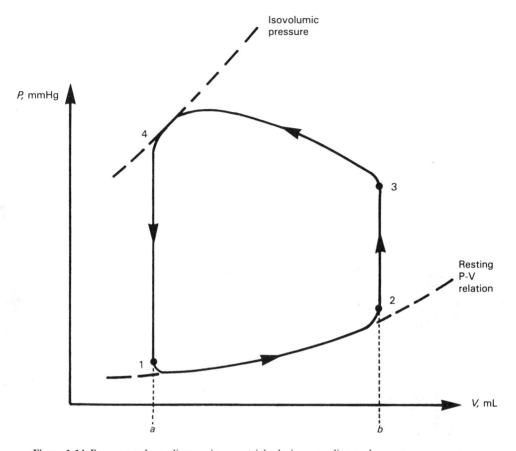

Figure 1.14 Pressure-volume diagram in a ventricle during a cardiac cycle.

clockwise, suggesting negative work (work done on blood). The difference between the two areas (e.g., area 1-2-3-4) represents the network done on the blood by the ventricular muscle. Since the loop 1-2-3-4-1 is counterclockwise, it is a negative work, called stroke work which is performed by the left ventricle on the blood as it contracts.

1.4 SUMMARY OF TRANSPORT-RATE EQUATIONS AND EQUATIONS OF STATE

The ''particular'' laws pertinent to transport phenomena include the transport-rate equations and the equations of state. Their mathematical expressions are summarized in Tables 1.1 and 1.2. Notice that Ohm's laws are incorporated into the table as they are the transport rate equations for electrical current. It

Table 1.1 Transport rate equations*

Transport process	Mass	Momentum	Heat	Electricity
Property transferred ϕ	Mass N_A (kg/s)	Shear stress τ (N/m²)	Heat q (W)	Current i (amp)
Potential Φ	Mass concentration (or density) C_A (kg/m³)	Momentum concentration ρv (kg/s m²) or velocity v (m/s)	Energy concentration ρe (kJ/m³) or temperature T(°C)	EC^* (volt farad/m) or voltage E (volt)
Diffusion type Diffusity or conductivity	Mass diffusivity D_{AB} (m²/s)	Kinematic viscosity ρ (m²/s) or absolute viscosity μ (kg/s m)	Thermal diffusivity α (m²) or thermal conductivity k (W/m s)	$\dfrac{1}{R^*C^*}$ (m²/Ω farad) $\dfrac{1}{R^*}$ (m/Ω)
Rate equation as a function of potential gradient	Fick's first law $\dfrac{N_A}{A} = -D_{AB}\dfrac{\partial C_A}{\partial y}$ (1.21)	Newton's law of viscosity $\tau = -v\,\dfrac{\partial(\rho v)}{\partial y}$ or $\tau = -\mu\,\dfrac{\partial v}{\partial y}$ (1.23)	Fourier's conduction equation $\dfrac{q_k}{A} = -\alpha\,\dfrac{\partial(\rho e)}{\partial y}$ or $\dfrac{q_k}{A} = -k\,\dfrac{\partial T}{\partial y}$	$i = -\dfrac{1}{R^*C^*}\dfrac{\partial(EC^*)}{\partial y}$ or $i = -\dfrac{1}{R^*}\dfrac{\partial E}{\partial y}$ (1.27) (1.25)

	Mass transfer coefficient h_D (m/s)	Fanning friction coefficient f (dimensionless)	Heat transfer coefficient h (W/m² °C)	Ohm's equation for electrical resistance
Convective type Coefficient			$h \begin{cases} h_k = \dfrac{k}{\ell} & \text{for conduction} \\[2mm] h_c & \text{for convection} \\[2mm] h_r = \dfrac{F\sigma(T_1^4 - T_2^4)}{T_1 - T_2} & \text{for radiation} \end{cases}$	$\dfrac{1}{R}\left(\dfrac{1}{\Omega}\right)$
Rate equation as a function of potential difference	$\dfrac{N_A}{A} = h_D \Delta C_A$ (1.22)	$\tau = f\dfrac{\rho v^2}{2g_c}$ for turbulent flow (1.24)	$\dfrac{q}{A} = h\Delta T$ (1.26) q_k, Fourier's conduction equation q_c, Newton's law of cooling q_r, Stefan-Boltzmann equation	Ohm's equation for electrical resistance $i = \dfrac{\Delta E}{R}$ (1.28) For a capacitor (C): $i = C\dfrac{dE_c}{dt}$ (1.29) For an inductor (L): $i = \dfrac{1}{L}\int E_t\, dt$ (1.30)

E, voltage; C, capacitance; R, resistance; C, capacitance/length; R*, resistance/length.

Table 1.2 Equations of state

(a) Thermodynamic equation of state (P-v-T) relationship*

Substance	Empirical parameters	Model or equation name	Expression	Equation no.
Ideal gas	None	Perfect gas law	$Pv = RT$	(1.31)
Real gas	Z		$Pv = ZRT$	(1.32)
	a, b	Van der Waals	$P = \dfrac{RT}{v - b} - \dfrac{a}{v^2}$	(1.33)
	A, B, ϵ	Beattie-Bridgeman	$P = \dfrac{RT(1 - \epsilon)}{v^2}(v + B) - \dfrac{A}{v^2}$	(1.34)

*Z, compressibility factor; R, gas constant; v = specific volume.

Table 1.2 Equations of state (*Cont.*)
(b) Rheological equation of state (shear stress-shear strain relationship)

Substance	Empirical parameters	Model or equation name	Expression	Comments
Newtonian	μ	Newton's law of viscosity	$\tau = -\mu\dot\gamma$	Common fluids behave as a Newtonian fluid
non-Newtonian	η, τ_0	Bingham-plastic	$\tau = \tau_0 - \eta\dot\gamma$	Very few materials possess the properties of an idealized Bingham plastic. The model is a crude approximation but the concept is simple and useful
	η, n	Power law or Ostwald de Waele	$\tau = -\eta\dot\gamma^n$	Suspensions of red blood cells often behave as power-law fluids with n of about 0.75. $n < 1.0$ called pseudoplastic $n > 1.0$ called dilatant
	η, τ_0	Casson	$\sqrt{\tau} = \sqrt{\tau_0} + \eta\sqrt{\dot\gamma}$	A modified Bingham model. Whole blood has been shown to follow this model
	$\eta, G,$ or λ	Maxwell or elasticoviscous	$\tau + \dfrac{n}{G}\dfrac{D\tau}{Dt} = \eta\dot\gamma$ $\lambda = \eta/G$	Under certain circumstances, blood behaves as an elasticoviscous fluid. The model is a superposition of the Hookean solid and Newtonian fluid
Solids	η, E	Kelvin or viscoelastic	$\tau = E\gamma + \eta\dot\gamma$	There is some evidence that the walls of the blood vessels and the surface membrane of red blood cells behave as Kelvin solids

is advantageous to utilize the analogy between the flows of heat, mass, and electricity in solving heat and mass transfer problems.

1.5 AN EXAMPLE

Consider the steady laminar flow of a fluid of constant density ρ in a long tube of internal radius R. One selects as the system a cylinder of the fluid of length l and radius r situated on the tube axis (Fig. 1.15). The difference in force between the ends of the cylinder is $\Delta P \pi r^2$. The opposing force comes from the viscous drag on the surface of the cylinder. The viscous drag force for the tangential stress τ_{rz} is $\tau_{rz} 2\pi r l$. For the velocity to be steady, it requires

$$\Delta P \pi r^2 = \tau_{rz} 2\pi r l \qquad \text{or} \qquad \tau_{rz} = \frac{\Delta P}{l} \frac{r}{2} \tag{1.35}$$

Hence, the momentum flux distribution τ_{rz} is linear, zero at the tube axis and maximum at the wall surface.

For a Newtonian fluid, the rheological equation of state reads

$$\tau_{rz} = -\mu \frac{dv_z}{dr} \tag{1.36}$$

The combination of the last two equations yields

$$\frac{dv_z}{dr} = -\frac{\Delta P}{l} \frac{r}{2\mu} \tag{1.37}$$

This is Eq. (B.20) for steady, one-dimensional, fully developed flow. Integration of this gives

$$v_z = -\frac{\Delta P}{l} \frac{r^2}{4\mu} + C_1$$

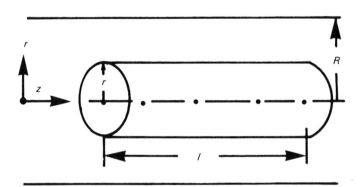

Figure 1.15 Fluid flowing in a cylindrical tube.

Because of the boundary condition that $v_z = 0$ at $r = R$, the constant C_1 has the value $(\Delta P/l)[R^2/(4\mu)]$. Hence, the velocity distribution is

$$v_z = \frac{\Delta P}{l}\frac{R^2}{4\mu}\left[1 - \left(\frac{r}{R}\right)^2\right] \tag{1.38}$$

This result indicates that the velocity distribution for laminar, incompressible flow in a tube is parabolic. The maximum velocity $v_{z,\max}$ occurs at $r = 0$ and has the value

$$v_{z,\max} = \frac{\Delta P}{l}\frac{R^2}{4\mu} \tag{1.39}$$

Now consider a cylindrical shell of fluid with an inner radius r and thickness dr (Fig. 1.16). The volumetric rate of fluid flowing down the annulus is given by

$$dQ = v_z 2\pi r\, dr \tag{1.40}$$

Substituting for v_z from Eq. (1.38) and integrating flow from $r = 0$ to $r = R$, one obtains

$$Q = \frac{\Delta P}{l}\frac{\pi R^4}{8\mu} \tag{1.41}$$

This result is called the Hagen-Poiseuille law which gives the relationship between the volumetric rate of flow and the force causing the flow. The average velocity is obtained by dividing the flow rate by the cross-sectional area as

$$\bar{v}_z = \frac{\Delta P}{l}\frac{R^2}{8\mu} \tag{1.42}$$

The Reynolds number of the flow is defined as

$$\mathrm{Re} = \frac{\bar{v}_z 2R\rho}{\mu} \tag{1.43}$$

The system in Fig. 1.16 is used to derive the momentum equation, where the rate of momentum across the cylindrical surface at r is $2\pi rl\tau_{rz}$.

The rate of momentum out across cylindrical surface at $r + dr$ is

$$2\pi(r + dr)l(\tau_{rz} + d\tau_{rz})$$

$$= 2\pi rl\tau_{rz} + 2\pi rl\tau_{rz}\, dr + 2\pi rl\, d\tau_{rz} + 2\pi l\, dr\, d\tau_{rz}$$

$$\cong 2\pi rl\tau_{rz} + 2\pi ld(r\tau_{rz})$$

The rate of momentum in across annular surface at entrance $z = 0$ is $2\pi r\, dr\, v_z(0)\rho v_z(0)$.

The rate of momentum out across annular surface at exit $z = l$ is $2\pi r\, dr\, v_z(l)\rho v_z(l)$.

The pressure force acting on annular surface at entrance is $(2\pi r\, dr)P(0)$.

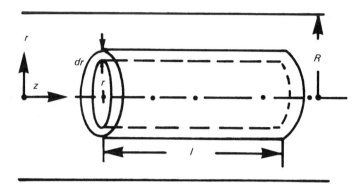

Figure 1.16 Fluid flowing in a cylindrical annulus.

The pressure force acting on annular surface at the exit is $(2\pi r\, dr)P(l)$. Neglecting the effect of gravity force, the momentum balance yields

$$2\pi r l \tau_{rz} - [2\pi r l \tau_{rz} + 2\pi l\, d(r\tau_{rz})] + 2\pi r\, dr\, \rho[v_z(0)]^2$$
$$- 2\pi r\, dr\, \rho[v_z(l)]^2 + 2\pi r\, dr[P(0) - P(l)] = 0$$

$v_z(0) = v_z(l)$ for an incompressible fluid. Dividing the equation by $2l\, dr$, one obtains

$$\frac{d(r\tau_{rz})}{r\, dr} - \frac{\Delta P}{l} = 0 \qquad (1.44)$$

This is Eq. (1.11) for steady, one-dimensional flow.

PROBLEMS

1.1 Some heart patients have their valve tissue calcified such that the open area for blood flow is reduced. When the area reaches a critical value A_c below which the patient's life is threatened, open heart surgery is needed to replace the patient's natural cardiac valves with prosthetic ones. The empirical expression to determine A_c is called the Gorlin equation which reads

$$A_c \text{ (cm}^2\text{)} = \frac{\text{flow rate through the valve } Q \text{ (ml/s)}}{K\sqrt{2g\Delta P \text{ (mmHg)}}}$$

where P is the pressure gradient (difference) across the valve and K is the empirical constant. Derive the expression from the Bernoulli equation:

$$\frac{P}{\rho} + \frac{v^2}{2} + gz = \text{constant} \qquad (1.45)$$

1.2 Derive the expressions for (a) the velocity profile, and (b) the volume rate of flow for blood flow through a circular tube with an inner diameter of $2R$. Assume that blood behaves like a Bingham plastic fluid (i.e., a shearing-core model)

$$\tau_{rz} = -\mu_0 \frac{dv_z}{dr} + \tau_0$$

here μ_0 is the viscosity and τ_0 denotes the yield stress.

1.3 In an invasive method of blood flow measurement, a hot wire (or film) probe fitted with a catheter is inserted into a blood vessel. An electrical current i is passed through the hot wire of diameter d and electrical resistance R_e to generate heat. The heat is then dissipated to the surrounding flowing blood by convection. The hot wire temperature is measured by T_w, while the blood temperature is known as T_b. Derive an expression for the blood flow velocity v, knowing that the heat transfer coefficient h_c is proportional to the square root of the velocity: $h_c = C_1\sqrt{v} + C_2$. Both C_1 and C_2 are constant.

1.4 Derive an expression for estimating the mechanical energy expenditure by the left ventricle with the measurement of blood velocity v. (*Hint:* This can be achieved by the rate of flow and the pressure difference ΔP between the aorta and the left ventricle prior to contraction.) If a measurement of the kinetic energy of the blood KE is made of a velocity measurement, what will be the expression (of the mechanical energy expenditure) instead? Use the Bernoulli equation in the latter case.

1.5 Blood is equilibrated with a gas mixture containing 14.5% oxygen at a barometric pressure of 752 mmHg and 38°C. How many volumes percent of oxygen are dissolved in the blood?

1.6 Dissolved oxygen and hemoglobin react to form a chemical compound:

$$O_2 + Hb \rightleftharpoons HbO_2$$

in which Hb denotes reduced hemoglobin and HbO_2 is oxyhemoglobin. One mole, or 32 g, of oxygen combines with 16,700 g (molecular weight) of hemoglobin. A sample of blood is equilibrated at 38°C with a gas mixture having a P_{O_2} of 200 mmHg at which the hemoglobin is fully saturated. It is found to contain 21.3 vol % oxygen. How much hemoglobin is in the sample?

1.7 The bladder is idealized as a thin-walled sphere of radius R. The pressure in the bladder, called intravesical pressure, is P_{ves}, while the pressure surrounding the bladder in the abdomen, called abdominal pressure, is P_{abd}. The intravesical pressure is recorded by a fine catheter passing through the urethra or through the abdominal and bladder walls. Similarly, the pressure in the rectum or stomach is measured and taken to be representative of the abdominal pressure. The difference between the intravesical and abdominal pressure is the pressure contribution from stresses in the bladder wall, called detrusor pressure P_{det}. Determine the expression for the radius of the bladder using static equilibrium of forces.

1.8 The Fick method for determining cardiac output $\dot{V}_{co}$ (ml/min of blood flow) is based on the formula

$$\dot{V}_{co} = \frac{\dot{V}_{CO_2}}{C_i - C_o}$$

Here, $\dot{V}_{CO_2}$ denotes the volumetric flow rate of CO_2 (ml/min) that is exhaled, C_i is the concentration of CO_2 in the venous blood entering the lung and C_o represents the concentration of CO_2 in the arterial blood leaving the lung. The unit of C_i and C_o is milliliters of dissolved CO_2 per milliliter of blood. One measure yields $\dot{V}_{CO_2} = 400$ ml/min, $C_i = 0.55$, and $C_o = 0.48$. If the measured values of $\dot{V}_{CO_2}$ and C_o are accurate while C_i has an error of 2%, determine the corresponding percentage error in the measured cardiac output.

REFERENCES

Fung, Y. C. (1981) *Biomechanics, Mechanical Properties of Living Tissues*, Springer-Verlag, New York.

Fung, Y. C. (1984) *Biodynamics: Circulation*, Springer-Verlag, New York.

Ghista, D. N., Van Vollenhooven, E., Yang, W.-J., and Reul, H. (1979) *Theoretical Foundations of Cardiovascular Processes, Advances in Cardiovascular Physics*, vol. 1, S. Karger, Basel, Switzerland.

Ghista, D. N., Van Vollenhooven, E., Yang, W.-J., Reul, H., and Bleifeld, W. (1983) *Cardiovas-*

cular Engineering Part IV, Prothesis, Assist and Artificial Organs, Advances in Cardiovascular Physics, vol. 5, S. Karger, Basel, Switzerland.

Griffith, D. J. (1980) *Urodynamics,* Adam Hilger Ltd., Bristol.

Guyton, A. C. (1971) *Medical Physiology,* W. B. Saunders, Philadelphia.

Houdas, H., and Ring, E. F. J. (1982) *Human Body Temperature,* Plenum, New York.

Kreith, F. (1965) *Principles of Heat Transfer,* 2d ed., International Textbook Co., Scranton, Penn.

Lighthill, M. J. (1972) Physiological fluid mechanics: A survey. *J. Fluid Mech. 52,* Pt. 3, 475–497.

Lih, M. M. S. (1975) *Transport Phenomena in Medicine and Biology,* Wiley Interscience, New York.

Newburgh, L. H. (ed.) (1968) *Physiology of Heat Regulation and the Science of Clothing,* Hafner, New York.

Seagrave, R. C. (1971) *Biomedical Applications of Heat and Mass Transfer,* Iowa State University, Ames, Iowa.

Shitzer, A., and Eberhart, R. C. (eds.) (1985) *Heat Transfer in Medicine and Biology, Analysis and Applications,* vols. 1 and 2, Plenum, New York.

Stolwijk, J. A., and Hardy, J. D. (1966) Temperature regulation in man—A theoretical study, *Pfluegers Arch. 291:*129–162.

Streeter, V. L., and Wylie, E. B. (1985) *Fluid Mechanics,* 8th ed. McGraw-Hill, New York.

Van Wylen, G. J., and Sonntag, R. E. (1986) *Fundamentals of Classical Thermodynamics,* 3d ed., Wiley, New York.

Wissler, E. H. (1961) Steady-state temperature distribution in man, *J. Appl. Physiol. 16:*734–740.

Wissler, E. H. (1966) A mathematical model of the human thermal system, *Chem. Eng. Progr. Symp. Ser. 62.*

Yang, W.-J. (1979a) Equations in Biotransport Mechanisms, in *Theoretical Foundations of Cardiovascular Processes, Advances in Cardiovascular Physics,* vol. 1, Karger, Basel, Switzerland, pp. 102–127.

Yang, W.-J. (1979b) Blood-Gas Interactions and Physiological Implications, in *Blood: Rheology, Hymolysis, Gas and Surface Interactions, Advances in Cardiovascular Physics,* vol. 3, Karger, Basel, Switzerland, pp. 45–99.

Yang, W.-J., and Yang, L. H. (1982) Heat and Mass Transfer Equipment with Medical Applications. Amer. Soc. Mech. Eng. Paper No. 82-WA/HT-87, presented at the 1982 ASME Winter Annual Meeting, Phoenix, Ariz.

TWO

FLUID DYNAMICS

2.1 RHEOLOGICAL EQUATIONS OF STATE

Physiological fluid dynamics deals with flow of biological fluids, for example, blood in a tube such as a blood vessel. The main objectives include the determination of velocity profiles, total flow rate, and pressure variations of flow of biological fluids in the circulation or secretion system. These physical quantities of interest depend on the nature of the fluid and its containing vessel which may vary with a change in the physiological state. It is, therefore, essential to understand the rheological properties of the fluid and the solid in the presence of internal forces. The basic difference between the solid and fluid states is that the former may exert a restraining force on the displaced plane inside the medium, while the latter is unable to sustain internal forces. The concepts of solidity and fluidity are idealizations that describe the behavior of real materials in certain limiting cases. Figure 2.1 summarizes the types of rheological materials.

In general, the behavior of real materials encompasses many intermediate properties in the figure. A rigid solid does not deform under shear or tangential force. An elastic medium deforms under stress but returns to its original state when the stress is removed. An inviscid fluid has zero viscosity which represents another extreme of a material. Bingham plastic remains rigid when the shear stress is of smaller magnitude than the yield value τ_0 but flows like a Newtonian fluid when the shear stress exceeds τ_0.

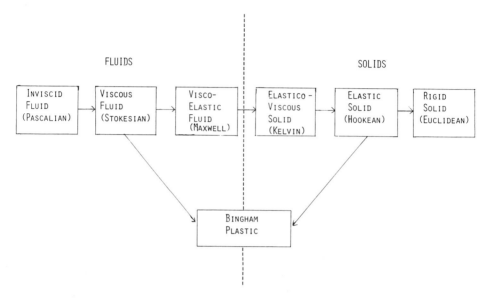

Figure 2.1 Rheological classification of materials.

2.1.1 Plasma and Whole Blood

Plasma may be separated from other constituents of the blood by sedimentation under the influence of gravitational or centrifugal force. Human plasma is a transparent, slightly yellowish fluid and its density is about 1.035 g/ml. It consists of a solution of plasma proteins in an aqueous medium. Proteins, consisting, in total, of about 7% of the total volume of the plasma, can be divided into three main groups: fibrinogen, globulins, and albumin. Plasma also contains emulsified fats (lipids), cholesterol, free fatty acids, adrenalin, dissolved oxygen, and dissolved carbon dioxide.

Whole blood consists of a suspension of red cells (or erythrocytes), white cells (or leukocytes), and platelets in an aqueous solution (plasma). While the density of erythrocytes is 1.10 g/cm^3, whole blood has a density of 1.05–1.06 g/cm^3. The red cells, numbering approximately 5 million per cubic millimeter are the dominant particulate matter in blood with about 40–45% by volume of the whole blood. The term "hematocrit" is normally used to specify the volume percentage of red cells and entrained plasma. The principal function of the red cells is to carry oxygen to living tissues and remove CO_2. Each 100 ml of whole blood contains about 15 g of hemoglobin which can transport 20 ml of oxygen at 37°C and 760 mmHg pressure. The cells have a density of about 1.08 g/ml and their shape is that of a biconcave discoid under normal conditions. Their size depends, to some extent, on the method of preparing the blood sample. Typical dimensions in humans are 7.8 μm in diameter, approximately 2 μm in thickness, and 88 μm^3 in volume. The cell shape is flexible enough to vary with the diameter of the capillaries in microcirculation.

White cells protect the body from disease. Their sizes vary from 16 to 22 μm for monocytes, and from 6 to 10 or 12 μm for lymphocytes and granulocytes. Their volume concentration in the blood is one white cell to every 1000 red cells, which is totally insignificant.

Platelets are much smaller than red or white cells, with a diameter of 2–3 μm. Their number is one-tenth of the red cells. Both leukocytes and platelets are ordinarily not numerous enough to influence the flow characteristics of blood. However, platelets play an important role in the formation of blood clots which may severely interfere with the flow.

An ideal fluid is one that is incompressible and has no viscosity. With zero viscosity, the fluid offers no resistance to shearing forces. Hence, all shear forces are zero during flow and deformation of the fluid. However, all real fluids have finite viscosity, and it is necessary to take into account the viscosity and the related shearing forces associated with deformation of the fluid. Real fluids are also called viscous fluids.

Real fluids are classified into two main categories. Newtonian fluids are characterized by the shear stress τ, which is proportional to the shear strain du/dy, as shown in Fig. 2.2. The slope of the line representing the viscosity coefficient μ is a constant at a given temperature and pressure. When a steady flow is achieved, a constant force called the shear force or viscous drag force is required to maintain the motion of the fluid. τ refers to the shear force per unit area, while the rate of shear strain $\dot{\gamma}$ signifies the velocity decrease in the distance perpendicular to the direction of flow. The shear-stress/shear-strain relationship may be described by

$$\tau = -\mu\dot{\gamma} \tag{2.1}$$

This is known as *Newton's law of viscosity,* and fluids that behave in this fashion are termed Newtonian fluids. All gases and homogeneous nonpolymeric liquids are described by this equation. There are, however, many fluids whose rheological behavior does not obey Eq. (2.1), and they are referred to as non-Newtonian fluids. While plasma is generally treated as a Newtonian fluid, the behavior of whole blood is non-Newtonian.

Non-Newtonian fluids are those in which the viscosity is a function of either $\dot{\gamma}$ or τ. A value calculated from the ratio of the applied tangential stress to the rate of shear produced is called apparent viscosity η, which reads

$$\eta = -\frac{\tau}{\dot{\gamma}} \tag{2.2}$$

η is equal to μ in the case of a Newtonian fluid. Figure 2.3a is a plot of apparent viscosity against the hematocrit concentration for various values of the diameter of the measuring capillary tube. It indicates not only an increase of viscosity with cell concentration but also shows a marked effect of the tube diameter. The latter is called the Fahraeus-Lindquist effect which results from the tendency of the cells to move away from the wall where they leave a layer of low viscosity. This effect becomes important for blood vessels having diameters smaller than about 1 mm, as illustrated in Fig. 2.3b.

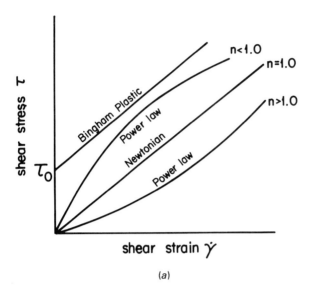

(a)

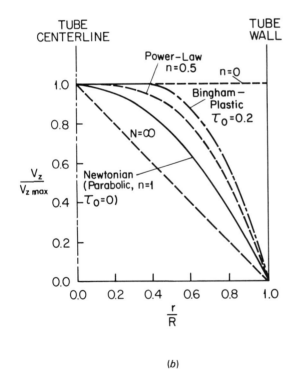

Figure 2.2 (a) Shear stress/shear strain relation, two-parameter models, (b) Comparison of velocity profiles of Newtonian, power-law, and Bingham-plastic fluids.

(b)

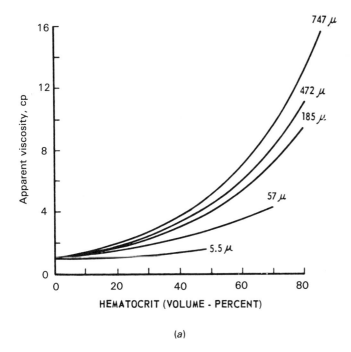

(a)

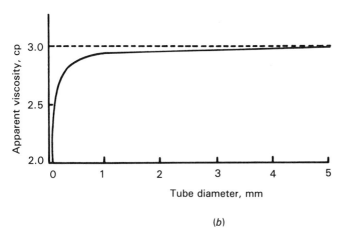

(b)

Figure 2.3 (a) Apparent viscosity of blood (in centipoise) as a function of hematocrit and capillary diameter (Haynes, 1961), (b) Effect of tube diameter on apparent viscosity of blood for a hematocrit of 40% and a temperature of 38°C (Haynes, 1960).

However, numerous empirical equations or models have been proposed to experss the steady-state relation between τ and $\dot{\gamma}$. Some are characterized by two parameters and others by three parameters. The equation most generally applicable to the case of blood assume that it behaves either as a power-law fluid or a Bingham plastic. Sometimes, blood is treated as a viscoelastic fluid of either the Maxwell or Oldroyd type.

The rheological behavior of a power-law fluid is expressed by

$$\tau = -\mu_p \dot{\gamma}^n \tag{2.3}$$

where μ_p denotes a measure of the consistency of the fluid and n is a constant. The behavior is pseudoplastic for values of n less than unity in which the apparent viscosity decreases with rate of shear. The viscosity becomes constant at high shear rates.

A Bingham plastic, sometimes called an ideal plastic, remains rigid when the shear stress is of smaller magnitude than the yield stress τ_0 but flows like a Newtonian fluid when the shear stress exceeds τ_0. The model reads

$$
\begin{aligned}
\tau &= \tau_0 - \mu_B \dot{\gamma} && \text{for } |\tau| > \tau_0 \\
\dot{\gamma} &= 0 && \text{for } |\tau| < \tau_0
\end{aligned}
\tag{2.4}
$$

where μ_B signifies the plastic viscosity. The yield stress in normal blood is less than 0.1 dyn/cm^2. When it is applied to whole blood, considerable errors may result from the model, particularly at low rates of shear. The Casson equation, a modified Bingham plastic model, has been empirically derived for a better fit at shear rates from 1 to 100,000 s^{-1}. The general expression for blood viscometry is

$$\sqrt{\tau} = K_c \sqrt{\dot{\gamma}} + \sqrt{\tau_0} \tag{2.5}$$

K_c is called Casson's viscosity. The average equation (Charm and Kurland, 1965) is

$$\sqrt{\tau} = 0.166\sqrt{\dot{\gamma}} + 0.33 \tag{2.6}$$

For theoretical study of the velocity and pressure fluctuations or blood in the clotting process, the flow models with a time factor are appropriate since whole blood, being a viscoelastic fluid, exhibits relaxation and memory. Two types of the viscoelastic model have been employed. The Maxwell (fluid) type describes the shear-stress/shear-strain relationship as

$$\tau + \lambda_1 \frac{D\tau}{Dt} = \eta \dot{\gamma} \tag{2.7}$$

where D/Dt is the substantial derivative; λ_1, the characteristic stress-relaxation time; and η, the shear viscosity.

The other type of Oldroyd, a three-parameter linear model, reads

$$\tau + \lambda_1 \frac{D\tau}{Dt} = \eta_0 \left(\dot{\gamma} + \lambda_2 \frac{D\dot{\gamma}}{Dt} \right) \tag{2.8}$$

Here, λ_2 denotes a characteristic strain-relaxation time, with $\lambda_1 \geqslant \lambda_2 \geqslant 0$. The special case in which $\lambda_1 = \lambda_2 = 0$ corresponds to a Newtonian fluid. For $\lambda_2 = 0$, Eq. (2.8) is reduced to Eq. (2.7).

2.1.2 Walls of Blood Vessels

An elastic solid is like a spring; the extension is a function of the applied force. The spring reverts immediately to its original position once the force is removed, according to

$$\tau = E\xi \tag{2.9}$$

where E is the elasticity modulus and ξ is the displacement or strain.

The Kelvin elasticoviscous solid is also called a Voigt body. The extension gradually increases up to a maximum value limited by the magnitude of the applied force. The model exhibits the phenomenon of creep. The shear stress at any time is described by

$$\tau = E_K\xi + \eta_K\dot{\gamma} \tag{2.10}$$

where E_K and η_K signify the elasticity and viscosity moduli, respectively. When the force is released, the material will gradually return to its initial form. There is some evidence that the walls of the blood vessels and the surface membrane of red cells behave as elasticoviscous solids.

One needs to know the properties of blood vessels in order to formulate the boundary conditions in the analysis of a flow process. The quantities of interest include average density, dimension and elastic properties of the wall materials. The density of soft tissue varies from 1.0 to 1.2 g/cm^3. One may therefore use the same value for the density of the vessel walls as for blood. Some typical dimensions are listed in Table 2.1.

Blood vessels consist of four types of tissue in varying proportions: endothelial lining, elastin fibers, collagen fibers, and smooth muscle. The endothelial lining provides a smooth wall and offers a selective permeability to various substances carried in the bloodstream. However, it is too soft to contribute to the elasticity of the wall. Elastin fibers can be easily stretched with an elastic modulus of 3×10^{16} to 6×10^6 dyn/cm^2. Collagen fibers possess an elastic modulus of 30×10^6 to 100×10^6 dyn/cm^3 and are arranged in the wall with a certain degree of slackness so as not to contribute to the elastic behavior until some stretching has occurred. The smooth muscles change the diameter of the vessels by exerting active tension through contraction under physiological control. Their elastic modulus ranges from 6×10^4 to 6×10^7 dyn/cm^2.

In general, elastic deformations of blood vessels take place at a substantially constant volume so that Poisson's ratio is close to 0.5. The effective elastic (Young's) modulus increases with blood pressure, as illustrated in Table 2.2.

Figure 2.4 demonstrates that an increase of stiffness of the arterial wall, equivalent to an increase in the mean blood pressure, results in an increase in the pulse-wave velocity. It also shows the effect of age.

Table 2.1 Typical data for some blood vessels

Vessel	Diameter, mm	Wall thickness, mm	Length, cm	Blood velocity, cm/s	Reynolds number
Dog					
Aorta	10	1.2	40	50	2500
Large arteries	3	0.5	20	13	200
Main arterial branches	1	0.2	10	8	40
Arterioles	0.02		0.2	0.3	0.03
Capillaries	0.004–0.008		0.1	0.07	0.003
Venules	0.03		0.2	0.07	0.01
Large veins	6.0		20	3.6	108
Vena cava	12.5		40	33	2100
Human					
Ascending aorta	32	1.6	4		
Thoracic aorta	20	1.2	16		
Abdominal aorta	19	0.9	16		
Femoral artery	8	0.5	32		
Carotid artery	9	0.75	18		
Radial artery	4	0.35	23		

Source: Rudinger (1966).

A further complication arises from the dependence of the elastic properties on the strain rate. For small strains ξ, the wall stress is equal to

$$\frac{1}{R}\left(E\xi + \eta\frac{d\xi}{dt} + M\frac{d^2\xi}{dt^2}\right)$$

Here, R denotes the unstressed tube radius. M is the mass modulus and is small enough to be neglected. Some values of E and E/η are given in Table 2.3.

Table 2.2 Pressure dependence of Young's modulus in 10^6 dyn/cm^2 for several arteries of a dog

Vessel	Pressure, mmHg			
	40	100	160	220
Thoracic aorta	1.2	4.3	9.9	18.1
Abdominal aorta	1.6	8.9	12.4	18.0
Femoral artery	1.2	6.9	12.1	20.4
Carotid artery	1.0	6.4	12.2	12.2

Source: Bergel (1964).

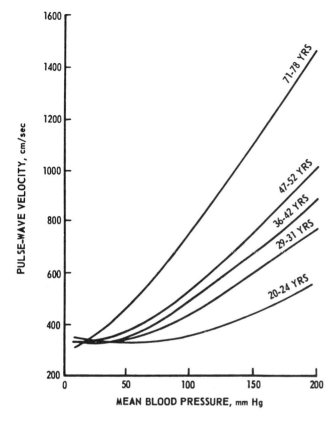

Figure 2.4 Velocity of the pulse wave in the human thoracic aorta as related to blood pressure and age (Rudinger, 1966).

Table 2.3 Elastic properties of several arteries in a dog

Artery	Diameter, mm	Wall thickness, mm	E, Elastic modulus, 10^6 dyn/cm^2	E/η, Ratio of elastic to viscous modulus, s^{-1}
Puppy				
Thoracic aorta	8.0	0.67	9.3	28
Abdominal aorta	7.0	0.41	15.5	36
Femoral artery	1.7	0.15	14.2	38
Carotid artery	2.0	0.19	15.7	40
Aged dog				
Abdominal aorta	11.7	0.42	34	24
Femoral artery	7.8	0.44	29	25
Carotid artery	8.2	0.44	30	27

Source: Rudinger (1966).

2.2 SYSTEMIC CIRCULATION

The systemic circulation refers to blood flow in larger vessels in which a detailed analysis of the performance of the underlying flow process may not lead to a deeper understanding of the corresponding biological mechanism. The primary interest in these large vessels is the rheological study of the resistance relationships and their underlying causes. Figure 2.5 lists various constraints imposed on arterial flow analyses. The simplest case corresponds to steady, fully developed, incompressible, laminar, Newtonian flow through a long, straight, rigid tube.

2.2.1 Steady, Fully Developed Flow

The study of the flow of fluids in tubes with application to the cardiovascular system can be traced back to Poiseuille in 1840 (in Prandtl and Tietjens, 1957). In an attempt to describe the flow of blood in capillaries, he investigated steady, incompressible, laminar, fully developed (constant velocity distribution in the axial direction) flows. Results are presented in Secton 1.5. The shear stress at the wall surface can be evaluated as

$$\tau_w = -\mu \left(\frac{\partial v_z}{\partial r} \right)_{r=0} = -\frac{4\bar{v}_z \mu}{R} \tag{2.11}$$

We can write the wall shear stress in terms of a nondimensional friction coefficient f by defining

$$\tau_w = f \frac{\rho \bar{v}_z^2}{2} \tag{2.12}$$

Then, employing Eq. (2.11) and considering the absolute value of the shear stress, one gets

$$f = \frac{16}{\text{Re}} \tag{2.13}$$

Example 2.1 Consider a laminar flow of an incompressible non-Newtonian fluid between two parallel plates of width $2h$. The channel is placed horizontally so that the effect of body force is neglected. In this case, the flow is one dimensional in the flow direction x. In other words, $v_x = v_x(y)$ and $v_y = v_z = 0$. The continuity equation [Eq. (B.3) in Appendix B] is thus satisfied, and the Navier-Stokes momentum equations (B.6), (B.7), and (B.8) are reduced, respectively, to

$$\frac{\partial P}{\partial x} = \mu \frac{d^2 v_x}{dy^2} \tag{2.14}$$

$$\frac{\partial P}{\partial y} = 0 \tag{2.15}$$

$$\frac{\partial P}{\partial z} = 0 \tag{2.16}$$

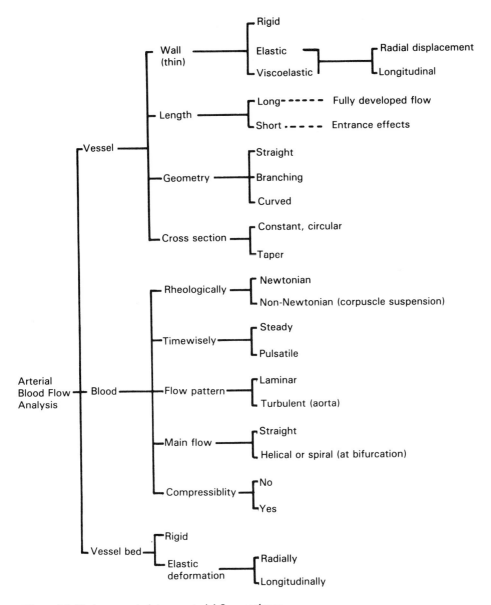

Figure 2.5 Various constraints on arterial flow analyses.

Equations (2.15) and (2.16) suggest that P is a function of x only. The differentiation of Eq. (2.10) with respect to x results in $\partial^2 P/\partial x^2 = 0$ since v_x is not a function of x. It implies that $\partial P/\partial x$ is constant. Integration of Eq. (2.14) with respect to y yields

$$v_x = \frac{y^2}{2\mu}\frac{dP}{dx} + C_1 y + C_2 \tag{2.17}$$

The boundary conditions due to the symmetry of the velocity profile with respect to the midplane are

$$\frac{dv_x(0)}{dy} = 0 \tag{2.18}$$

and due to no-slip conditions at the walls

$$v_x(h) = 0 \tag{2.19}$$

One gets

$$v_x = -\frac{1}{2\mu}(h^2 - y^2)\frac{dP}{dx} \tag{2.20}$$

which is a parabolic profile like a laminar flow in a pipe.

The volumetric flow rate can be obtained by the integration as

$$Q = 2\int_0^h bv_x \, dy = -\frac{4bh^3}{3\mu}\frac{dP}{dx} \tag{2.21}$$

where b denotes the width of the channel. The mean velocity is obtained by dividing the flow rate Q by the cross-sectional area of the channel $2bh$. It yields

$$\bar{v}_x = -\frac{h^2}{3\mu}\frac{dP}{dx} \tag{2.22}$$

The shear stress at the wall τ_w is given by $-\mu(dv_x/dy)$ at $y = h$. The ratio of τ_w to the mean dynamic head $\rho\bar{u}^2/2$ is defined as the friction coefficient f. One obtains

$$f = \frac{\text{Wall shear stress}}{\text{Mean dynamic head}} = \frac{\tau_w}{\rho\bar{v}_x^2/2} = \frac{24}{\text{Re}} \tag{2.23}$$

where the Reynolds number Re is defined as

$$\text{Re} = \frac{4h\rho\bar{v}_x}{\mu} \tag{2.24}$$

The limit of laminar flow is Re = 2300.

Example 2.2 The occurrence of turbulent flow (Re > 2300 in steady flow) in the circulation system is limited to a few locations, such as in the aorta near the exit of the left ventricle and the flow separation region of some stenoses in the artery. Turbulence dissipates energy and causes a large resistance to flow. The velocity distribution in fully developed turbulent flow in a vessel of radius R is

$$v_x = v_{x\max}\left(1 - \frac{r}{R}\right)^{1/7}$$

Using the coordination transformation of $1 - r/R = x$, determine (1) the volumetric flow rate through the vessel, (2) the relationship between the average and maximum velocities, (3) the radius at which the actual velocity is equal to the average velocity, and (4) the shear stress at the vessel wall.

The elementary discharge through an annulus dr is given by

$$dQ = 2\pi r v_x \, dr = 2\pi r v_{xmax} \left(1 - \frac{r}{R}\right)^{1/7} dr$$

and the discharge is through the vessel by

$$Q = 2\pi v_{xmax} \int_0^R r \left(1 - \frac{r}{R}\right)^{1/7} dr$$

Let $1 - r/R = x$, then

$$v_x = v_{xmax} \, x^{1/7}$$

$$\frac{dx}{dr} = -\frac{1}{R} \quad \text{and} \quad dr = -\frac{R}{dx}$$

Therefore, one gets

$$Q = 2\pi v_{xmax} R^2 \int_0^1 (1 - x) x^{1/7} \, dx$$

$$= \frac{49\pi R^2 v_{xmax}}{60}$$

$$\bar{v}_x = \frac{Q}{\pi R^2} = \frac{49 v_{xmax}}{60}$$

For $v_x = \bar{v}_x$,

$$v_{xmax} \left(1 - \frac{r}{R}\right)^{1/7} = \frac{49 v_{xmax}}{60}$$

Therefore, $r/R = 1 - (49/60)^2 = 0.758$. Hence $r = 0.758\,R$.

The balance between the force due to shear stress τ_w on the vessel wall and the pressure drop ΔP on the end faces yields

$$\tau_w = \frac{\Delta P}{l} \frac{R}{2}$$

Thus, one can determine the shear stress τ_w directly by measuring the pressure gradient $\Delta P/l$ along the vessel. Alternately, the law of friction can be utilized to determine τ_w as

$$\tau_w = f \frac{\rho v_x^2}{2}$$

where the friction factor for turbulent flow $f = 0.023(2Rv_x\rho/\mu)^{-0.2}$.

Example 2.3 In many mammals, the blood flow would become turbulent if the Reynolds number relating to the flow exceeded the value

$$Re = a \ln r - br$$

Here, r is the radius of the aorta and a and b are positive constants. Find the radius r that makes the critical Reynolds number a maximum.

Differentiation of Re with respect to r yields

$$\frac{d(Re)}{dr} = \frac{a}{r} - b$$

The expression is set equal to zero, producing

$$r = \frac{a}{b}$$

It gives a maximum value of Re because

$$\frac{d^2(Re)}{dr^2} = -\frac{a}{r^2}$$

is negative at $r = a/b$.

Example 2.4 A general law, called Weber-Fechner law, relates the magnitudes of stimulus and response. It states that the relationship between the level of response X to a certain type of stimulus at the level Y can be expressed as

$$X = a \ln Y + b$$

where a and b are empirical constants. Alternately, one can write this as

$$Y = BA^x$$

in which $\ln A = 1/a$ and $\ln B = -b/a$.

An experiment is conducted to determine the effect of a certain drug on the adrenalin flow of dogs. Y is the amount of drug necessary to produce a flow of adrenalin X. The following results were obtained:

X (cm/s):	1	2	3	4	5	6
Y (mg):	3.2	6.1	11.4	23.0	49.0	95.0

Determine the constant A and B (or a and b).

The experimental points are correlated using a semilog plot, $\ln Y$ versus X. They lie close to a straight line whose slope is $\ln A$ and Y intercept is $\ln B$. By measuring the slope and intercept, the two constants A and B are determined.

Other rheological models that yield different results are obtained for the velocity and flow rate to replace Eqs. (1.38) and (1.41).

For a power-law fluid:

$$v_x = \frac{n}{n+1} \left(\frac{\Delta P}{2\mu_p l} \right)^{1/n} (R^{n+1/n} - r^{n+1/n}) \tag{2.25}$$

$$Q = \frac{n\pi R^3}{3n+1} \left(\frac{R\Delta P}{2\mu_p l} \right)^{1/n} \tag{2.26}$$

For a Bingham plastic:

$$v_z = \frac{1}{\mu_B} \left[\frac{(R^2 - r^2)\Delta P}{4l} - \tau_0(R - r) \right] \tag{2.27}$$

$$Q = \frac{\pi R^4 \Delta P}{8\mu_B l} \left[1 - \frac{4}{3} \frac{2\tau_0 l}{R\Delta P} + \frac{1}{3} \left(\frac{2\tau_0 l}{R\Delta P} \right)^4 \right] \tag{2.28}$$

For Casson's model:

$$v_2 = \frac{\Delta P}{4K_c l}(R^2 - r^2) - \frac{4}{3} \frac{\sqrt{\tau_0}}{K_c^2} \left(\frac{\Delta P}{2l} \right)^{1/2} (R^{3/2} - r^{3/2}) + \frac{\tau_0}{K_c^2}(R - r) \tag{2.29}$$

$$Q = \frac{\pi R^4 \Delta P}{8K_c^2 l} \left[1 - \frac{16}{7} \left(\frac{2\tau_0 l}{R\Delta P} \right)^{1/2} + \frac{4}{3} \frac{2\tau_0 l}{R\Delta P} - \frac{1}{21} \left(\frac{2\tau_0 l}{R\Delta P} \right)^4 \right] \tag{2.30}$$

Their velocity profiles are compared in Fig. 2.2b.

Several studies of incompressible, laminar, viscous flows in elastic tubes have been made. Morgan (1952) investigated steady, fully developed flow in elastic pipes, while the propagation of small amplitude, large wavelength pressure waves in thin-walled elastic tubes was treated by Morgan and Keily (1954), Morgan and Ferrante (1955), and Womersley (1957). Other references include Fry et al. (1980), Holenstein et al. (1980), Pedley (1980), Patel and Vaishnav (1980), Stettler et al. (1980), Lyon et al. (1980), Yen and Foppiano (1981), Yen et al. (1980), Nerem (1981), Fung (1984), and Sugawara et al. (1985).

A fully developed velocity profile is established only after the fluid has flowed over a long tube. In the case of flow through a short tube, the velocity profile continues to vary from the inlet to the exit. The flow in the aorta provides a good example. The velocity distribution of the blood entering the aorta (which depends on the nature of the ventricular contraction and the valves at the aortic entrance) differs from that at downstream because of the action of fluid viscosity. This phenomenon is called the entrance effect, which is characterized by higher friction coefficients. Because of continuous branchings in the systemic circulation, the short lengths of many vessels may preclude the flow in them from ever becoming fully developed. Similarly, entrance effects may also play important roles in the flows in some vessels in microcirculation. This is due to the velocity profiles at the entrances to these vessels (which depend on the flow in the upstream vessels and geometric factors) being modified by viscosity as the fluid proceeds down the vessel.

To study entrance effects, Nikuradse (in Prandtl and Tietjens, 1957) and Langhaar (1942), among others, extended Poiseuille's work to include a description of the steady, laminar flow development in the entrance regions of rigid tubes, while Kuchar and Ostrach (1966) dealt with the problem of flow development in elastic tubes. Langhaar solved the momentum equations, linearizing them by assuming that the acceleration of the fluid along the tube was a function of the axial distance z alone:

$$\frac{\partial v_r}{\partial r} + \frac{v_r}{r} + \frac{\partial v_z}{\partial z} = 0 \qquad \text{[from Eq. (B.4)]} \qquad (2.31)$$

$$v_{z0}\frac{\partial v_z}{\partial z} = -\frac{1}{\rho}\frac{\partial P}{\partial z} + \frac{\mu}{\rho}\left(\frac{\partial^2 v_z}{\partial r^2} + \frac{1}{r}\frac{\partial v_z}{\partial r}\right) \qquad \text{[From Eq. (B.20)]} \qquad (2.32)$$

At inlet $z = 0$: $\qquad\qquad\qquad\qquad v_z = v_{z0} \qquad P = P_0 \qquad (2.33)$

At tube wall $r = R$: $\qquad\qquad\qquad v_r = 0 \qquad v_z = 0 \qquad (2.34)$

At tube center $r = 0$: $\qquad\qquad\quad v_r = 0 \qquad \dfrac{\partial v_z}{\partial r} = 0 \qquad (2.35)$

Here v_{z0} and P_0 signify the axial flow velocity and fluid pressure, respectively, at the entrance, and R is the unstressed tube radius. The solutions are

$$\frac{P - P_0}{\rho v_{z0}^2} = -\frac{1}{3} - \frac{8z}{R\,\text{Re}} - 4\sum_{n=1}^{\infty}\frac{1}{\alpha_n^2}\exp\left(-\frac{\alpha_n^2 z}{R\,\text{Re}}\right) \qquad (2.36)$$

$$\frac{v_z}{v_{z0}} = 2 - 2\left(\frac{r}{R}\right)^2 - 4\sum_{n=1}^{\infty}\frac{1}{\alpha_n^2}\left[1 - \frac{J_0(\alpha_n r/R)}{J_0(\alpha_n)}\right]\exp\left(-\frac{\alpha_n^2 z}{R\,\text{Re}}\right) \qquad (2.37)$$

$$\frac{v_r}{v_{z0}} = \frac{4}{\text{Re}}\sum_{n=1}^{\infty}\frac{J_1(\alpha_n r/R) - (r/R_0)J_1(\alpha_n)}{\alpha_n J_0(\alpha_n)}\exp\left(-\frac{\alpha_n^2 z}{R\,\text{Re}}\right) \qquad (2.38)$$

where Re is defined as $\rho R v_{z0}/\mu$ and α_n are the roots of J_2, the Bessel function of the first kind and of the second order.

Figures 2.6 and 2.7, based on Eq. (2.37), portray the behavior of the axial

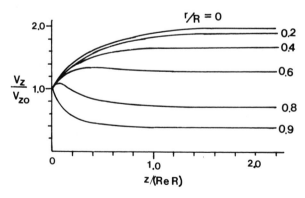

Figure 2.6 Longitudinal fluid velocity distributions in the entrance region for various values of r/R.

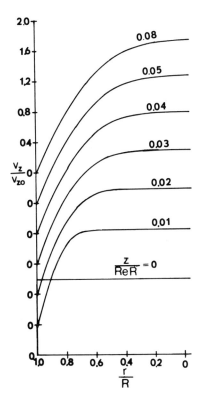

Figure 2.7 Longitudinal velocity distributions in the entrance region of the tube.

velocity component v_z as a function of the parameters $z/\text{Re}R$ and r/R. It is apparent that at small values of $z/\text{Re}R$, the profiles have a strong boundary-layer character consisting of a relatively thin region of retarded fluid near the wall and a relatively unsheared core of fluid at the center. The boundary layer grows rapidly so that at a $z/\text{Re}R$ of about 0.05, the layer has almost penetrated to the center. The core flow continually accelerates as it proceeds down the tube due to the influx of fluid from the retarded boundary-layer region. The process of flow development continues even after the boundary layer has penetrated to the tube center, with the velocity profile proceeding symetrically to the fully developed parabolic shape. At a distance Le from the tube entrance, the fully developed profile is essentially attained. Within the hydrodynamic entrance regime, the Poiseuille's flow does not exist and the pressure drop is higher than that occurs in the fully developed flow [Eq. (2.36)]. Langhaar shows that

$$\frac{Le}{D} = 0.115 \, \text{Re} \tag{2.39}$$

where Le is called the hydrodynamic entrance length required for the center-line velocity to reach 99% of its fully developed value, and $D = 2R$.

The behavior of the radial velocity component v_r [Eq. (2.38)] is depicted in Fig. 2.8. It is seen that v_r is always negative, indicating outflow from the retarded boundary-layer region. The location r/R of the maximum value of v_r shifts toward the tube center with increasing z/RRe. Eventually the magnitude of v_r decreases to zero as the flow becomes fully developed.

Equation (2.36) is plotted in Fig. 2.9. The pressure gradient near the entrance is larger than that downstream, owing to the acceleration of the core flow which results from the buildup of the boundary layer. Because of the coupling of the pressure with the elastic tube equations, a knowledge of the pressure distribution is much more important in the case of elastic tubes than in the rigid tube case.

Kucher and Ostrach (1966) obtained the analytic solution for the radial wall displacement which displays a "boundary layer" type behavior. The elastic entrance length Le', based on the standard 99% criterion, is about 0.01 ReR for $G = (\delta/R\text{Re}^2)^{1/2} = 0.003$, where δ is the tube wall thickness. It increases with an increase in G and becomes 0.02 ReR for $G = 0.006$. Kucher and Ostrach found that

$$Le = 2.67(\delta R)^{1/2},$$

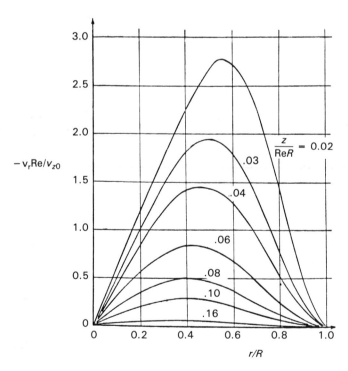

Figure 2.8 Radial fluid velocity component for various values of $z/$ReR.

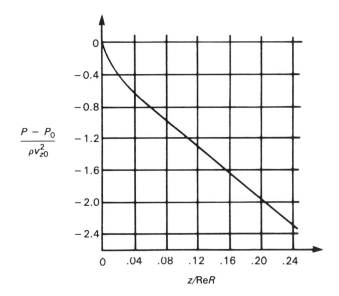

Figure 2.9 Pressure distribution in the entrance region of the tube.

which is much smaller than the hydrodynamic entrance length Le in Eq. (2.39). Within the elastic entrance region, bending stresses are high. Consequently, the usual membrane model of the tube is invalid and a model that includes bending effects must be employed. Outside of the length, the radial displacement component of the tube wall decreases slowly but almost linearly with z as the fluid pressure falls due to viscosity.

2.2.2 Pulsatile Flows

In general, the flow of blood in the cardiovascular system in the human or animal body is characterized as (1) being time dependent and (2) flowing through a distensible vessel. The latter is especially unique as compared to nonbiological flows through rigid channels that are encountered in all engineering practice. Under physiological conditions, blood may be treated as an incompressible fluid of density ρ. The assumption of cylindrical symmetry of the flow is generally valid. It leads to the use of cylindrical coordinates (r, z, t), where r is the radial coordinate; z, the axial distance; and t, time. Let the radial and axial velocity components be denoted by v_r and v_z, respectively, and the pressure by P.

Two theoretical models are treated in the following: (1) the simplest case of unsteady flow through a long, rigid pipe, and (2) a more realistic case of a distensible vessel.

2.2.2.1 Long, Rigid Pipes. Pulsatile flow in a long pipe under the influence of a periodic pressure gradient can be expressed as, from Eq. (1.20),

$$\frac{\partial v_z}{\partial t} = -\frac{1}{\rho}\frac{\partial P}{\partial z} + \nu \left(\frac{\partial^2 v_z}{\partial r^2} + \frac{1}{r}\frac{\partial v_z}{\partial r} \right) \tag{2.40}$$

It is subject to the boundary conditions

$$r = 0: \qquad \frac{\partial v_z}{\partial r} = 0$$

$$r = R: \qquad v_z = 0$$

The pressure gradient is harmonic and is given by

$$-\frac{1}{\rho}\frac{\partial P}{\partial z} = k \cos \omega t \tag{2.41}$$

where k denotes a constant, namely the pulsating amplitude, and ω is the pulsating frequency. At steady-periodic state, the velocity has the form

$$v_z = f(r) \cos \omega t$$

in which $f(r)$ depends on the Bessel functions of the first kind and of zero order. Expanding the Bessel function in a series and retaining only the quadratic terms, one obtains the following expression, which is valid for the case of very slow oscillations, that is, very small values of $\alpha = (\omega/\nu)^{1/2} R$:

$$v_z = \frac{KR^2}{4\nu} \left[1 - \left(\frac{r}{R} \right)^2 \right] \cos \omega t \tag{2.42}$$

The velocity distribution is in phase with the pulsatile pressure, while the amplitude is a parabolic function of the tube radius as was in the case of steady flow.

2.2.2.2 Distensible Vessels. The equations describing both the flow and the movements of the vessel walls must be solved simultaneously. The basic flow equations describe the behavior of the fluid and are thus given by the continuity equation Eq. (B.4)

$$\frac{\partial v_r}{\partial r} + \frac{v_r}{r} + \frac{\partial v_z}{\partial z} = 0$$

and the Navier-Stokes equations for a Newtonian fluid are, from Eqs. (B.18) and (B.20),

$$\frac{\partial v_r}{\partial t} + v_r \frac{\partial v_r}{\partial r} + v_z \frac{\partial v_r}{\partial z} = -\frac{1}{\rho}\frac{\partial P}{\partial r} + \frac{\mu}{\rho}\left(\frac{\partial^2 v_r}{\partial r^2} + \frac{1}{r}\frac{\partial v_r}{\partial r} + \frac{\partial^2 v_r}{\partial z^2} - \frac{v_r}{r^2} \right) \tag{2.43}$$

$$\frac{\partial v_z}{\partial t} + v_r \frac{\partial v_z}{\partial r} + v_z \frac{\partial v_z}{\partial z} = -\frac{1}{\rho}\frac{\partial P}{\partial z} + \frac{\mu}{\rho}\left(\frac{\partial^2 v_z}{\partial r^2} + \frac{1}{r}\frac{\partial v_z}{\partial r} + \frac{\partial^2 v_z}{\partial z^2} \right) \tag{2.44}$$

where μ represents the apparent or effective viscosity of blood. The assumption of a Newtonian fluid is valid for limited regions of the vascular system.

The differential equations are subject to the appropriate initial and boundary conditions describing the behavior of the vessel walls and their surrounding tissue and the condition at the fluid-vessel wall interface. It is most difficult to properly explain the behavior of the vessel walls. Depending on the purpose of the analysis and the extent of sophistication desired, such descriptions could range from the simplest model of a rigid tube of constant cross section (Womersley, 1955) to the most elaborate formulation for a thick-walled tube with nonisotropic and nonuniform viscoelastic properties subject to prescribed external boundary conditions. Womersley's model (1958) for a tethered elastic tube is most frequently cited in the literature as it is sophisticated enough and mathematically tractable. In this model, the blood vessel is considered a thin-walled, isotropic, elastic tube surrounded by tissue which adds to the inertia of the tube without contributing to the resistance to elastic deformations. It yields two linear equations describing the balance between the inertia, pressure, and elastic and viscous forces:

$$\rho_w \delta_e \frac{\partial^2 \zeta_r}{\partial t^2} - P + \frac{E_c \delta_e}{1 - \sigma_c^2} \left(\frac{\sigma_c}{R} \frac{\partial \zeta_z}{\partial z} + \frac{\xi_r}{R^2} \right) = 0 \qquad (2.45)$$

$$\rho_w \delta_e \frac{\partial^2 \zeta_z}{\partial t^2} + \rho_w \delta_e \Omega^2 \zeta_z + \mu \left(\frac{\partial v_r}{\partial r} + \frac{\partial v_z}{\partial r} \right) - \frac{E_c \delta}{1 - \sigma_c^2} \left(\frac{\partial^2 \zeta_z}{\partial z^2} + \frac{\sigma_c}{R} \frac{\partial \zeta_r}{\partial z} \right] = 0 \qquad (2.46)$$

Here, ρ_w denotes the vessel density; δ_e, the effective wall thickness; ζ_r and ζ_z, the radial and longitudinal displacements of the wall, respectively; E_c and σ_c, the complex elastic modules and Poisson's ratio, respectively; R, the vessel radius; Ω, the natural frequency of longitudinal elastic constraint of the wall; and δ, the wall thickness.

The equations for the fluid and those for the wall are coupled by the nonslip flow condition at the fluid-wall interface, $r = R + \zeta$:

$$v_r = \frac{\partial \zeta_r}{\partial t} \qquad v_z = \frac{\partial \zeta_z}{\partial t} \qquad (2.47)$$

Even for this reasonably simplified model, a general solution of the above equations is not available. Two major methods are in use that are based either on a linearization of the equations or on the method of characteristics. In the first case, the flow is considered as axisymmetric, while in the second, the flow is treated as one dimensional with the nonlinear terms of the equations retained.

Using the first method based on linearized equations, Womersley (1957) obtained the flow velocity averaged over the cross section of the duct $\bar{v}_z$ as

$$\bar{v}_z = \frac{A}{\rho_c} K \exp \left[i\omega \left(t - \frac{z}{c} \right) \right] \qquad (2.48)$$

in which A is the amplitude of pressure oscillations; c the pulse wave velocity; K the complex constant given in Womersley (1957); ω, the circular frequency of the pulse waves propagating along the blood vessel; and $i = (-1)^{1/2}$. Actually, reflected waves are also present and any pressure measurements record the superposition of advancing and receding waves. The solution for the receding wave is obtained simply by replacing c by $-c$. If A_1 and A_2 are the pressure amplitudes of the advancing and receding waves, the pressure can be expressed as

$$P = A_1 \exp\left[i\omega\left(t - \frac{z}{c}\right)\right] + A_2 \exp\left[i\omega\left(t + \frac{z}{c}\right)\right] \qquad (2.49)$$

Then, the corresponding superposition of the velocities yields

$$\bar{v}_z = \frac{A_1 K}{\rho c} \exp\left[i\omega\left(t - \frac{z}{c}\right)\right] - \frac{A_2 K}{\rho c} \exp\left[i\omega\left(t + \frac{z}{c}\right)\right] = \frac{K}{i\omega\rho}\left(-\frac{\partial P}{\partial z}\right) \qquad (2.50)$$

Therefore, if the pressure gradient $\partial P/\partial z$ can be measured at some point as a function of time, the corresponding instantaneous flow velocity $\bar{v}_z$ can be calculated. As a linear system, the wave form of the measured pressure gradient may be described by a number of Fourier series'. If A' is the amplitude of a given harmonic of the pressure gradient, then the amplitude of the corresponding harmonic of the velocity would be $A'K/i\omega\rho$.

Figure 2.10 is a plot of the ratio Q_{max}/Q_{steady} as a function of the parameter α for two values of Poisson's ratio $\sigma = 0$ and $\sigma = 0.5$. Q_{max} signifies the maximum flow rate during a simple harmonic cycle, while Q_{steady} is the Poiseuille flow that would result if the pressure gradient were constant at the maximum value during the cycle. α is defined as $R(\rho\omega/\mu)^{1/2}$. It can be seen that the maximum amplitude of the flow pulsations decreases with increasing fre-

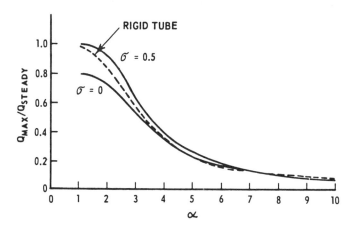

Figure 2.10 Comparison of nonsteady and steady flow rates for a rigid and two elastic tubes (Rudinger, 1966).

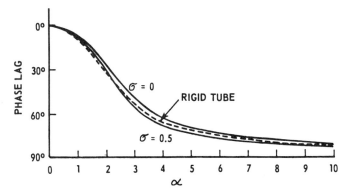

Figure 2.11 Phase lag of the flow rate behind the pressure gradient for a rigid and two elastic tubes (Rudinger, 1966).

quency. The results for a rigid tube lie between those for $\sigma = 0$ and $\sigma = 0.5$. Figure 2.11 shows the corresponding phase angles by which the flow rate lags behind the pressure gradient. One observes that the pulsations of the velocity are nearly in phase with those of the pressure gradient at very low frequencies which lag almost 90° at high frequencies. Since the above-mentioned Fourier series' are comprised of only periodic terms, it is necessary to superimpose the mean flow to obtain the total flow. Theory agrees fairly well with the experimentally determined velocity distribution.

The second technique (method of characteristics) is used to build up periodic solutions from arbitrary initial and prescribed boundary conditions by following individual wavelets in the position-time plane. The equation of motion for one-dimensional flow and the continuity equation in terms of the cross-sectional area of the duct are solved by the method of characteristics.

2.2.3 Pulsatile Blood Flow in the Aorta

Fich et al. (1966) analyzed pulsatile blood flow in the aorta based on the assumptions of (1) an exponentially tapered arterial section, (2) elastic, tapered wall, (3) validity of Poiseuille's law (nonuniform hydrodynamic resistance R), (4) nonlinearity of the elastic modulus with pressure, and (5) the inertance considered as a lumped parameter. Then, with the use of the definition of hydrodynamic capacitance, the flow problem is reduced to the analogous exponentially nonuniform resistance-capacitance electric transmission line with the lumped loading. Figure 2.12 is a schematic diagram of an arterial section

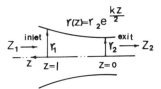

Figure 2.12 Exponentially tapered arterial section.

of length l, with the origin fixed at the exit and z measuring the distance in the direction opposite to the blood flow. The formulation leads to the governing equations

$$-\frac{\partial Q}{\partial z} = C\frac{\partial P}{\partial t} \qquad (2.51)$$

$$-\frac{\partial P}{\partial z} = L\frac{\partial Q}{\partial t} + RQ \qquad (2.52)$$

Here, Q denotes the blood flow rate at a point; t, time; C, capacitance; L, inertance or inductance; and R, resistance. It is defined that

$$C = \frac{A}{\beta_e} \qquad \text{per unit flow length}$$

$$\frac{1}{\beta_e} = \frac{1}{\beta} + \frac{2}{E[(D_o/D_i) - 1]}$$

$$L = \frac{\rho}{A_c} \qquad \text{per unit flow length} \qquad \left(\text{inertance} = \frac{\rho l}{A_c}\right)$$

$$R = \frac{8\pi\nu}{A_c^2} \qquad \text{if Poiseuille's law holds}$$

where A_c = cross-sectional area of the blood vessel
D_i, D_o = inside and outside diameters of the blood vessel, respectively
E = Young's modulus of the vessel wall ($= E_2/r$)
E_2 = E at the exit of the aorta
β = bulk modulus of elasticity of the blood
r = mean radius of the vessel [$= r_2 \exp(kz/2)$]
r_2 = mean radius at the exit of the aorta
$k/2$ = taper coefficient

Equations (2.51) and (2.52) are analogous to those for an electrical transmission line. If the distributed inertance is replaced by a lumped one, the distributed system would be without inertance and the above equations become

$$-\frac{\partial Q}{\partial z} = C\frac{\partial P}{\partial t}$$

$$-\frac{\partial P}{\partial z} = RQ$$

They can be solved using the Laplace transformation and matrix techniques to yield the pressure and flow transfer functions

$$\frac{P_2}{P_1} = \frac{\exp{(kl)}}{\cosh{\gamma l} + (1/\gamma)(k + R_2/Z_2)\sinh{\gamma l}}$$

$$\frac{Q_2}{Q_1} = \frac{\exp{(-kl)}}{\cosh{\gamma l} + (1/\gamma)[(\gamma^2 - k^2)(Z_2/R_2) - k]\sinh{\gamma l}}$$

in which Z_2 is called the exit impedance being equal to $R_2/(\gamma - k)$, $\gamma = (k^2 + sR_2C_2)$ and s is the Laplace variable. The other quantities of interest include the inlet and exit impedance defined as $Z_2 = P_2/Q_2$ and $Z_1 = P_1/Q_1$, respectively.

Numerical results were obtained by Fich et al. (1966) for the human aorta, with $l = 30$ cm, $A_1 = 4$ cm^2, $A_2 = 2$ cm^2, $\rho = 1.005$ g/cm^3, $\mu = 0.04$ Ns/m^2, $\beta_{el} = 16{,}000$ N/m^2, and heart rate of 75 beats per minute. Results are shown in Figs. 2.13 and 2.14.

Typical pressure histograms at the inlet ($z = 0$) and exit (at $z = 30$ cm) of the aorta are illustrated in Fig. 2.13. One observes the well-known peak rise, increase in dichrotic notch, and growth of the dichrotic wave (Rushmer, 1961). Figure 2.14 shows the inlet impedance calculated for a nonreflective exit. The results compare favorably with the test data (McDonald, 1964).

2.2.4 Similarity Analysis of Pulsatile Blood Flow

The flow in the circulatory system is, in general, pulsatile and nonuniform, the fluid is non-Newtonian, and the walls of the vessels are flexible. There is no analytic method to solve such a complex problem in its full generality. It has

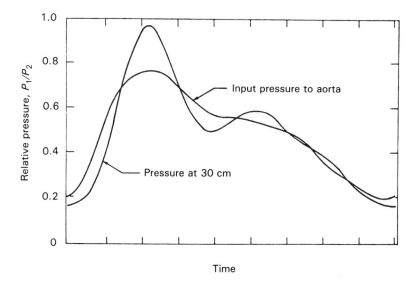

Figure 2.13 Calculated pressure wave form (Fich et al., 1966).

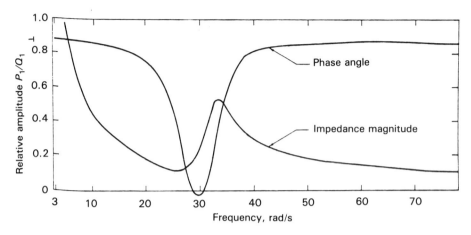

Figure 2.14 Calculated input impedance (Fich et al., 1966).

been necessary to make simplifying assumptions to obtain useful, theoretical solutions so that experiments must be conducted to check their validity and to develop criteria for determining the range of validity. Thus the need for experimental results is apparent.

The principle of similitude is widely used in the planning of experiments and in correlating test data. Using the method of dimensional analysis, Young and Cholvin (1966) obtained similarity criteria for the in vivo study of pulsatile blood flow in a segment of the circulatory system. Considering the simultaneous mean velocity $\bar{v}$ as the dependent variable, the governing physical quantities and associated primary dimensions are listed in Table 2.4. Application of the Pi theorem on the variables yields a set of dimensionless parameters listed in Table 2.5. Here, $\bar{v}_0$ denotes the mean velocity in a steady, uniform flow induced by a pressure gradient p_0', and c_0 is the propagation velocity of a pressure pulse of circular frequency ω in a thin-walled elastic tube containing a nonviscous incompressible fluid. They are defined as

$$\bar{v}_0 = \frac{p_0' R^2}{8\mu} \qquad c_0 = \left(\frac{E\delta}{2R\rho}\right)^{1/2}$$

Four velocities, $\bar{v}$, $\bar{v}_0$, c_0, and R, are involved in pulsatile flows in which ωR is the group velocity. One finds that the instantaneous mean velocity in dimensionless form $\bar{v}/\bar{v}_0$ is governed by two force parameters, three geometric parameters, three factors pertinent to the pressure gradient in the wave form, four wall properties, and two parameters accounting for the surrounding tissue characteristics. The second dimensionless parameter is obviously a Reynolds number, while the square root of the third dimensionless group is called the α parameter. That is,

$$\text{Re} = \frac{\rho \bar{v}_0 R}{\mu} \qquad \alpha = R\left(\frac{\rho\omega}{\mu}\right)^{1/2}$$

Table 2.4 Physical quantities and associated primary dimensions

Physical quantity	Symbol	Dimensions	Remarks
Mean velocity	$\bar{v}$	$L\theta^{-1}$	Instantaneous flow rate divided by cross-sectional area of vessel.
Vessel radius	R	L	Factors required to describe the basic geometry of the system
Characteristic length	l	L	such as radius of curvature, taper, location of measuring site
Other geometric factors	l_i	L	relative to major branches, etc., must be included. Wall thick-
Wall thickness	δ	L	ness will be significant if flexibility of wall is considered.
Steady pressure gradient	P'_0	$ML^{-2}\theta^{-2}$	Assume that pressure gradient is periodic and can be described
Modulus of pressure gradient	P'_n	$ML^{-2}\theta^{-2}$	in terms of Fourier series
Phase angle	ϕ_n	θ^{-1}	$\partial P/\partial x = P' = P'_0 + \sum_1^n P_n \cos(n\omega t - \phi_n).$
Fundamental frequency	ω	θ^{-1}	
Time	t	θ	
Fluid viscosity	μ_a	$M\theta^{-1}L^{-1}$	Assume that blood can be treated as a Newtonian fluid of con-
Fluid density	ρ	ML^{-3}	stant apparent viscosity and density.
Modulus of elasticity of wall	E	$ML^{-1}\theta^{-2}$	For a flexible vessel wall, material properties of wall must be
Poisson's ratio	σ		included. These are major unknowns. Only typically different
Wall damping factor	η	$M\theta^{-1}L^{-1}$	classes of properties are included.
Density of wall material	ρ_w	ML^{-3}	In a living system, the various blood vessels are not isolated but
Effective mass of surrounding tissue per unit length	m	ML^{-1}	are surrounded by other tissue. Assume that this effect can be
Natural frequency of effective mass	Ω	θ^{-1}	adequately accounted for by the addition of an added mass having a longitudinal stiffness $m\beta^2$.

57

Table 2.5 Dimensionless parameters governing pulsatile blood flow in the circulatory system

Dimensionless parameters	Group
$\dfrac{\bar{v}}{v_0}$	Velocity ratio
$\dfrac{\rho \bar{v}_0 R}{\mu}, \dfrac{\rho(R\omega)R}{\mu}$	Forces
$\dfrac{l}{R}, \dfrac{l_i}{R}, \dfrac{\delta}{R}$	Wall geometry
$\dfrac{P_n'}{P_0'}, \phi_n, \omega t$	Pressure gradient in wave form
$\dfrac{c_0}{R\omega}, \sigma, \dfrac{\eta}{\mu}, \dfrac{\rho_w}{\rho}$	Wall properties
$\dfrac{m}{\rho_w \delta R}, \dfrac{\Omega}{\omega}$	Surrounding tissue characteristics

A large amount of biological data was analyzed by Stahl (1963). He arrived at the conclusion that many physical and physiological parameters X_i of living systems can be related to the body mass m through the expression

$$X_i = a_i m^n$$

where a_i denotes the proportionality constant, and n is the exponent. X_i includes the heart rate (Clark, 1927), basal heart rate (Altman, 1959; Brody, 1945), aorta radius (Clark, 1927), and α parameter (Young and Cholvin, 1966) as illustrated in Fig. 2.15. For many geometries, the surface area is proportional to $m^{2/3}$ (Davson, 1970). Even the basal metabolic rate q_m''' as well as the specific metabolic rate q_m'''/m are found to be proportional to m to the $\frac{3}{4}$ and $-\frac{1}{4}$ powers, respectively (Davson, 1970).

It is known that the similarity parameters in Table 2.5 can also be derived through the analysis of the governing equations of the system or that of force ratios (Klein, 1965).

2.3 MICROCIRCULATION

From the fluid mechanics viewpoint, microcirculation may be defined as the circulation occurring in vessels where the ratio of erythrocyte to vessel size is so large as to demand consideration of blood as a nonhomogeneous two-phase fluid. Blood by definition is such a fluid. However, for the purposes of analysis, it is usually not essential to consider it nonhomogeneous for an understanding of the phenomena occurring in the larger vessels. Vessels ranging from, say, 50 erythrocyte diameters downward may be considered to be microcirculation. For human blood, this corresponds to vessel diameters ranging from 350 μm to the true capillaries, 7 μm or less in diameter. The larger vessels in this range continue to fulfill the transport function performed by the upstream arteries. Thus the rheological study of the resistant relationships and their underlying cause is of primary interest.

On the other hand, the smallest vessels are the sites of the exchanges between blood and surrounding tissue and also of the energy dissipation in the circulatory system. Therefore, they should be investigated both from the rheological viewpoint and from that of the characteristics of the heat and mass transfer processes. The transition from the larger to the smaller vessels occurs through repeated branchings whose performance governs the repartition of the erythrocyte concentration and its stability. The progression from the larger to the smaller vessels in vivo is one from a flow of essentially constant erythrocyte concentration to flow characterized by large changes in concentration both in space and in time.

The capillaries are the smallest blood vessels in the human body, with an inner diameter of about 8 μm, approximately the same as the diameter of red blood cells. Being lined with a layer of endothelial cells, they are closely integrated with the surrounding tissues. The length of the capillary vessels ranges from 0.01 to 0.1 cm, in which the blood flows at a speed of about 1–2 mm/s. Despite such a short length, each capillary provides a large resistance to flow because of its very small diameter. It takes up approximately 27% of the total resistance to blood flow in the entire circulatory system. In other words, one-quarter of the blood pressure pumped out from the heart, amounting to about 120 mmHg, is consumed in forcing the blood through that short distance. In contrast, the resistance to flow in larger blood vessels of the same length is much lower.

The capillaries branch off from arterioles of roughly 0.002 cm diameter and 0.2 cm length and empty into venules of about 0.003 cm diameter and 0.2 cm length. A substantial fraction, about 41%, of the available pressure gradient is used up in the arterioles, while only a small fraction, about 4%, is consumed in the venules. At the junction of the capillary with the arteriole is a smooth muscle called the precapillary sphincter whose function is presumed to control the blood flow into the capillaries by opening or closing the entrance port. Only a fraction of an enormous number of capillaries, estimated to be 1.2×10^9, are opened at a time.

The main interests in microcirculation include: (1) pressure-flow relationship; (2) mechanism of flow regulation; (3) movement of red blood cells in the capillaires; (4) stress distribution in red blood cells and the capillaries; (5) exchange of water and solute molecules across the capillary walls; (6) regulation of temperature; (7) specific features of microcirculation in different organs; and (8) aging, pathology, etc.

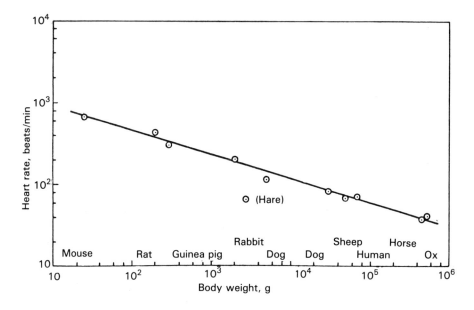

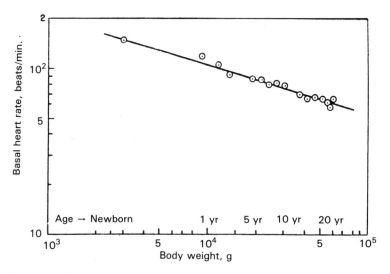

Figure 2.15 Dependence of heart rate, α parameter, and aorta radius on body weight (Young et al., 1966).

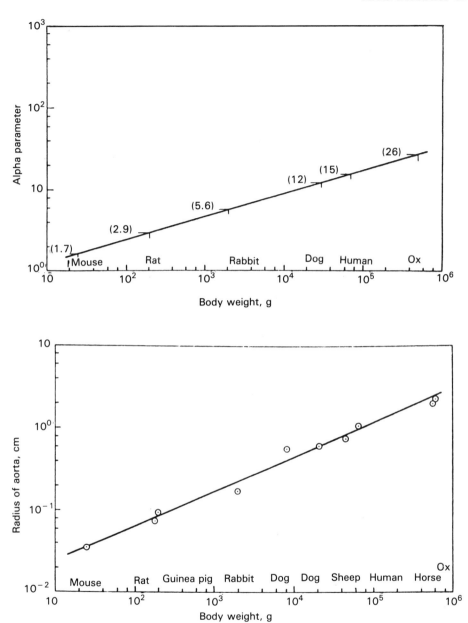

Figure 2.15 (cont.)

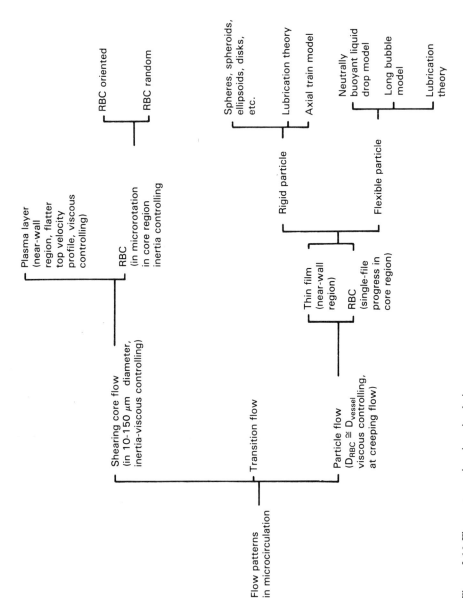

Figure 2.16 Flow patterns in microcirculation.

In microcirculation the flow is so slow that the viscous effect of the blood becomes extremely important. On the other hand, the virtual disappearance of the pulsatory characteristics greatly reduces the importance of the viscoelastic properties of the vessel walls. As the diameter of the vessel diminishes, the particulate nature of blood plays an increasingly important role in determining its flow properties, and eventually blood can no longer be treated as a fluid because the red cells are restricted in the orientations and positions adjacent to the vessel wall. The blood flow patterns in microcirculation can be classified into three kinds: "shearing-core" flow, transition flow, and "particle" flow (Fig. 2.16).

2.3.1 Shearing-Core Flows

The microvelocity field is the detailed velocity in the vicinity of a single particle, while macrovelocity is averaged over a volume that contains many particles. Thus, it is reasonable to define macrocontinuum field variables in terms of averages over microcontinuum variables. One such application is the so-called micropolar fluid theories, which are continuum theories incorporating the local macroscopic fluid rate of rotation into the mean speed of rotation of the particles. The use of micropolar theories for blood flow (Ariman, 1969) revealed that the mean velocity profile for a polar fluid in a tube flow is flatter than the parabolic distribution for a Newtonian fluid. The prediction was confirmed by experimental data on blood flow in tubes 30–130 μm in diameter (Gaehtgens et al., 1970). The measurements by Sevilla-Larrea (1968) indicated that the rate of rotation of the red cells flowing in glass tubes 40–70 μm in diameter is generally less than that of the blood as a whole. That means that the red cell rotation is impeded by neighboring cells and, thus, a relative microrotation exists. The observations of a flatter-top velocity profile and a relative microrotation support the "shearing-core" flow model for microcirculation in larger arterioles and venules of 10–500 μm in diameter.

Velocity measurements (Gaehtgens et al., 1970) in the mesenteric microvessels of a cat show pulsations in both arterioles and venules, while pressure pulsation was detected only in arterioles (Intaglietta et al., 1970). The compliance of the vessels in microcirculation is small. It can be neglected when considering the details of the flow (Fung, 1966) but must remain nonzero with regard to the pulsatile part of blood flow. Inertia can be neglected and the flow network may be regarded as having capacitance and resistance in analogical simulation.

In larger arterioles and venules where the vessel diameter exceeds 8 or 10 cell diameters, the red cells arrrange themselves in the main core of the vessels with the leukocytes and platelets accompanying the peripheral cells. A plasma layer of between one-half and one cell diameter exists at the wall. Hence, both the inertia effect in the core and the viscous effect in the plasma layer play important roles in the flow. Relative movement between the red cells in the core appears in two patterns: oriented and random. In the oriented shearing-core flow, the red cells arrange themselves in roughly concentric laminae with their discoidal surfaces oriented predominately parallel to planes passing through the vessel axis. In the second pattern, there is a random orientation of the red cells and every small portion of the blood mass oscillates across the vessel in an irregular manner, the cells bending, rotating, tumbling, and exchanging places in a chaotic fashion. They make frequent contact with the walls, and the plasma spaces frequently extend transversely across one-third or more of the vessel cross section.

2.3.2 Particle Flows

In the narrowest arterioles and capillaries, where the vessel diameter approximates the diameter of the red cells, only single-file progress is possible (Fig. 2.17). In a capillary of about 7 μm diameter, most of the cells show the

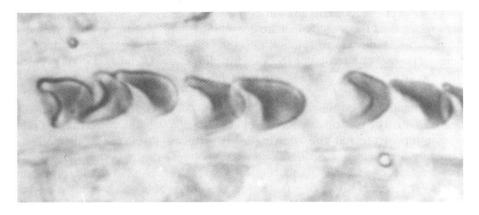

(a)

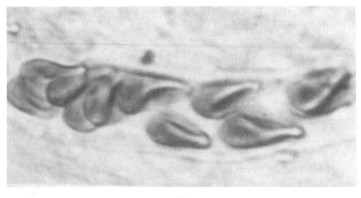

(b)

Figure 2.17 Shapes of human red blood cells in vivo in a capillary (Skalak and Branemark, 1969).

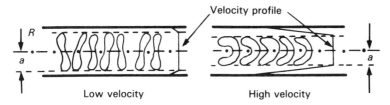

Figure 2.18 The axial-train model.

parachute-like shape, but the cell on the left takes a tail-flap appearance. The flattening of red blood cells at the rear is observed in a vessel of about 12 μm diameter. The longest axis may be oriented in the direction of flow, particularly if the cells take an ellipsoidal shape. When their shape is biconcave discoid, the cells move axially with discoidal surfaces perpendicular to the direction of flow either individually or in short trains (Fig. 2.18). As they move in the capillaries, the cells may deform into a great variety of shapes, such as tear drop (Monro, 1964), parabolic (Palmer, 1959; Guest et al., 1963), bullet (Stalker et al., 1966), and parachute (Branemark, 1971). With the exception of the smallest vessels, the flexibility of the red cells leads to the creation of a thin film of plasma at the wall (Branemark, 1965). The flow pattern is called particle flow, plug flow, or axial-train flow, since each cell practically fills the cross section of the vessel along which it is in motion.

The various theories and models that have been developed to study the capillary blood flow can be classified into two categories: rigid particles and flexible particles. Since the Reynolds number is of the order of 0.01 or less, creeping (or Stokes) flow equations can be used. The red cells are treated as neutrally buoyant, since the effect of gravity on capillary blood flow can be neglected (with the specific gravity of red cells at about 1.10 and that of plasma at about 1.03). For a neutrally buoyant particle in Stokes flow, the resultant force and moment on the particle due to pressures and viscous stresses exerted by the fluid must be zero.

2.3.2.1 Rigid Particles. Most theoretical studies have treated red cells as spheres, spheroids, ellipsoids, or disks moving axisymmetrically in a cylindrical tube. For each assumed shape, the velocity of the cell U and pressure drop ΔP can be determined. Solutions have been obtained for a line of spherical particles by Wang and Skalak (1969) and for a line of spheroidal particles by Chen and Skalak (1970) (Figs. 2.19–2.21). In Figs. 2.18–2.21 a and b denote the major and minor axes of spheroids, respectively; R, vessel radius; s, particle spacing between the centers of two adjacent cells; $\bar{v}$, mean velocity of the suspension including both the liquid and solid phases; $\lambda = b/R$; $\beta = s/R$; and G and K, the additional pressure drop coefficients. The total pressure drop ΔP over length l is expressed as

$$\Delta P = \frac{8\nu\bar{v}l(1 + G)}{R^2}$$

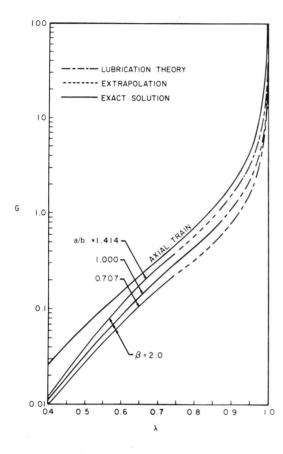

Figure 2.19 Pressure gradient coefficient G versus λ ($= b/R$) (Wang and Skalak, 1969).

A special case for $G = 0$ results in Poiseuille's law. The factor $(1 + G)$ is the relative apparent viscosity of the suspension compared to the suspending fluid. G is a strong function of b/R as illustrated in Fig. 2.19. For particle spacing exceeding $2R$, the additional pressure drop coefficient G varies approximately as R/s. Hence, the additional pressure drop per particle ΔP_a is practically constant and is given by

$$\Delta P_a = \frac{2v\bar{v}K}{R} \tag{2.53}$$

where $K = 4sG/R$. The additional pressure drop coefficient K is plotted versus s/R in Fig. 2.20. The left end of each curve corresponds to the case of particles touching, like a string of beads. Little variation of K with s/R suggests that results for a single particle in an infinite vessel can be used to estimate results for a line of particles by multiplying the pressure drop for one particle by the number of particles in a given length l, if s exceeds $2R$. Figure 2.21 is a plot of $U/\bar{v}$ versus b/R for spheres at spacing $s = 2R$. The curve for spheroids deviates from that shown for spheres by less than 2%. The curves in Figs. 2.19 and 2.21 consist of exact series solutions for b/R less than 0.9 and approximate results based on lubrication theory for large b/R.

Based on the lubrication theory, Hochmuth and Sutera (1970) derived an asymptotic equation for close-fitting spheres ($b = R$) as

$$K = 2\pi \left(\frac{2}{1 - b/R}\right)^{1/2} - 15.75$$

Whitmore (1967) developed the so-called axial-train of stacked coins model which consists of a cylinder of radius a (or a_1) made up of coins or disks (Fig. 2.18). Here, a denotes the static radius of a red blood cell, while a_1 is the effective radius of the red blood cell. The suspending fluid occupies the annular space between a and R with the velocity profile being independent of axial

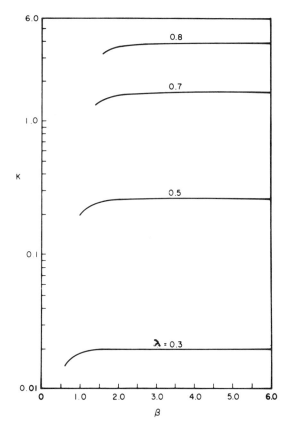

Figure 2.20 Pressure drop (per particle) coefficient K versus β ($= s/R$) (Wang and Skalak, 1969).

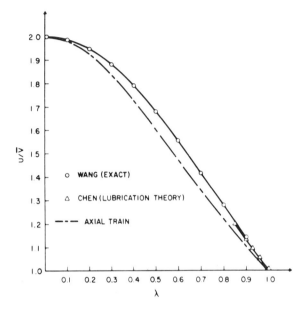

Figure 2.21 $U/\bar{v}$ versus λ ($= b/R$) for β ($= s/R$) $= 2$.

distance. He obtained

$$G = \frac{(b/R)^4}{1 - (b/R)^4}$$

which is superimposed in Fig. 2.19 forming an upper limit for any train of particles. The $U/\bar{v}$ ratio is found to be

$$\frac{U}{\bar{v}} = \frac{2}{1 + (b/R)^2}$$

which is plotted in Fig. 2.21.

The model for a line of axisymmetric disks (Bloor, 1968) yields the flow in the gap which is the same as the axial-train solution (Whitmore, 1967). The train of cells and interspersed fluid travel as a single cylindrical unit possessing an infinite viscosity. This model makes no allowance for the possibility of secondary or bolus flow developing in the gaps between the cells. However, the viscous drag of the fluid in the annulus on the fluid in the space between the red cells leads to the development of the bolus flow in the axial-train model. The blood cells proceed in bolus flow, either isolated or in groups, separated by gaps containing only the plasma (Fig. 2.22). To reach a detailed quantitative understanding of the phenomenon, a two-dimensional model (Fig. 2.23) is developed to determine the characteristics of the flow field in the plasmatic gaps by integrating the Navier-Stokes equations in the Stokes flow range. A parabolic velocity profile is assumed to have been developed between the solid plugs. Figure 2.24 shows an example of the configuration of the absolute and relative

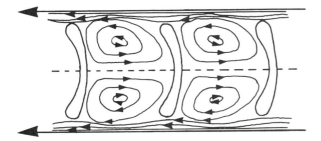

Figure 2.22 Bolus flow (streamlines shown relative to the red cells).

streamlines for half of the channel width, obtained by integrating the Navier-Stokes equations in the Stokes flow range. Here, L and D denote the spacing between two consecutive erythrocytes and capillary diameter, respectively; ψ_A and ψ are the absolute and relative stream functions.

In the bolus flow between two disks with $b = R$ (Bugliarello and Hsiao, 1967; Lew and Fung, 1969; Aroesty and Gross, 1970), the velocity distribution changes from a uniform profile required on the disk face to within 1% of a Poiseuille flow in a distance of 1.3 tube radii from the disk. This coincides with the lower limit of the entrance length as the Reynolds number approaches zero. The additional pressure drop for one disk face is

$$\Delta P_a = \frac{16\nu\overline{V}}{3R}$$

The total pressure drop over Poiseuille flow between the particles (disks) decreases when they are spaced closely. Aroesty and Gross (1970) calculated the toroidal vortex motion in the plasma between two particles in order to determine the effect of this circulation on mass transfer across the capillary wall and through the plasma to the red cells by a combination of diffusion and convection. The circulation was found to enhance mass transfer to the downstream cells but reduces it for the upstream cell. Consequently, the net effect is small.

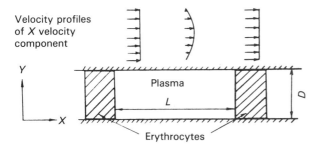

Velocity profiles of X velocity component

Y

X

Plasma

L

D

Erythrocytes

Figure 2.23 Idealized model of a capillary.

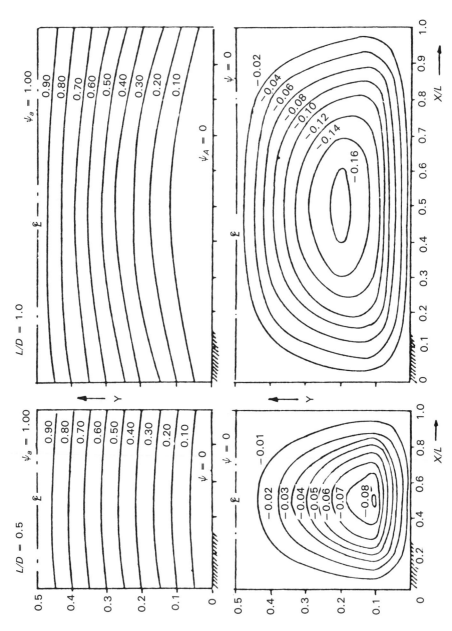

Figure 2.24 Absolute and relative stream functions ($\psi_A + \psi$).

Test results confirmed the prediction of various theories that particles spaced more than one or two tube diameters apart do not interact and that closely spaced disks suffer less additional pressure drop than those spaced apart. The predictions of $U/\bar{v}$ versus b/R in Fig. 2.21 are also confirmed by the data which indicate a smaller dependence of $U/\bar{v}$ on the shape of the particle. Spherical caps (planoconvex disks) with $b/R = 0.732–0.9982$ were found to be self-centering and stable in the transverse orientation (with the plane of the disk perpendicular to the axis of the tube) with the flat side facing upstream. Biconcave disks resembling red cells with $b/R = 0.8$ were always in the edge-on orientation (with the plane of the disk including the tube axis). Disks with larger values of b/R appeared in stable transverse orientations. Hemispheres were less stable and tended to rotate slowly. A symmetric shape such as a sphere may be stable in translation when centered on the tube axis. A body with a predominant convex surface facing downstream is stable.

2.3.2.2 Flexible Particles. Red cells with diameters larger than capillaries must be deformed in translation. Even cells smaller than the blood vessels may change shape during flow. The extent of deformation depends on the magnitude of the viscous stress and hence on the velocity and pressure gradient. Since the additional pressure drop is a function of shape which is velocity dependent, the apparent viscosity of a suspension of deformable particles is a function of mean velocity and pressure gradient as well as tube diameter.

Using the lubrication theory, the models of Lighthill (1968) and Fitzgerald (1969a,b) yield a velocity-dependent resistance and predict the shape of flexible particles. As shown in Fig. 2.25, the gap width between the red cells and the capillary wall narrows gradually from the leading end of the cell and tapers to a minimum near its tail end. The necking of the gap at the rear end of the particle would produce a fishtail shape at its back end; this has been observed on red blood cells in narrow capillaries by Branemark and Lindstrom (1963). The distance between the deformed and reference shape in Fig. 2.25 is a measure of the pressure distribution which is characterized by a local minimum

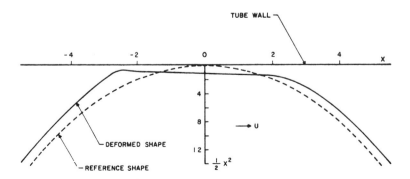

Figure 2.25 Particle shape in the vicinity of the tube wall (Lighthill, 1968).

near the rear end of the cell at the point of maximum necking. Other studies using the neutrally buoyant liquid drop model (Hyman and Skalak, 1970) and the long bubble model (Goldsmith and Mason, 1962) also predict the same qualitative features such as concavity of the rear end and a gap width increasing with velocity.

Observations of red cells in flow through glass capillaries confirmed that red cells in tubes below 9 μm in diameter were usually in the edge-on orientation (Hochmuth et al., 1970). As the velocity increases, the rear of the cell develops the typical concavity and the cells become almost axisymmetric. The shapes of thin rubber models filled with oil and red blood cells in glass capillaries were found to be quite similar (Sheshadri et al., 1970).

2.3.3 Transition Flow

In capillary blood vessels, the Reynolds numbers based on either the cell diameter or the vessel diameter are small, 0.01 or less, and thus creeping flow prevails. For the larger vessels in microcirculation, the Reynolds number based on the individual red cell may still be small but the Reynolds number based on the vessel diameter can become sufficiently large to make the effects of inertia appreciable. The transition from wave propagation in larger arterioles and ven- ules (or shearing-core flow) to the creeping flow in capillaries (or particle flow) takes place in the microvessels of several cell diameters. There the passage of one individual cell past another can be achieved only by deformation of both cells. Under these conditions, an important characteristic of the flow is the variety of shapes red cells can assume. They may elongate or flex, sometimes symmetrically, sometimes not, and rotate about an axis which forms an angle with the axis of elongation of flection (Branemark, 1965).

2.3.4 Flow-Through Functions

The vascular system is typified by repeated branchings (diverging junctions or bifurcations) and joinings (converging junctions) of vascular vessels. At di- verging junctions in the microcirculation, unequal repartition of red cell con- centrations occurs. There are two reasons: timewise variation of erythrocyte concentrations in the feeding vessel and plasma skimming. When a minor vessel draws blood from a larger vessel having a layer of plasma at the wall, the side vessel tends to draw a preponderance of plasma. As a result, the fluid in the side vessel is practically devoid of cells in some cases. However, flow at a junction does not seriously disturb the preferred orientation of the red cells.

Bugliarello and Hsiao (1964) conducted experiments on macrobifurcations of rigid plastic tubes with a fluid consisting of rigid spherical particles made neutrally buoyant in a suspension of glycerine and water. Figure 2.26 plots the ratio of the concentration C_2 in the side branch to the concentration C_1 in the continuation of the main branch versus the upstream concentration C_0 for different ratios Q_2/Q_1 of the flows in the two branches. The ratio D_2/D_1 of the

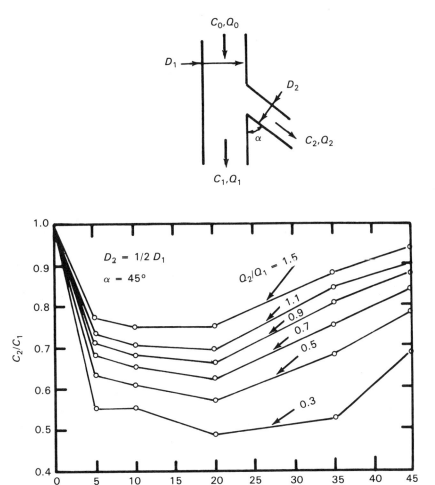

Figure 2.26 Concentration repartition in a model of a bifurcation (Bugliarello and Hsiao, 1964).

side to the main branch diameter is $\frac{1}{2}$. C_2 is considerably less than C_1 even when Q_2 is greater than Q_1. The phenomenon becomes more pronounced as D_2 increases, and less pronounced as C_2 increases or the upstream flow becomes turbulent. A low value of C_2 is due to the deflection into the branch of layers of lower momentum and concentration when the velocity and concentration profiles upstream of the bifurcation are nonuniform.

The flow behavior at converging junctions depends, among other factors, on the relative sizes of the vessels. When two capillaries converge, the cells suffer considerable distortion as the two streams merge to form a single flow. The flow rates in the capillary beds show a rhythmic intermittency due to vasomotion. Flow pulsates at a lower rate than the heart beat. In the case of venules of similar size, two streams merge so smoothly that the plasma layers come together to form a region of reduced concentration in the center of the collecting vessel. This region spreads downstream for many vessel diameters (Stehbens, 1967). When a small branch joins a main vessel, the side stream often maintains its integrity at the edge of the mainstream.

2.4 MODELING OF FLOW REGULATION

2.4.1 Compartmental Systems

Transport phenomena for any substance or energy in physiological systems at a macroscopic level can be simulated by lumped-parameter or compartmental analysis. A compartment refers to an organ or a tissue in any body region that can be distinguished from others with regard to the transport phenomena of substance. Physiological parameters contained in the mathematical expressions are determined by simulation studies using analog computers, and the results are compared with actual measurements.

Dyes or compounds labeled with radioisotopes are used as tracers for the compartmental analysis of transport process in the circulatory system. Consider a compartment with a fixed volume V and a single inlet and outlet, as shown in Fig. 2.27a. Let the inlet and outlet flow rates of blood be the same at Q. It is commonly postulated that the substance C_i entering the compartment is instantaneously mixed well and the substances leaving or accumulated in the compartment are identical in quantity to the substance $C(t)$. The balance of tracer in the compartment, known as the Fick principle, requires that

$$V \frac{dC}{dt} = Q(C_i - C)$$

which is subject to the initial condition of $C(0) = 0$.

If the inlet concentration changes stepwise, with time from 0 to C_i, the solution reads

$$C(t) = C_i(1 - e^{-t/\tau})$$

where $\tau = V/Q$ is called the time constant of the compartment.

There are many cases in the circulatory system where blood circulation must be simulated by a model consisting of a set of subcompartments in parallel and/or series. Figure 2.27b and c show two compartments arranged in parallel and in series, respectively. Let the two compartments have volumes V_1 and V_2.

For two parallel compartments, let blood flow rates through V_1 and V_2 be Q_1 and Q_2, respectively. The application of Fick's principle yields

$$V_1 \frac{dC_1}{dt} = Q_1(C_i - C_1) \qquad V_2 \frac{dC_2}{dt} = Q_2(C_i - C_2)$$

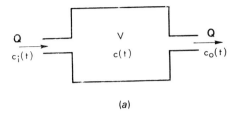

(a)

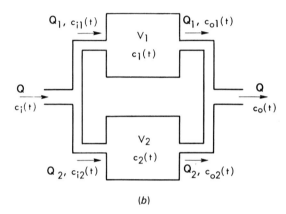

(b)

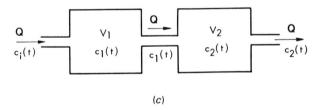

(c)

Figure 2.27 Compartmental systems. (a) Single, (b) two parallel, and (c) two series compartments.

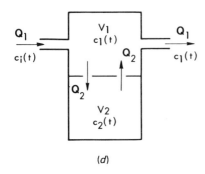

(d)

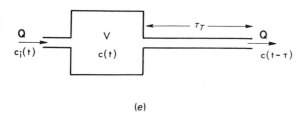

(e)

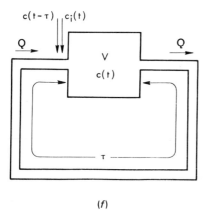

(f)

Figure 2.27 (Cont.) (d) Two interacting compartments; (e) single compartment and a transport delay.

in which C_1 and C_2 correspond to the outlets of the compartments V_1 and V_2. The mass balance equation is

$$QC = Q_1 C_1 + Q_2 C_2$$

where $Q = Q_1 + Q_2$ and C denote the outlet concentration of the two compartments combined. C is useful for clinical studies since the combined concentration in the vein is monitored rather than the regional concentrations.

When two subcompartments are connected in series, one obtains

$$V_1 \frac{dC_1}{dt} = Q(C_i - C_1) \qquad V_2 \frac{dC_2}{dt} = Q(C_1 - C_2)$$

The equations can be rewritten as

$$\tau_1 \frac{dC_1}{dt} + C_1 = C_i$$

$$\tau_1 \tau_2 \frac{d^2 C_2}{dt} + (\tau_1 + \tau_2) \frac{dC_2}{dt} + C_2 = C_i$$

where $\tau_1 = V_1/Q$ and $\tau_2 = V_2/Q$ are the time constants of the two compartments.

Figure 2.27d illustrates a system consisting of two compartments with volumes V_1 and V_2 that are interacting with each other. The injected substance will be transplanted from one compartment to the other. Blood flow rates (Q values) and concentrations of substance (C values) have subscripts 1 and 2 for the compartments. C_i is the inlet concentration of compartment 1. The governing equations for this system are

$$V_1 \frac{dC_1}{dt} = Q_1(C_i - C_1) + Q_2(C_2 - C_1)$$

$$V_2 \frac{dC_2}{dt} = Q_2(C_1 - C_2)$$

They can be rewritten as

$$\tau_1 \tau_2 \frac{d^2 C_1}{dt^2} + \left(\tau_1 + \tau_2 + \frac{Q_2}{Q_1} \tau_2 \right) \frac{dC_1}{dt} + C_1 = \tau_2 \frac{dC_i}{dt} + C_i$$

$$\tau_1 \tau_2 \frac{d^2 C_2}{dt^2} + \left(\tau_1 + \tau_2 + \frac{Q_2}{Q_1} \tau_2 \right) \frac{dC_2}{dt} + C_2 = C_i$$

where $\tau_1 = V_1/Q_1$
$\tau_2 = V_2/Q_2$

Comparing the last equation with that for two series compartments, it is obvious that the term

$$\frac{Q_2}{Q_1} \tau_2 \frac{dC_2}{dt}$$

for compartment 2 represents the interaction effect between the two compartments.

Finally, consider a system of a single compartment of volume V and a transport lag τ_T shown in Fig. 2.27e. This model of the transport process is applicable to the case of a complex vascular bed for obtaining its overall performance. It represents an approximation of the two series compartments when V and τ_T are related to C_2, V_1, and V_2 of the two series compartments as follows:

$$C(t - \tau_T) = C_2 \qquad V + Q\tau_T = V_1 + V_2$$

where V/Q represents the time constant of the compartment τ (Fig. 2.27e).

2.4.2 Determination of Cardiac Output in Human Beings

2.4.2.1 Dye Dilution Method. The standard technique for determining cardiac output is the indicator dilution method developed by Hamilton and his co-workers (Hamilton, 1962). One end of a small polyethylene catheter tube is inserted into the radial artery and the other end is connected to a cuvette densitometer, which can automatically record the concentration of the dye in the blood. Another catheter is inserted into the pulmonary artery through an antecubital vein, the right atrium, and the right ventricle. A known amount of a dye, such as Coomassie blue or indocyanine green, is injected rapidly into the pulmonary artery through the catheter. Immediately following the injection, the arterial blood, withdrawn at a constant rate through the polyethylene tube, is fed to the cuvette densitometer. A typical dye dilution curve is shown in Fig. 2.28a. The injected dye appears in the radial artery after a delay of a few seconds. The concentration reaches a maximum at about 15 s and then falls off, followed by a rise due to the recirculation of the dye. The curve is replotted on semilog paper as shown in Fig. 2.28b in which the descending limb of the dilution curve is approximated by a straight line. The cardiac output can be calculated from the replotted curve as

$$\text{Cardiac output} = \frac{I \times 60}{\displaystyle\int_0^\infty C(t)\,dt} = \frac{I \times 60}{A} \qquad \text{(L/min)}$$

where I denotes the total amount of the injected dye in milligrams; $C(t)$, the concentration of the dye in milligrams per liter; and A, the area under the C-t curve in mg s/L.

2.4.2.2 Radiocardiogram (RCG). The procedure of radiocardiography consists of (1) the injection of a nondiffusible γ-emitting radioisotope such as ^{131}I-labeled human serum albumin (RIHSA) into a peripheral vein, (2) the recording of the time-radioactivity curve by a scintillation counter placed over the precordial region, and (3) the calibration of the concentration of radioisotope in the heart

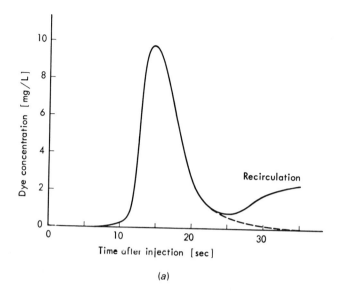

(a)

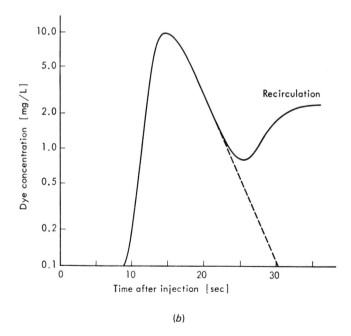

(b)

Figure 2.28 Dye dilution curves. (a) Single injection dye dilution model; (b) semilog plot of the dye dilution curve.

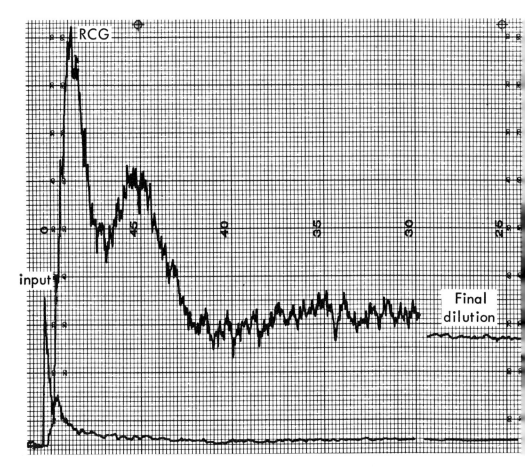

Figure 2.29 Typical radiocardiogram.

after its complete mixing in the circulatory system, by sampling the venous blood. A typical double-peaked RCG is shown in Fig. 2.29. The first peak is caused by the passage of the radioisotope through the right side of the heart and the second one is due to the radioisotope appearing at the left side of the heart in normal circulation.

The cardiac output is $(C_f/\text{CBV}/60)/A$, expressed in liters per minute. C_f denotes the final concentration recorded by the scintillation counter in counts per minute (cpm); CBV is the circulating blood volume in liters; and A is the area under the circulation curve in cpm.

It is important to note that if the detector is placed in a narrow limited field such as the aortic arch, the RCG becomes a one-peaked curve as in the case of the dye dilution curve. However, if the detector views the entire heart region through a wide collimation, the RCG consists of a complex of several dilution curves such as those of the right side of the heart, the pulmonary circulation, the left side of the heart, and others.

2.4.3 Circulatory System

The whole circulatory system can be simulated by a closed loop of four compartments consisting of the right side of the heart, the lungs, the left side of the heart, and the body, as illustrated in Fig. 2.30. The right and left sides of the heart are represented by a single mixing chamber including the atrium and ventricle, while the lungs and body are represented by a mixing chamber with a transport delay. The subscripts r, p, l, and b in the figure denote the right side, lungs, left side, and body, respectively. V (in milliliters) is the equivalent volume of each mixing chamber and $C(t)$ (in μCi/ml) is the concentration of isotope in each compartment in time t. Transport delays in the lungs and body are representd by τ_{Tp} and τ_{Tb} (in seconds), respectively. An RCG can be considered as a time-radioactivity curve of radioisotope in the right and left sides of the heart.

A radioisotope is injected into the peripheral vein. Therefore, the injection and transport processes of the radioisotope from the injection site to the right side of the heart must be taken into account. Figure 2.31 is a block diagram simulating the transport processes of the injected radioisotope in the circulatory system of a normal subject. Figure 2.32 shows the RCG on which the simulated curve is superimposed. Similarly, Figs. 2.33–2.35 are the RCGs and simulation results of a patient with mitral stenosis and congestive heart failure, of a patient with atrial septal defect, and of a patient with aortic insufficiency, respectively. Simulated data agree very well with catheterization data in Figs. 2.32 through 2.35.

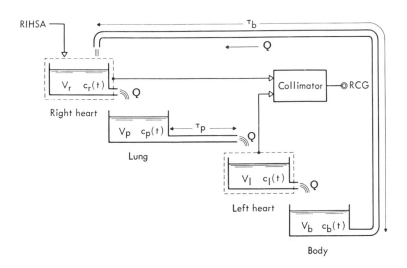

Figure 2.30 Simplified model of the circulatory system.

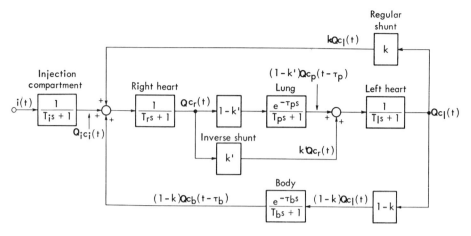

Figure 2.31 Transport process of injected radioisotope in the circulatory system taking into account the intracardiac shunts.

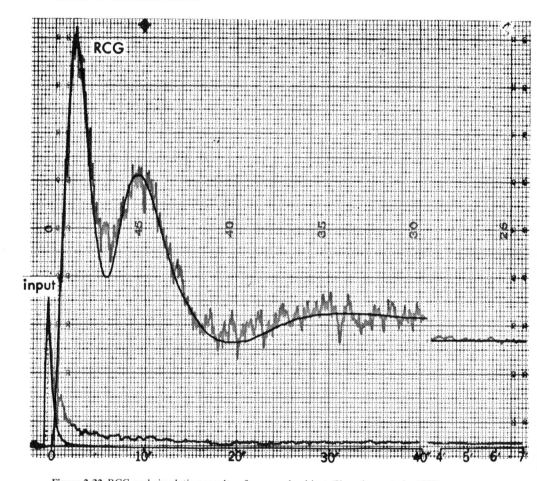

Figure 2.32 RCG and simulation results of a normal subject (Kuwahara et al., 1979).

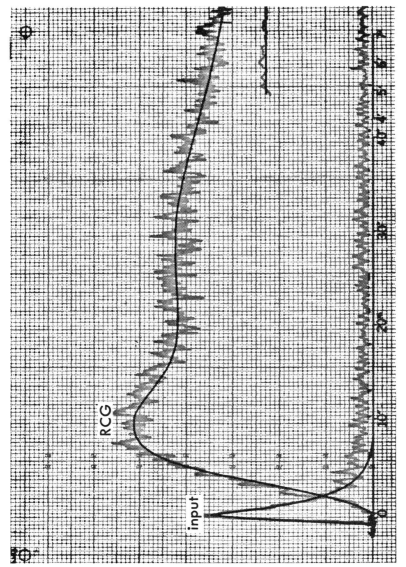

Figure 2.33 RCG and simulation of a patient with mitral stenosis and congestive heart failure (Kuwahara et al., 1979).

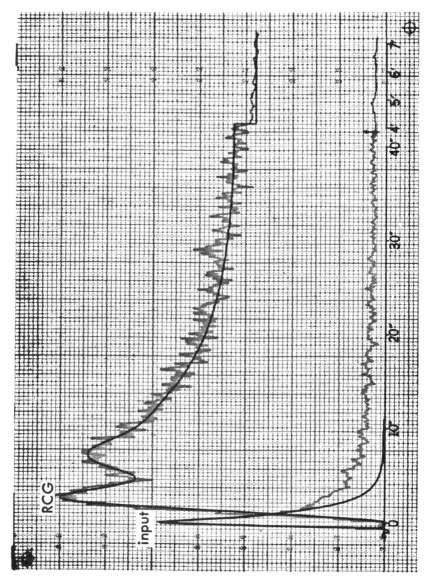

Figure 2.34 RCG and simulation results of a patient with atrial septal defect (Kuwahara et al., 1979).

84

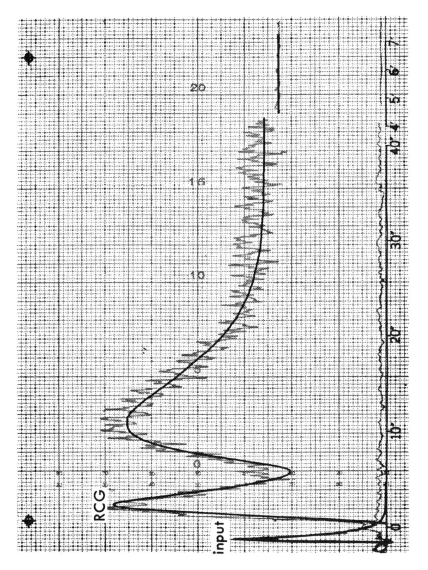

Figure 2.35 RCG and simulation results of a patient with aortic insufficiency (Kuwahara et al., 1979).

Table 2.6 Normal values of the circulatory parameters obtained by analog simulation of radiocardiograms (mean ± SD)[1]

Parameter	Right heart	Left heart	Heart in total	Lungs	Body
Mean transit time (s)	2.26 ± 0.50	2.25 ± 0.48	4.52 ± 0.90	4.53 ± 0.68	1032.8 ± 5.0
Mean blood volume (ml/m²)	135 ± 24	134 ± 24	268 ± 43	272 ± 45	1.955 ± 199
Percent to circulating blood volume	1.4 ± 0.9	5.4 ± 1.1	10.8 ± 1.8	10.9 ± 1.1	78.3 ± 2.3
Circulating blood volume (L/m²)	2.50 ± 0.24				
Cardiac index (L/min/m²)	3.63 ± 0.49				

*Total number, 34 (male 22, female 12).
Source: Kuwahara et al. (1979).

Table 2.6 illustrates the normal functional values (mean ± SD) of the circulatory system obtained by the simulation method in 34 normal subjects. The circulating blood volume and cardiac index (per body surface area) are comparable with those obtained by other standard techniques. The simulation method is unique in that it can determine the mean transit times and the blood volume distributions in the four blood chambers.

2.4.4 Cerebral Blood Circulation

A radioisotope can be used for external measurement of the blood flow rate in a cerebral circulation system, which is located away from other parts of the body, and the amount of radioisotope in it can be easily monitored by external collimation. A radioisotope dilution curve recorded by a scintillation counter placed over the head is called a radioencephalogram (REG). If a simulation circuit for the transport process of the radioisotope in the cerebral circulation is added to the analog circuit for the RCG as a part of the body circulation, one can simulate an REG obtained through peripheral venous injection of the radioisotope.

The cerebral circulation system may be approximated by a single compartment of volume V_c and a transport delay τ_{Tc}. r_c denotes the ratio of cerebral blow flow rate to systemic blood flow rate. Since the REG is greatly affected by the inflow process of radioisotope through the cardiopulmonary circulation into the cerebral compartment, one records and analyzes the RCG and REG simultaneously as a radiocardioencephalogram (RCEG).

Figure 2.36 depicts an RCEG and simulation results of a patient with hypertension. The obtained circulatory values are: cardiac index, 4.61 L/min/m²; cerebral blood flow rate fraction, 17.8%; cerebral blood flow rate, 1270 ml/min; and cerebral blood volume, 199 ml. Normal values of cerebral circulation obtained by this method are listed in Table 2.7. They agree well with those obtained by other standard techniques.

2.4.5 Renal Blood Circulation

A fast and easy measurement of an individual kidney function can be obtained by radioisotope renography. The information on essential renal functions such as effective renal plasma flow rate (RPF) and glomerular filtration rate (GFR) can be obtained from the radiorenogram (RRG). A simulation model of the renogram is presented later in the section. Figure 2.37 depicts a normal ¹³¹I-hippuran renogram which exhibits three features:

1. A rapid upstroke of inflow which is followed by a slower upstroke of "accumulation" by renal active and/or passive transfer of radioisotope into the kidney.
2. A peak at which inflow and/or accumulation are equivalent to outflow.

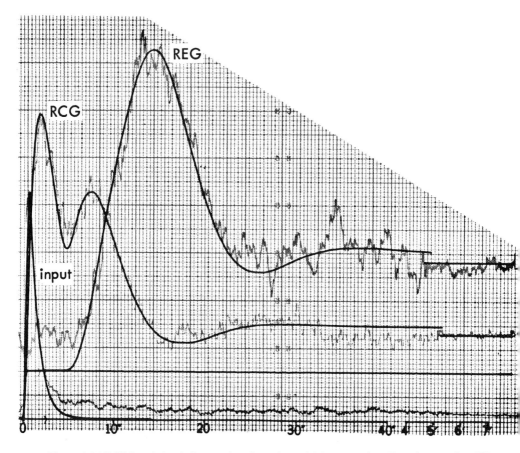

Figure 2.36 RCEG and simulation results of a patient with hypertension (Kuwahara et al., 1979).

Table 2.7 Normal values obtained by the analog simulation of radiocardioencephalograms

Parameter	Value
Mean transit time	7.55 ± 0.91 s
Cerebral blood flow rate	63.9 ± 12.5 ml/min/100 g
Cerebral blood volume	104 ± 21 ml
Cerebral blood flow rate fraction	12.0 ± 3.7%

*n = 14; age = 43.8 ± 12.1 yr.
Source: Kuwahara et al. (1979).

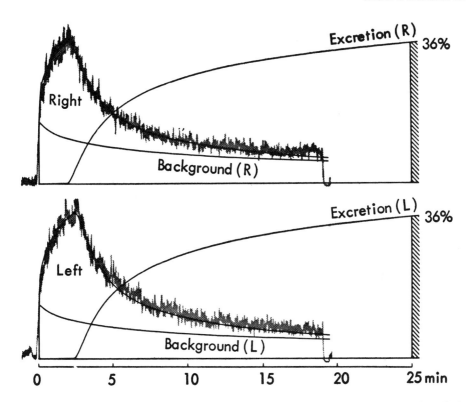

Figure 2.37 Normal ^{131}I-hippuran renogram with the observed excretion rate of 72% in 25 min. superimposed by the calculated renogram, background and excretion indicating 36% (right) + 36% (left) in 25 min. (Kuwahara et al., 1979).

3. A downslope that is normally rapid in an early period, reflecting excretion of radioisotope concentration in the kidney as urine, followed by a slower decay period which indicates a falling plasma concentration caused by the excretion of radioiostope through the kidney.

Figure 2.38 illustrates the input-output relationship of an RCG which consists of (1) the total amount of radioactivity entering the kidney; (2) the total amount of radioactivity leaving the kidney to the bladder; and (3) the background radioactivity from blood circulating through the tissues within the field of the scintillator.

Figure 2.39 is a schematic diagram of the transport process of a radioisotope with the renal circulation system. The renal system consists of a single compartment with time constant τ and a transport delay τ_T. C_p in μCi/ml signifies the radioisotope concentration in plasma, $V_{pe}(t)$ in ml, the time-varying hip-

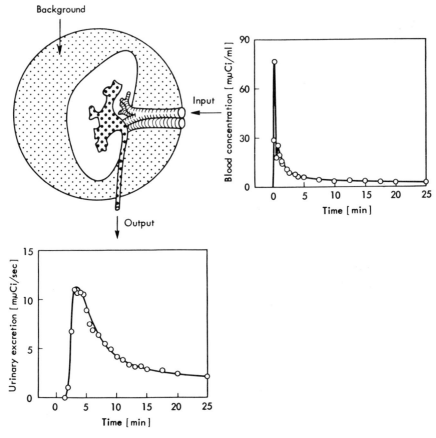

Figure 2.38 Input and output relationship of renogram, measured with ^{131}I-hippuran in dog (well-hydrated state). (Kuwahara et al., 1979).

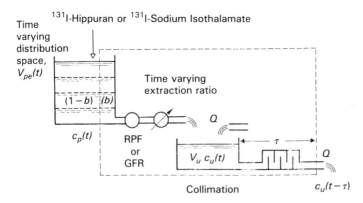

Figure 2.39 A model of the renogram (Kuwahara et al., 1979).

puran distribution space; RPF in ml/min, the effective renal plasma flow rate; $e'(t)$, the time-decaying extraction ratio of hippuran; V_{ui} in ml, the kidney volume; $C_{ui}(t)$ in μCi/ml, the radioisotope concentration in urine; and Q_i, the urine flow rate.

A typical ^{131}I-hippuran renogram is shown in Fig. 2.37 which is superposed by the simulated results. The sum of the two calculated excretions at 25 min of the right and left kidneys agrees well with clinical data.

2.5 BOUNDARY LAYER FLOWS AND FLOW BIFURCATION

2.5.1 Boundary Layer Flows

For fluids having relatively low viscosity, the effect of internal friction in a fluid is appreciable only in a narrow region surrounding the fluid-solid inter-faces. The flow outside the narrow region may be considered as ideal (or potential) flow. The narrow region is called the boundary layer in which the velocity is affected by shear stress. The concept of the boundary layer was proposed by L. Prandtl in 1904. It can be applied to internal flows (inside blood vessels) as well as external flows (over body surfaces).

For a flow over smooth boundaries, the boundary layer upstream starts out as a laminar boundary layer in which the fluid particles move in smooth layers. One layer glides smoothly over an adjacent layer with only a molecular interchange of momentum. As the thickness of the laminar boundary layer increases, it becomes unstable and finally transforms into a turbulent boundary layer in which the fluid particles move in a random manner with a violent transverse interchange of momentum. The velocity of the fluid particles is reduced by the action of viscosity at the boundary. When the boundary layer has become turbulent, there is still a very thin layer next to the solid boundary that has laminar motion. It is called the laminar sublayer (see Fig. 2.40).

The nature of the flow, whether laminar or turbulent, and its relative po-sition along a scale, indicating the relative importance of turbulent to laminar tendencies, are indicated by the Reynolds number. The Reynolds number is a dimensionless parameter expressing a ratio of the inertial forces (forces set up by acceleration or deceleration of the fluid) to the viscous shear forces:

$$\text{Re} = \frac{Lv\rho}{\mu}$$

where L denotes the characteristic length. For flows over a smooth surface, L is the distance from the leading edge. L represents the diameter for crossflows over a cylinder or a sphere. In internal flow cases, L is the hydraulic diameter D_h defined as

$$D_h = \frac{4 \times \text{cross-sectional area of flow}}{\text{wetted perimeter}}$$

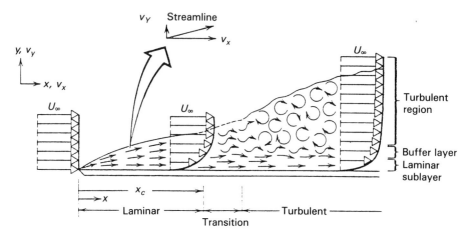

Figure 2.40 Velocity boundary layer development on a flat plate.

D_h is exactly the inside diameter for a flow inside a circular tube, for example blood vessels. The internal flow is laminar for Re less than 2100 and becomes turbulent when Re exceeds 10,000. When Re for the plate is between 5×10^5 and 10^6, the boundary layer becomes turbulent.

The basic geometry in boundary layer flows is a flat plate. In the case of steady, two-dimensional flow (Fig. 2.40), the continuity and momentum equations [Eqs. (B.3) and (B.15) in Appendix B] are reduced to

$$\frac{\partial v_x}{\partial x} + \frac{\partial v_y}{\partial y} = 0$$

$$v_x \frac{\partial v_x}{\partial x} + v_y \frac{\partial v_x}{\partial y} = -\frac{1}{\rho} \frac{dP}{dx} + \nu \frac{\partial^2 v_x}{\partial y^2}$$

respectively. The appropriate boundary conditions are

$$y = 0: \qquad v_x = v_y = 0 \qquad \text{(no-slip)}$$

$$y = \infty: \qquad v_x = U_\infty \qquad \text{(free-stream velocity)}$$

We now introduce a similarity variable:

$$\eta = y \sqrt{\frac{U_\infty}{\nu x}}$$

and a stream function:

$$\psi(x, y) = \sqrt{\nu x U_\infty} \, f(\eta)$$

$f(\eta)$ denotes the dimensionless stream function. The velocity components become:

$$v_x = \frac{\partial \psi}{\partial y} = \frac{\partial \psi}{\partial \eta}\frac{\partial \eta}{\partial y} = U_\infty f'$$

$$v_y = -\frac{\partial \psi}{\partial x} = \frac{1}{2}\frac{\nu U_\infty}{x}(\eta f' - f)$$

The prime denotes differentiation with respect to η. The continuity equation is satisfied, while the momentum equation is reduced to the Blasius equation

$$ff'' + 2f'' = 0$$

The boundary conditions become

$$\eta = 0: f = f' = 0$$

$$\eta = \infty: f' = 1$$

The Blasius equation can be solved in terms of a power series expansion or can be numerically integrated by the Runge-Kutta method. The viscous drag, F_D, for one side of the plate is

$$F_D = b \int_0^l \tau_s \, dx$$

Here, b is the width and l is the length of the plate. The local shear stress at the wall is given by

$$\tau_s(x) = \mu \left(\frac{\partial v_x}{\partial y}\right)_{y=0} = \mu U_\infty \sqrt{\frac{U_\infty}{\nu x}} f''(0)$$

with $f''(0) = 0.332$. Hence,

$$\frac{\tau_s}{\rho U_\infty^2} = \frac{0.332}{\sqrt{Re_x}} \qquad F_D = 0.664 b U_\infty \sqrt{\mu \rho l U_\infty}$$

wherein Re_x is $U_\infty x/\nu$. With the introduction of a dimensionless drag coefficient defined as

$$C_f \equiv \frac{2D}{\rho A U_\infty^2/2}$$

Since $A = 2bl$, one obtains

$$C_f = \frac{1.328}{\sqrt{Re_l}}$$

Here, $Re_l = U_\infty l/\nu$ denotes the Reynolds number formed with the plate length and the free stream velocity. This law of friction on a plate is valid only in the laminar flow region, that is, for $Re_l < 5 \times 10^5$ to 10^6. In the turbulent flow region for $Re_l > 10^6$, the drag becomes considerably greater than that given in the above equation.

Results for flows over other smooth surfaces and in the entrance region of ducts and cylinders can be obtained by performing a series expansion on the solution of a plate.

2.5.2 Flow Bifurcation in the Aorta

The aorta rises from the base of the left venticle, arches over to the left side of the vertebral column and passes downward along the vertebral column to the level of the fourth lumbar vertebra. It terminates at this level by dividing into the common iliac arteries, as shown in Fig. 2.41. The aorta may be divided into four parts: (1) the ascending aorta, whose branches are the coronary arteries; (2) the aortic arch, which extends to the left side of the fourth thoracic vertebra and whose branches include the innominate, the left common carotid, and the left subclavian arteries; (3) the thoracic aorta, which extends from the arch to the diaphragm and gives off visceral and parietal branches; and (4) the abdominal aorta, which extends from the diaphragm to the fourth lumbar vertebra and gives off both visceral and parietal branches (Greisheimer, 1955).

Typical branches of the aorta are either unpaired, for example the celiac, the superior mesenteric, and the inferior mesentric, or paired, like the renal

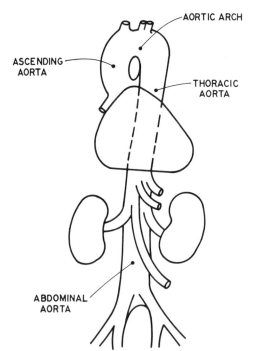

Figure 2.41 Aorta.

and the internal spermatic. Clinical and postmortem studies indicate that both primary atherosclerotic lesions and the deposition of platelet thrombi appear to occur most frequently at certain sites in the arterial tree, such as bifurcations, branchings and curved segments of the aorta. There, flow is disturbed, and separation of streamlines and formation of eddies may occur. Flow patterns inside the aortic branchings have been studied both empirically in vitro and theoretically, using isolated transparent natural blood vessels or artificial vessel molds exact as well as simulated molds. Experimental studies include both velocity measurements and flow visualization. The aortic bifurcation models commonly studied include the T, Y, and double T junctions (Fig. 2.42). The Y bifurcation is a crude analog of the junction of the abdominal aorta and the iliac arteries. The T junction having the side (called the "daughter" branch) at a right angle or inclined with the trunk (called the "parent" vessel) models many aortic branchings, for example, the carotid and celiac branchings. A double T junction is an idealization of the renal branches.

Secondary flows and flow separation are commonly observed in arterial branches and bifurcations (Fig. 2.43). They are most prominent during the systole and diminish appreciably or disappear during the diastolic phase of the cycle. Secondary flows may appear in a spiral or double helical form. The zone of flow separation (recirculating flow) appears at sites of arterial branching and bifurcations along the outer wall. The reverse flow in the separation zone is characterized by vortex formation, adjacent low shear zones, high pressure, and changes in local mass transfer between the wall and the mainstream. Theoretical results for two-dimensional flow models are presently available, whereas theoretical studies using more realistic three-dimensional flow models need to be undertaken.

2.6 FLOW MEASUREMENTS

The cardiac output of an adult at rest is about 5 L/min. This is the amount of blood that must pass through the systemic and pulmonary circulation.

Flow measurements commonly include the determination of cardiac output, blood velocity, and blood pressure. The techniques are divided into two categories: steady-state measurements, where equilibrium must be achieved prior to the measurement, and transient measurements, where the measurement is made while the system is varying with time. There are noninvasive methods and invasive techniques.

2.6.1 Flow Rate

Typical invasive flow measurement devices include the rotameter, the electromagnetic flowmeter, and the ultrasonic flowmeter. The blood vessel must be cut and the blood passed through the rotameter. An anticoagulant such as heparin must be used to prevent blood clotting. The rotameter is a variable-

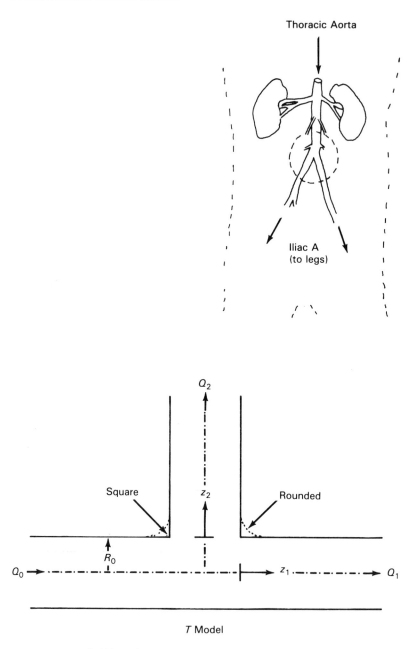

Figure 2.42 Aortic bifurcation models.

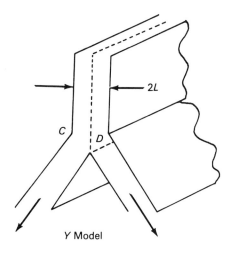

Y Model

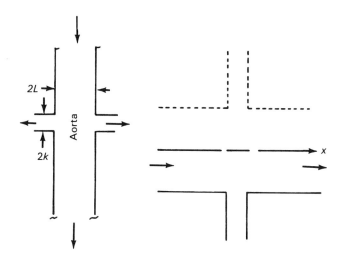

Double *T* Model

Figure 2.42 (cont.) Aortic bifurcation models.

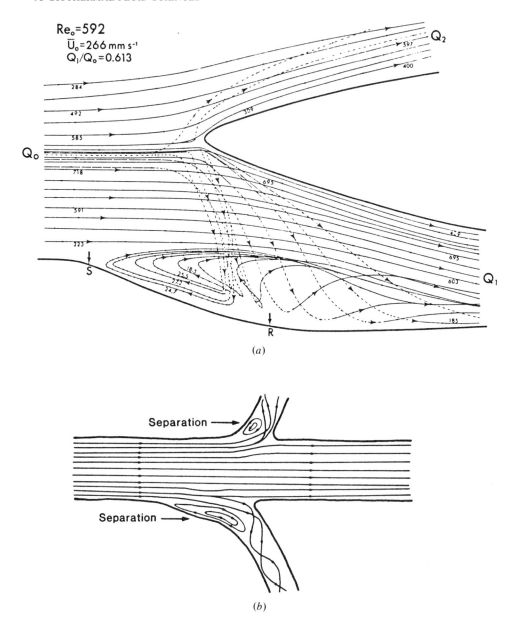

Figure 2.43 (*a*) Detailed flow patterns in a human carotid artery bifurcation, showing the formation of a recirculation zone and a counter-rotating double helicoidal flow (symmetrical about the common median plane) in the carotid sinus. The solid lines are the paths of particles in or close to the common median plane and the dashed lines are paths that are far away from the common median plane (projection of the particle paths on the common median plane). The arrows at S and R denote the respective separation and stagnation points. The numbers on the streamlines indicate the particle velocities in mm s^{-1} (Karino et al., 1983). (*b*) Renal artery (Sabbah et al., 1984).

area meter that consists of an enlarging transparent tube and a metering float that is displaced upward by the upward flow of blood through the tube. The tube is graduated to read the flow rate directly. Bernoulli's equation and the continuity equation are applied to calibrate the device. With the use of the appropriate electronic apparatus, the blood flow can be recorded continuously.

Both the electromagnetic and ultrasonic flowmeters are implanted around the blood vessel. In the former unit, an electromagnetic force is generated in electrodes on a blood vessel when the vessel is placed in a strong magnetic field and blood flows through the vessel. The probe contains both the strong magnet and the electrodes. A minute piezoelectric crystal is mounted in the walls of each half of an ultrasonic flowmeter. An electronic unit alternate the direction of sound transmission several hundred times per second. The sound is transmitted first downstream and then upstream. Sound waves (100,000–4,000,000 cyles per second) travel downstream faster than upstream. An electronic apparatus measures the difference between these two velocities, which is a measure of blood flow.

The laser Doppler velocimeter, ultrasonic Doppler velocimeter, and a plethysmograph are examples of noninvasive methods. Laser Doppler velocimetry (LDV) is superior for its high spatial resolution, fewer optical constants (needed for calculating the object velocity from the detected signal), ability to measure very low velocity (in the order of 1 μm/s) and easy identification of the probing area. It is suitable for the velocity measurement of blood flow in microvessels. For example, in a dual scatter LDV, a moving scatter center (a particle) is illuminated by two focused equal intensity laser beams (produced by a beam splitter) (Fig. 2.44). A set of stationary fringe planes are formed in the crossover region at the scatter center. As a scatter center passes through the fringe planes, it scatters light whose intensity is modulated at a frequency

$$f_D = \frac{2v}{\lambda} \sin \frac{\theta}{2}$$

where f_D is called Doppler frequency, v is the velocity of the scatter center, θ denotes the angle of beam crossing, and λ is the wavelength. v can be determined by measuring f_D.

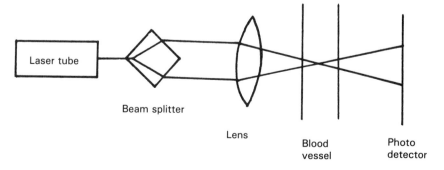

Figure 2.44 A schematic of a double-beam laser Doppler velocimeter.

Unfortunately, the application of LDV to blood velocity measurements is restricted to small vessels with thin walls because of relatively low transparency of blood and vessel walls to laser light. Recently, a new fiberoptic (graded index multimode, 0.125 mm) LDV (Tomonaga et al., 1983) has been developed to measure the velocity profile in coronary arteries. Its schematic diagram is shown in Fig. 2.45. The object beam from the fiber tip is scattered by the flowing red blood cells. The back-scattered light with the Doppler shift signals is transmitted back through the same fiber and is superposed with the reference beam. The photocurrent from a photodetector is fed into a spectrum analyzer to determine the Doppler shift frequency of the scattered light Δf. The blood velocity v is related to Δf by

$$\Delta f = \frac{2nv}{\lambda} \cos \theta$$

where n is the refractive index of blood, approximately 1.33, and θ is the angle of the beam incident into the bloodstream. Figure 2.46 compares the velocity profiles in steady flow through a transparent straight tube of 1.8 m length, 5.5 mm inside diameter, and Re = 800. A very good agreement confirms the validity of the fiberoptic LDV method which is applied to coronary flow measurements. Figure 2.47 shows a set of velocity profiles in the proximal portion of the circumflex coronary artery of an adult mongrel dog at the peak velocity of 28 cm/s and the cardiac cycle length of 0.44 s. The diastolic prominant pattern, which is characteristic of the left coronary artery flow, is observed. A set of velocity profiles in the distal portion of the same dog at the peak velocity of 10 cm/s and the cardiac cycle length of 0.40 s is illustrated in Fig. 2.48. The velocity profile across the vascular lumen is developed and skewed toward the inner wall.

The plethysmograph is used for recording blood flow in a forearm, leg, toe,

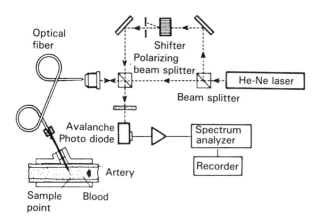

Figure 2.45 LDV that uses an optical fiber (Tomonaga et al., 1983).

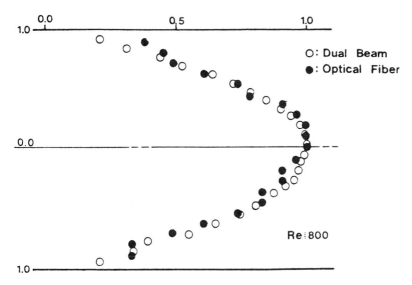

Figure 2.46 Comparison between the velocity profiles measured by the fiberoptic LDV and by the dual-beam mode LDV (Tomonaga et al., 1983).

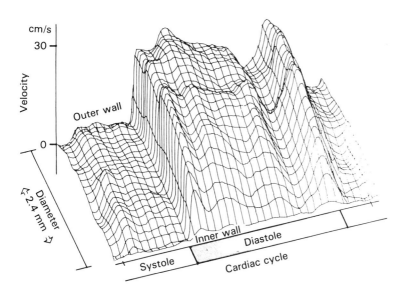

Figure 2.47 Set of velocity profiles in the proximal portion of the circumflex coronary artery of a dog at peak velocity of 28 cm/s and cardiac cycle length of 0.44 s (Tomonaga et al., 1983).

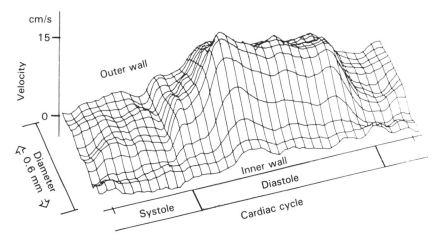

Figure 2.48 Velocity profiles in distal portion of same dog artery as in Fig. 2.47 at peak velocity of 10 cm/s and cardiac length of 0.40 s (Tomonaga et al., 1983).

finger, or a number of internal organs, such as spleen, kidney, liver, various glands, or even the heart. The method is to place the forearm, for example, in a chamber that is connected to a tambour recorder so that an increase in volume of the arm inside the airtight chamber causes the membrane of the tambour to rise. A blood pressure cuff is used to inflate the plethysmograph.

2.6.2 Velocity

Hot-wire anemometry is a technique based on forced convective heat transfer from a hot wire or thin film (a thin, electrically conducting film being deposited on an insulator catheter which is inserted into the vessel). To determine the blood velocity, the probe is either maintained at a constant temperature by adjusting the current i and the blood speed determined from the measured value of the current, or the gauge is heated by a constant current and the velocity is deduced from a measurement of the electrical resistance or the voltage drop in the probe.

The heat dissipated to the flowing blood is equal to the heat generated in the probe. In laminar flow over a single wire, the heat transfer coefficient is proportional to the square root of the velocity. Thus,

$$i^2R = (C_1 v^{1/2} + C_2)(T_s - T_b)$$

where C_1 and C_2 are constants. For the constant temperature type, the equation can be expressed alternately as

$$i^2 = K_1 v^{1/2} + K_2$$

with K_1 and K_2 being obtained in the calibration. Seed (1972) described the use of hot film anemometry in human aorta.

2.6.3 Pressure

The mercury manometer has been used as the standard reference for measuring blood pressure for many years. A cannula is inserted into an artery, a vein, or the heart. The pressure from the cannula is transmitted to one side of the manometer. The difference between the two levels of mercury ΔZ is equated to the pressure in the circulation: $\Delta p = \rho g \Delta z$.

The mercury manometer is used for recording steady pressures. For rapid pressure changes, say one cycle every 2–3 s, electronic pressure transducers are preferred. They convert pressure into electrical signals which are recorded on a high-speed electrical recorder. Pressure cycles of up to 500 cycles per second can be accurately recorded using high-fidelity recording systems.

2.6.4 Cardiac Output

In humans, cardiac output is measured through indirect methods that do not require surgery. Commonly used are the oxygen Fick method and the indicator dilution method. The latter is treated in detail in Section 2.4. The Fick procedure calculates the cardiac output by

$$\text{Cardiac output (ml/min)} = \frac{\text{oxygen absorbed by the lungs (ml/min)}}{\text{arteriovenous oxygen difference (ml/ml of blood)}}$$

The rate of oxygen absorption by the lungs is usually measured by a respirometer. The mean concentration of oxygen in the venous blood is obtained from the blood sample directly from the right ventricle or pulmonary artery (using a catheter), whereas the arterial oxygen concentration can be obtained from any artery in the body.

2.7 FLUID MECHANICAL ASPECT OF SOME DISEASES

2.7.1 Air Embolism

Cavitation refers to the formation of a cavity (gas-filled bubble) within a liquid, for example, in the blood circulatory system, caused by a pressure reduction outside the bubble nuclei due to hydrodynamic forces. Its effect in the blood is illustrated by destructive action to the fluid, that is, the red blood cells, rather than to the containing vessel walls (in sharp contrast to cavitation damages on the containing walls in the engineering field). Actually all three types of blood cells (red cells, white cells, and platelets) are readily destroyed by adverse combinations of internal pressure and flow shear stresses, but the red cells are particularly sensitive to cavitation effects because of their gas content. The membrane of the red blood cells can sustain only a slight reduction in local pressure (bursting pressure gradient). Thus, fast decompression in divers or aviators can destroy red blood cells, in addition to initiating a number of critical-size cavity nuclei which can grow into gas-filled bubbles. Hemolysis, or destruction of the blood cells during decompression, is considered the act of the excess nitrogen forming gas bubbles, since the oxygen and carbon dioxide content are amenable to the regulatory variations in blood flow and pulse rate.

Normally, the inception of cavitation in the vascular system is expected primarily in the vena cava leading to the right atrium. There, the local diastolic blood pressure decreases below atmospheric pressure during the suction stroke as the heart is filling. Cavity nuclei can form but are redissolved during the pressure stroke of the heart. Consequently, no gas bubbles are circulated.

However, during decompression, cavity nuclei that form from excess nitrogen are large enough to persist through the pressure cycle. They continue to grow during circulation in the arteries due to the reduction in blood pressure, resulting in the occurrence of dybarism or ebullism. Dybarism, commonly called the bends, is the outgassing of nitrogen in the blood during the pressure release from a higher pressure than the atmospheric pressure, and occurs in divers and caisson workers. Ebullism refers to local boiling at ordinary temperatures at a reduced ambient pressure or a dynamic simulated reduced atmospheric pressure due to accelerating forces in airplane dives or satellite maneuvering.

The destruction of red blood cells by cavitation effects can trigger coagulation mechanisms that produce blood clots; a thrombus is a stationary clot and an embolus is a moving clot. One should realize that the human heart has a pulmonary filter which removes a limited number of gas bubbles and blood degradation products in the right side of the heart lung system. Tamponade (partial blockage) of the pulmonary circulation results in a rise in the pulmonary blood pressure which is reflected in the right side of the heart, enabling blood-laden gas bubbles to traverse the foramen ovale (a bypass orifice) and enter the systemic circulation. Therefore, it is possible that gas bubbles may circulate to the brain where they contribute to a sudden circulatory collapse.

The degree of cavitation or the tendency to cavitate at a reference point within a flowing liquid is indicated by the cavitation number of the Thoma coefficient K. If the gas pressure is considered the sole important liquid property, then

$$K = \frac{P_\infty - P_i}{\rho v^2/2}$$

where P_∞ denotes the absolute static pressure in the liquid; P_i, the bubble pressure ($= P_v + P_g$); P_v and P_g, the vapor and gas pressures, respectively, inside the bubble; ρ, the liquid density, and v, the reference velocity. The numerator $P_\infty - P_i$) denotes the net pressure acting to collapse the bubble, with the denominator $\rho v^2/2$ is the velocity head to induce nucleation and growth of the bubble.

The critical liquid pressure for cavitation can be determined as follows. The relatively slow diffusion of dissolved foreign gas out of the liquid into the cavity nuclei can induce the formation of a gas-filled bubble. The process is called gaseous cavitation and its critical pressure is given by

$$P_c = P_v + P_g - \frac{2\sigma}{R_0}$$

where R_0 is the initial radius of the bubble nucleus and σ denotes the surface tension coefficient. The formation of vapor-filled bubbles by the rapid expansion of the vapor phase out of the liquid into the cavity nuclei is termed vaporous cavitation. The static equation for the critical pressure in vaporous cavitation reads

$$P_c = P_v - \frac{2\sqrt{3}}{9} \frac{2\sigma}{R_0} \left[1 + (P_0 - P_v) \frac{R_0}{2\sigma} \right]^{1/2}$$

where P_0 denotes the initial pressure of the surrounding liquid.

Figure 2.49 (Strasberg, 1956) shows the critical pressure for cavitation as a function of the air bubble nucleus in water (close to plasma). R_0 corresponds to the size at atmospheric water pressure. The indicated values of critical pressure are relative to vapor pressure so that zero corresponds to vapor pressure. The equations for both vaporous and gaseous cavitation are indicated in

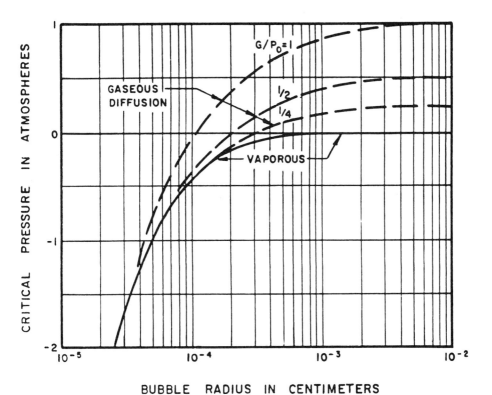

BUBBLE RADIUS IN CENTIMETERS

G = GAS SATURATION PRESSURE

P_0 = AMBIENT PRESSURE

Figure 2.49 Critical pressure for cavitation versus bubble nucleation size. G = gas saturation pressure; P_0 = ambient pressure.

the figure. It is seen that when the bubbles are small enough, R_0 being less than about 1.6×10^{-4} cm, both gaseous and vaporous cavitation require negative pressure (tensile stresses) of equal magnitude. However, for the larger bubbles, as usually encountered in water, gaseous cavitation can occur at pressures above the vapor pressure, provided that there is time for the slow diffusion process to occur. Vaporous cavitation also occurs at pressures below the vapor pressure (tensile stresses) but the cavity growth is rapid, even explosive.

Under certain circumstances, decompression sickness can be fatal. During rapid decompression, fat deposits in the human body become supersaturated, causing gas nuclei to form in the fat cells. The growth of gas bubbles yields a pressure difference so that fat cell membranes rupture. As a result, gas bubbles and fat emboli enter the venous blood flowing into the right side of the heart. The pulmonary filter removes some of them but a clogging of this filter induces a slight elevation in the pulmonary blood pressure in the right side of the heart which lets blood-laden gas bubbles traverse the foramen ovale into the left side of the heart. Thus, the gas bubbles are carried in the systemic circulation directly to the brain where they cause a sudden circulatory collapse.

Boycott and Damant (1908) reported the influence of fat on the susceptibility to caisson disease; fat acts as a nitrogen reservoir which becomes easily supersaturated during rapid decompression. The presence of cavities in the liver of victims of fatal decompression was discovered by Hill (1912) during the decompression from increased cassion pressures and by Trowell (1943) during the decompression to altitude. Fatty tissue is treated as the important site of nitrogen supersaturation and bubble formation during decompression (Behnke, 1945; Harvey, 1945). Haymaker et al. (1956) reported on neurocirculatory collapse at altitude during jet aircraft flight. Chan and Yang (1969a) surveyed the literature pertinent to the problems of gas embolism in human body. The dynamic characteristics of gas emboli in blood were theoretically and experimentally studied by Chan and Yang (1969b, c), Tansawa et al. (1969, 1970), Yang (1971), and Yang et al. (1971a, b, 1972).

Clinical treatment for decompression sickness includes denucleation by overpressure (Downey, 1963; Downey et al., 1963), addition of wetting agents such as heparin in the blood, and decompression with ultrasonic vibration.

2.7.2 Hemolysis

Hemolysis refers to the loss, that is, damage of red blood cells. It is a measure of cell trauma and occurs significantly in pumps and oxygenators, especially of the nonmembrane type (Peirce, 1969). Normally, the motion and resulting forces acting on blood components do not destroy them but may even facilitate their function. These blood components are produced in an orderly fashion and are utilized or recycled. They are only mildly affected by the stress produced by motion. However, in some diseased states or in the presence of prosthetic devices, the destruction of blood components occurs and the orderly recycling process is altered. For example, the red blood cell is produced in the bone marrow and lives for approximately 4 months. It is then scavenged by the reticuloendothelium system. With the insertion of prosthetic devices, its physical properties and morphology may be so altered that the red blood cell expires prematurely and dumps its constituents into circulating blood. A similar shortening of the life-span and altering of material pathways can occur in other blood components, for example, in protein denaturation.

Hemolysis has been thought to be of mechanical origin. It occurs in virtually every recipient of a prosthetic valve. The occurrence of intravascular hemolysis (vasculature) is evidenced by elevated serum hemoglobin, virtual absence of haptaglobin, reduction in the life-span of red blood cells, and the appearance of fragmented cells in the circulating blood.

Three classes of hemolysis have been observed and hypotheses have been proposed to explain the mechanism (Blackshear, 1972; Yang, 1974):

1. *Wall–red blood cell interaction.* Cells that come in physical contact with surfaces are mechanically torn or collapse upon themselves. This hemolysis, which is observed at every level of shear rate, is dependent on the surface condition and contact duration.
2. *Prolonged intermediate shear stress (1000–2000 dyn/cm^2).* The cells become distorted in such a way that the membranes collapse upon themselves yielding osmotically fragile hemoglobin-filled cell fragments. Such cells are actually observed in the residue from tests.
3. *Short-duration high shear stress (in excess of $40,000$ dyn/cm^2).* This hemolysis is caused by the yield stress of the cell membrane being exceeded. Either ghosts or membrane fragments are observed in tests.

2.7.3 Thrombosis

Thrombosis refers to the formation of a clot or thrombus (an abnormal clot) in any part of the vascular or lymphatic system. The thrombus lump or clot is a coagulum of blood elements or a growth of cells, like growth of tumor cells, formed in the heart, a blood vessel, or a lymphatic. Once a clot has developed, continued flow of blood past the clot may break it away from its attachment. Such freely flowing clots are called emboli. In general, emboli do not stop flowing until they come to a narrow point in a circulatory system. Thus, emboli originating in large arteries or in the left side of the heart eventually plug either smaller arteries or arterioles. Those emboli originating in the venous system or in the right side of the heart flow into the vessels of the lungs to cause pulmonary embolism.

There are usually two causes of thromboembolic (formation of thrombus and embolus) conditions that initiate the clotting process: (1) any roughened endothelial surface of a vascular or lymphatic vessel which may be caused by arteriosclerosis, infection, or trauma, and (2) very slow blood flow.

2.7.4 Atherosclerosis

Atherosclerosis is a disease of the large arteries in which lipid deposits, called atheromatous plaques, appear in the subintimal layer of the arteries. Since these plaques contain a particularly large amount of cholesterol, they are often called cholesterol deposits. They are also associated with degenerative changes in the arterial wall. Calcium often precipitates with the lipids to form calcified plaques. In a later stage, progressive sclerosis of the arteries occurs due to the infiltration of fibroblasts in the degenerative area. When both reactions take place, the arteries become extremely hard. The disease is then called arteriosclerosis or simply "hardening of the arteries."

Arteriosclerotic arteries lose most of their distensibility and are easily ruptured due to the degenerative areas. The atheromatous plaques often protrude through the intima into the flowing blood. The roughness of their surfaces then causes clots to develop. When a small clot has developed, platelets become entrapped and cause more clot to develop (thrombus) or the clot breaks away (embolus) and plugs a smaller vessel further downstream. This mechanism causes most coronary occlusions.

In an autopsy study of 500 cases, ranging in age from birth to 80 years, evidence of atherosclerosis in the renal arteries was first found in the first decade of life in the form of fatty streaks. By the second decade of life, 43% of cases showed renal atherosclerosis which increased to 82% of cases by the sixth decade.

It is generally recognized that there are three factors involved in atherosclerosis in human large elastic arteries: (1) the vessel wall, especially with injury to the endothelium, (2) the blood, including the lipids within the plasma, and (3) the hydrodynamics of the blood flow. The presence of flow separation at branching points and bifurcations is considered to be one form of hydrodynamic disturbance that contributes to atherogenesis at these sites (Fox and Hugh, 1966). The formation of atheroma in zones where local stasis is produced due to flow separation supports the view that atherogenesis is due to a process of deposition or thrombosis from the circulating blood. Many studies have been conducted to investigate the behavior of flow near branching points using in vitro models that simulate the arterial geometry, as mentioned in Section 2.5.2.

2.7.5 Hemodialysis

Blood is transported to the kidneys through the renal (arcuate) artery. It returns to the heart through the renal (arcuate) vein and the lower vena cava. Two small tubes called ureters carry the waste matter from the kidneys to the bladder, which expels the urine through the urethra. Figure 1.9 is a schematic diagram of a human kidney and a nephron.

Each kidney contains about 1 million nephrons, each nephron being capable of forming urine by itself. The nephron is composed of a glomerulus (a capillary bed) through which fluid is filtered out of the blood and a long tubule in which the filtered fluid is converted into urine on its way to the pelvis of the kidney. Blood enters the glomerulus through the afferent arteriole and exits through the efferent arteriole. Most of the blood then flows through the peritubular capillaries and empties into the cortical veins. Only a small fraction of the total renal blood flow, about 1–2%, passes through the vasa recta, a network of capillaries descending around the lower portions of the loops of Henle. About 1200 ml/min of blood, equivalent to 21% of the normal cardiac output of a 70-kg man, flow through both kidneys. The pressures in the different parts of the renal circulation are given in the Fig. 1.9.

The two major areas of resistance to blood flow through the nephron are the afferent arteriole and the efferent arteriole. Thus, the glomerulus operates as a high-pressure capillary bed, while the peritubular capillary system is a low-pressure bed. The glomerular pressure regulates the filtration rate.

The glomerular filtrate (filtered at a rate of approximately 125 ml/min), entering the tubules of the nephron, flows through the proximal tubule, the loop of Henle, the distal tubule, and the collecting tubule into the pelvis of the kidney. Along this course, substances are selectively reabsorbed or secreted by the tubular epithelium. The resultant fluid entering the pelvis is urine.

Artificial kidneys have been in use for about 35 years to treat patients with (1) severe renal insufficiency, in which so many nephrons are destroyed that

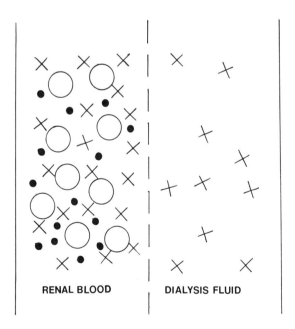

RENAL BLOOD   DIALYSIS FLUID

Figure 2.50 Basic principle of artificial kidney. The large white circles represent compounds too large to pass through the membrane (plasma protein). The ×'s represent the compounds on both sides of the membrane (some electrolytes and glucose) and the small black circles represent impurities that are filtered out by dialysis.

the kidneys cannot perform all the necessary functions or (2) acute renal shutdown, in which the kidneys stop working almost entirely. The idea of hemodialysis, that is, the dialization of blood, was introduced by Abel et al. in 1914. They thought that toxic materials could be removed by external dialysis of the blood. The basic principle of the artificial kidney is to pass blood through very thin flow channels bounded by thin membranes (see Fig. 2.50). On the other side of the membranes is a dialyzing fluid into which urea and other waste products in the blood pass by diffusion. A membrane material such as cellophane is porous enough to allow diffusion of all constituents of the plasma, that is, from plasma into dialyzing fluid and vice versa.

In normal operation of the artificial kidney, blood is continually removed from an artery (e.g., the radial artery), pumped through the kidney, and returned to a vein (such as the saphenous vein).

Various types of artificial kidney systems, sometimes referred to as hemodialyzers, are now being employed clinically. Some systems require pumps for both blood and dialysis solution, while others are designed for low flow resistance and can be used without pumps (Esmond and Clark, 1966).

2.8 ARTIFICIAL FLOW–REGULATION DEVICES

2.8.1 Some Consequences of Nonpulsatile Flow

Why must the flow of blood in a human body be pulsatile? The consequences of nonpulsatile (namely steady) flow over a long term result in various changes: (1) changes in the kidney function, including a significant depression in the renal function despite a high blood flow rate, changes in renal arteriograms such as narrowing, straightening, and loss of the normal configuration, reduced renal perfusion rate, changes in urine volume and contents, etc.; (2) capillary circulation (a marked slowing of red cell movement and an intravascular aggregation of erythrocytes); (3) cellular metabolism (no flow of the interstitial fluid and lymph); (4) lower oxygen consumption; (5) brain damage (swelling of Purkinje's cells); and (6) higher peripheral resistance resulting in a significant increase in mean pressure in both the systemic and pulmonary circulation and a lower flow rate.

2.8.2 Human Heart

Figure 2.51 is a schematic diagram of the human heart. It is essentially a four-chambered pump. The atria are priming chambers, while the ventricles are the main pumping chambers. The heart's contractions are synchronized by electrical impulses originating in the sinus node. The node has a regular intrinsic electrical discharge and serves as the pacemaker for the heart as a whole. Its impulse initiates the contraction in the atria directly. This impulse is also relayed to the atrioventricular node and from there it is conveyed to the muscular

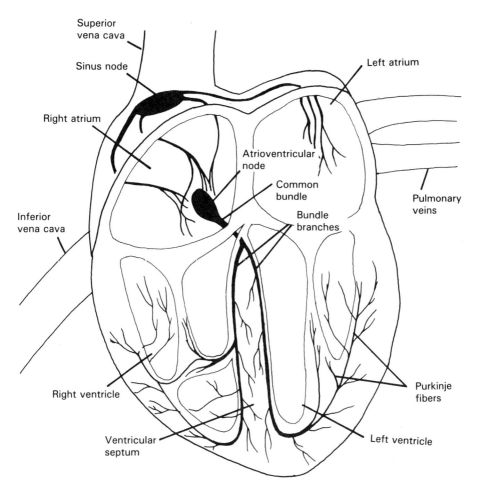

Figure 2.51 Human heart.

ventricles by specialized conducting fibers, among them Purkinje's fibers. Under normal conditions, the heart contracts almost as a single mass.

2.8.3 Cardiac Functions

Figure 2.52 depicts the location and shape of valves in the heart. The parentheses indicates the direction of flow through each valve.

Figure 2.53 shows the pressure and volume of the heart. The superimposed electrocardiogram (ECG) shows the characteristic P wave of atrial electric activity, the QRS complex denoting ventricular repolarization, and the T wave designating the ventricular repolarization. The systolic and diastolic phases are indicated together with the isovolumic contraction and relaxation phases. The opening and closing of the aortic (AO, AC) and mitral (MO, MC) valves are indicated in the figure. The timings of closing of the tricuspid valve (TC) and opening of the pulmonic valve (PO) fall in between the timings of MC and AO with TC first followed by PO. Similarly, closing of the pulmonic valve (PC) and opening of the tricuspid valve (TO) occur between AC and MO, with PC first followed by TO. The circled numbers correspond to the state on the pressure-volume diagram in a ventricle during a cardiac cycle shown in Fig. 2.54. In Figure 2.54, 1–2, 2–3, 3–4, and 4–1 correspond to the ventricular filling, isovolumic contraction, ejection, and isovolumic relaxation processes, respectively. The opening and closing of the tricuspid and pulmonic valves of the right ventricle are indicated in parenthesis. Processes 2–3 and 3–4 form the systolic phase and processes 4–1 and 1–2 make up the diastolic phase in the operation of a ventricle.

Figure 2.55 shows a technical model simulating the cardiac circulation system. The fluid (simulating venous blood) in a tube (blood flow system, 8) enters a reservoir (atrium, 1) and flows through an inlet valve (ventricular valve, 2) into a pump (heart or ventricle, 4). The fluid pressure is raised through the motion of a piston (cardiac wall, 5) operated by a motor (cardiac muscle, 6) while both the inlet and outlet valves are closed. The fluid then forces the outlet

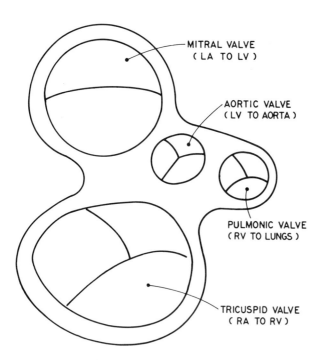

Figure 2.52 Location and shape of valves in the heart. LA, left atrium; LV, left ventricle; RA, right atrium; RV, right ventricle.

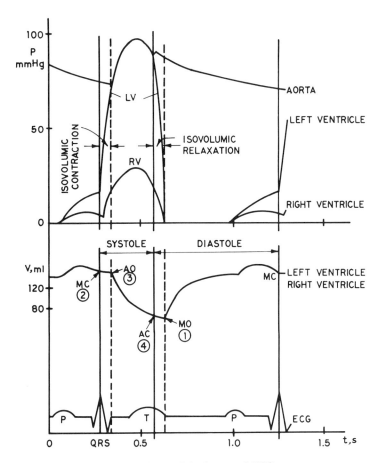

Figure 2.53 Pressure and volume of the heart and ECG.

valve (arterial valve, 3) to open and flows into a tube (blood flow system, 8) equipped with an air chamber (elastic property of large arteries, 7). The fluid (arterial blood) flows through a throttle (peripheral flow resistance, 9) to the other tube to complete a cycle (cardiac cycle).

Example 2.5 Figure 2.56 illustrates the systemic circulation system consisting of heart, aorta, and circulatory network. During the systolic phase of the heartbeat cycle, blood is pumped from the heart into one end of the aorta, and the walls of the aorta stretch to accommodate the blood. Let V be the volume of the aorta and P the pressure of blood within it at time t. Then, since V increases under P, it is reasonable to postulate that V is a linear function of P:

$$V = V_0 + aP$$

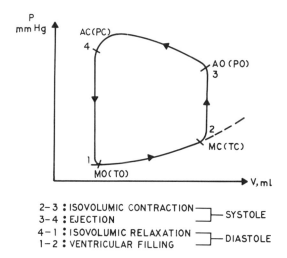

2-3 : ISOVOLUMIC CONTRACTION —⌉
3-4 : EJECTION —⌉ — SYSTOLE
4-1 : ISOVOLUMIC RELAXATION —⌉
1-2 : VENTRICULAR FILLING —⌉ — DIASTOLE

Figure 2.54 Pressure-volume diagram in a ventricle during a cardiac cycle.

where V_0 is the initial volume at $t = 0$ and a is the proportionality constant. The conservation of mass requires that the rate of change of volume, dV/dt, is equal to the difference between the rate at which the heart pumps blood into the aorta $Q_i(t)$ and the rate at which blood flows out of the aorta into the circulatory system $Q_e(t)$. The compartmental model assumes that Q_e is proportional to the pressure driving the blood. Hence, one can write

$$\frac{dV}{dt} = Q_i(t) - bP$$

wherein b is a constant. For simplicity, Q_i can be reasonably taken to $A \sin \omega t$ where A is the amplitude and ω is the frequency of blood flow rate. During the diastolic phase, there is no blood flow into the aorta ($Q_i = 0$). The walls of the aorta contract, thus squeezing the blood out of the aorta around the circulatory network of the body.

The combination of the two equations yields

$$a \frac{dP}{dt} + bP = Q_i(t) = \begin{cases} 0 & \text{during diastolic phase} \\ A \sin \omega t & \text{during systolic phase} \end{cases}$$

The initial condition is

$$P(0) = P_0 \qquad P(t_i) = P_1$$

Here, P_0 is the aortic pressure at the beginning of the systolic phase, $t = 0$, and P_1 is the pressure at the end of the systolic phase, $t_1 = \pi/\omega$. The solution for P during the systolic phase, $0 < t < \pi/\omega$, is

$$P = P_0 \exp\left(-\frac{bt}{a}\right) + \frac{A[a\omega \exp(-bt/a) + b \sin \omega t - a\omega \cos \omega t]}{b^2 + a^2\omega^2}$$

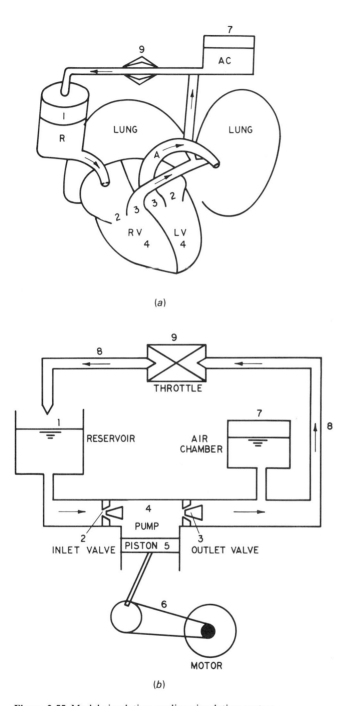

(a)

(b)

Figure 2.55 Model simulating cardiac circulation system.

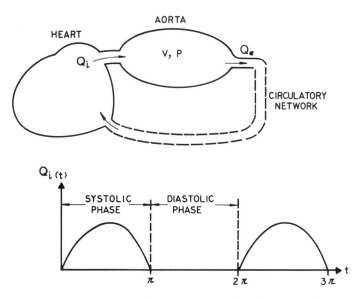

Figure 2.56 Systemic circulation system and its idealized volume flow rate–time diagram.

The solution for P during the systolic phase is a pure experimental decay:

$$P = P_1 \exp\left[-\frac{b(t - t_1)}{a}\right]$$

2.8.4 Artificial Heart

The purpose of cardiac assist devices is to relieve the natural heart of some of its work load until the muscle tissue can recover. If the heart is damaged beyond repair, either a new human heart must be transplanted or a totally prosthetic device has to be implanted. The latter consists of pumping chambers, an actuator, a power supply, and a feedback control system. The interest of thermal sciences is focused on the problem of pumping chambers.

The main problems in the use of artificial hearts are destruction of red blood cells (hemolysis) and clotting (thrombosis). The flow must be pulsatile for long-term operations. A convenient way to classify artificial hearts is by the type of flow: pulsatile or nonpulsatile. The nonpulsatile pumps in common use are finger pumps and roller pumps. They are used during the short periods of externally assisted circulation of open heart operations. Most of the pulsatile pumps work by flexing a diaphragm of some sort or by using pistons. While the compression of the tube (functioning as an artificial ventricle) in the nonpulsatile pumps is mechanical, compressing the ventricle in the pulsatile pumps has been mechanical (Medical Monitor pump), pneumatic (Army pump), and hydraulic (Davol pump). Most of the pumps for implantation are diaphragm type. The diaphragm pump is a flexible-walled container and is activated by liquid or gas which is constrained on one surface by an inflexible chamber.

Most diaphram-type artificial hearts fall into one of four categories:

1. A flexible, blood-filled tube in a rigid vessel (Fig. 2.57a).
2. A flexible bladder extending into a rigid blood-filled vessel (Fig. 2.57b).
3. The converse of (2) (Fig. 2.57c).
4. A conventional diaphragm type pump with blood on one side of the diaphragm and the activating fluid on the other (Fig. 2.57d).

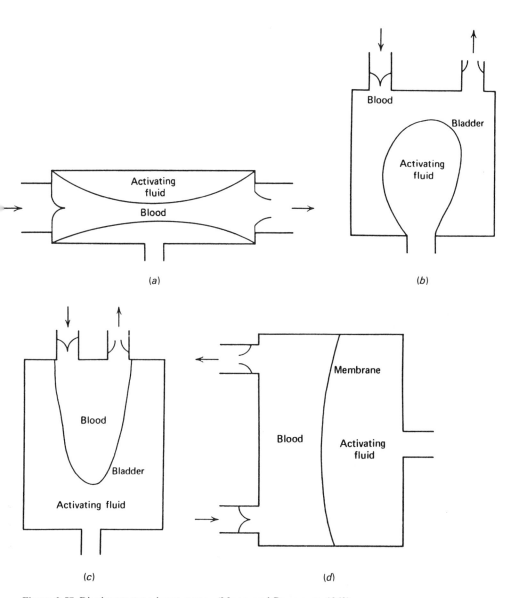

Figure 2.57 Diaphragm-type heart pumps (Myers and Parsonnet, 1969).

All diaphragm pumps inherently produce a pulsatile flow whose wave form can be regulated to be almost identical to that of a natural heart. The wave form is adjusted by controlling the activating fluid. The implantable artificial hearts by Kolff, DeBakey, Jarvik and Pennsylvania State University are diaphragm types (Fig. 2.58).

Two separate pumps are required for total heart replacement, since the systemic circulation delivers 5–14 L/min of blood against a normal pressure load of 120/80 mmHg, while the pulmonary circulation delivers the same flow against that of 25/12 mmHg. The inlet pressure available to both pumps is about 0–12 mmHg. Due to physiological constraints, the artificial heart must be close to the natural heart in appearance and configuration. The minimum size must be able to deliver the smallest end diastolic volume Q of 5–16 L/min at a reasonable beat rate. That means,

$$Q = V_e \dot{b} \eta \tag{2.54}$$

in which V_e represents the volume of blood ejected per stroke; $\dot{b}$, the frequency in beats per minute; and η, the ejection efficiency. With η of 80%, a maximum $\dot{b}$ of 100 beats per minute, and Q of 16 L/min, Eq. (2.54) gives the design volume of each ventricle to be 200 cm^3. The material of the entirely implantable artificial heart must be selected with consideration to light weight for suspension, compatibility with blood, and strength for a long fatigue life (on the order of 10^8 cycles in a 5-year duration at a rate of 60 beats per minute). Table 2.8 lists the specifications of the Pennsylvania State University artificial heart.

Figure 2.58 Pennsylvania State University diaphragm heart showing inlet and outlet valves (Phillips, 1983).

Table 2.8 Sac artificial heart specifications

Size	10.2 cm diameter
	4.45 cm thick
Weight	233 g (1 ventricle)
Stroke volume (nominal)	85–100 ml
Total pump volume	200 ml
Sac material	Segmented polyurethane (Biomer)*
Case material	Epoxy
Right heart outlet valve	# 13—A Starr-Edwards (18-mm diameter orifice)
Right heart inlet valve	# 29 Pyrolite Bjork-Shiley (24-mm diameter orifice)
Left heart outlet valve	# 13—A Starr-Edwards (18-mm diameter orifice)
Left heart inlet valve	Pennsylvania State University caged ball type (20-mm diameter orifice)
Aortic graft	26.5-mm diameter by 30-mm long Debakey Dacron graft
Pulmonary Artery Graft	26.5-mm diameter by 25-mm long Debakey Dacron graft
Nominal cardiac output	5–15 L/min

*Ethinor Corporation
Source: Davey (1983).

The artificial heart power requirement covering various types of daily activities can be derived from the hydraulic energy needed for cardiac output, using aortic and pulmonary artery pressures of 100 and 15 mmHg, respectively.

The total efficiency of the artificial heart is assumed to be 60%. For the male, the average power is 3.23 W at a cardiac output of 7.6 L/min with a maximum of 4.9 W at 11.5 L/min and a minimum of 2.55 W at 6 L/min. It is lower for the female, 2.9 W average at 6.9 L/min, 3.98 W maximum at 9.3 L/min and 2.53 W minimum at 6.0 L/min.

2.8.5 Artificial Heart Valves

Human heart valves function as check valves causing blood to flow in one direction through the heart. There are four valves: pulmonic, aortic, mitral, and tricuspid (Fig. 2.59). Defects in heart valves are of two primary types: (1) stenosis, where the valve tissue becomes calcified and restricts the opening of the valve, thus causing a reduction in blood flow, and (2) insufficiency, where the valve fails to close properly causing excessive regurgitant flow. Defects can occur in any of the four valves. However, the most serious problems occur at the left side of the heart, involving either the aortic or the mitral valve, or both. Table 2.9 shows the extent of reduction in the flow area of diseased natural valves as compared to that of healthy natural valves (Davey et al., 1983).

Cutler and Souttar (Netter, 1969) were the first to perform corrective surgery on diseased valves in the 1920s. Thereafter, the operations were limited to correction for aortic or mitral stenosis. The advent of open heart surgery in 1954 permitted the treatment of defective valves by direct surgical approach to all parts of the heart in order to restore valve function while retaining natural valve structures. Since only a few long-term operations succeeded, efforts were then directed toward the development of prosthetic valves.

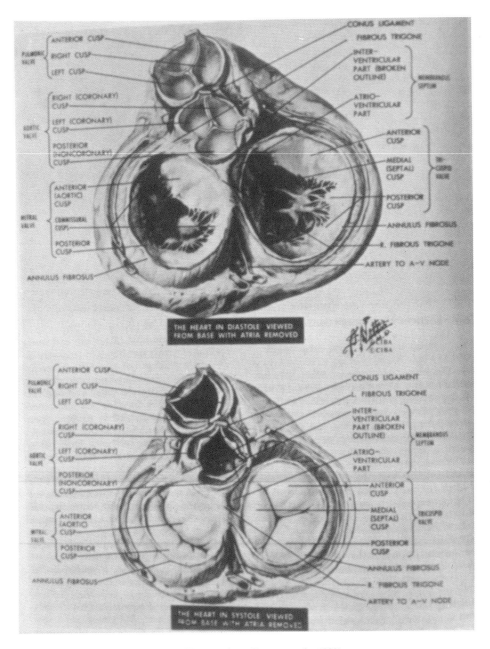

Figure 2.59 Location and shape of heart valves (Davey et al., 1983).

**Table 2.9 Typical orifice dimensions and pressure gradients
in natural, diseased, and prosthetic heart valves**

	Natural	Diseased natural	Prosthetic
Typical valve areas (cm²)			
Aortic	3.0	0.5	1.2
Mitral	4.5	1.0	2.0
Typical pressure gradients, peak to peak (mmHg)			
Aortic	0		20
Mitral	2–3		10
Typical blood flow rates 2.7 L/min/m² to 10 L/min/m² of body surface			
Average male	1.7 m²		
Average female	0.9 m²		

Source: Phillips (1983).

The first artificial heart valve was implanted by Hufnagel in 1952. At present, more than 40 different types of valves are in use. The most popular are the ball and disk types. There are numerous medical, material, and hydrodynamic requirements for artificial heart valves (Reul, 1983a). Materials used for valve cages include titanium, stellite 21, and stainless steel. For the ball type, the best available material is silicon rubber, but pyrolytic carbon, stellite,

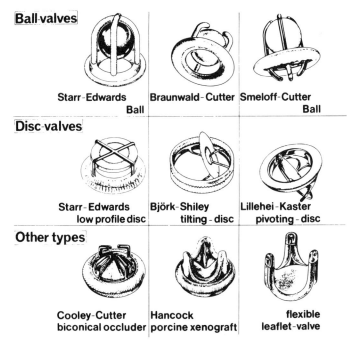

Ball-valves

Starr-Edwards Ball | Braunwald-Cutter | Smeloff-Cutter Ball

Disc-valves

Starr-Edwards low profile disc | Björk-Shiley tilting-disc | Lillehei-Kaster pivoting-disc

Other types

Cooley-Cutter biconical occluder | Hancock porcine xenograft | flexible leaflet-valve

Figure 2.60 Characteristic types of artifical heart valves (Reul, 1983a).

titanium, and fluorosilicone are also in use. Pyrolytic carbon occluders have proved satisfactory for disk valves. Before an artificial heart valve is implanted in humans, it is necessary to perform in vitro tests, including steady flow, pulsatile flow, and fatigue tests, and in vivo animal tests. Figure 2.60 shows some typical heart valve designs. Typical orifice dimensions and pressure gradients are listed in Table 2.9.

The major problems with long-term use of prosthetic valves have been thrombus formation, ball variance due to lipid absorption, and severe hemolysis (in a small number of cases). Thrombosis can be reduced by coating the valves with antithrombogenic substances. Ball and disk type valves have profiles that are known to develop turbulence. The potential hazards of turbulence include (1) triggering the clogging mechanism of blood, (2) collecting thrombi at stagnation areas, (3) causing increased drag to dislodge the valve, (4) producing higher fluid shear stresses to cause hemolysis. Thus, a streamlined valve would be advantageous (Vorhauer and Byars, 1966). Proper valve designs are a trade-off of various factors. Today, the art of cardiac surgery and valve design allow routine replacement of defective natural valves.

2.8.6 Circulatory Assist Devices

Mechanical devices designed to assist diseased natural hearts in circulating blood are called circulatory assist devices. They are intended to provide time for the correction of cardiac defects. These devices may be grouped into three categories: blood pumps, intraaortic balloons, and external body compression. Of engineering interest among the three groups are the blood pumps.

Blood pumps are used either intracorporeally or extracorporeally to assist a diseased heart in circulating blood on a short-term intermittent or permanent basis. The modes of their applications include venoarterial pumping, right ventricular bypass, left ventricular bypass, venovenous oxygenation, arteriovenous oxygenation, and the heart-lung machine, as depicted in Fig. 2.61. Blood is led to and from the organisms by tubes and cannulas. All modes are conveniently carried out with extracorporeal roller pumps.

There exist various designs of pumps with different pumping mechanisms, geometry, and driving devices. Design considerations must include hemodynamic requirements for minimizing hemolysis and thrombus formulation, physiological need of pulsatile flow for long-term perfusion, and material selection. Materials being used for blood pumps include stainless steel, titanium, stellite 21, pyrolytic carbon, natural rubber, and various polymers.

In summary, the main purposes of biofluid dynamics are to determine (1) the velocity distribution, flow rate, and pressure drop in the vessel, (2) the hydrodynamic aspects of circulatory diseases, and (3) guidelines for the design of prosthetic and assist devices in the circulatory system.

Other pertinent references include Bayliss (1960), Bird et al. (1960), Bugliarello (1969), Burton (1965), Fredrickson (1964), Kandarpa and Davids (1976), Karino et al. (1979), Lambert (1958), Lighthill (1972), Moravec and Liepsch

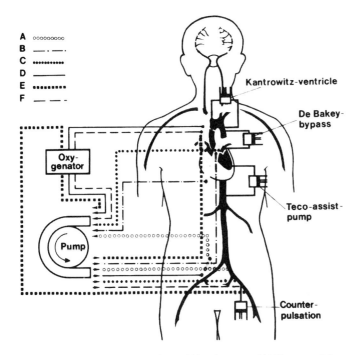

A oooooooooo
B _ . _ .
C ••••••••••
D
E •••••••••
F _ _ -

Kantrowitz-ventricle

De Bakey-
bypass

Oxy-
genator

Teco-assist-
pump

Pump

Counter-
pulsation

Figure 2.61 Different applications of blood pumps. (A) Venoarterial pumping, (B) right ventricular bypass, (C) left ventricular bypass, (D) venovenous oxygenation, (E) Arteriovenous oxygenation, and (F) heart-lung machine (Reul, 1983a).

(1983), Morgan and Ferrante (1955), O'Brien et al. (1976), Siouffi et al. (1984), Skalak (1972), Streeter et al. (1963, 1964), Whitmore (1968), and Yang and Liang (1972).

PROBLEMS

2.1 Show the mechanical models analogous to (a) the Maxwell elasticoviscous fluid, (b) the Kelvin viscoelastic solid, (c) an idealized Bingham plastic, and (d) the Oldroyd elasticoviscous fluid.

2.2 Derive the expression for the Young's modulus of a thin elastically walled tube ($R_i, R_0 =$ inner and outer radii) if the internal pressure is increased by an amount ΔP_i above the outer pressure so that the outer radius increases by ΔR_0. Young's modulus is the ratio of the tensional stress applied to the displacement or strain achieved.

2.3 Derive the expressions for both velocity profiles and flow rates of the following materials in laminar flow in a cylindrical tube: (a) a power-line fluid; (b) an idealized Bingham plastic; and (c) whole blood using the empirical Casson equation.

2.4 Consider all formed elements in whole blood to be rigid and geometrically shaped as (a) spheres, (b) ellipsoids, (c) rods, and (d) disks.

A dilute suspension is confined between two parallel flat plates. The rigid particles are initially randomly oriented. The shearing (the two plates move in opposite directions) of the suspension rearrange themselves so that they spend the majority of their time rotating. What are the orientations that are the most suitable and stable for rotation?

2.5 Determine the pressure drop in each of the different vascular systems listed in Table 2.1 for the systemic circulation of a dog. It is assumed that whole blood is a Newtonian fluid. What is the conclusion on the pressure drops?

2.6 Determine the total pressure drop, per unit length, additional pressure drop per particle, and ratio of particle velocity to mean velocity of the suspension flow in the microcirculation based on (a) the spheres model, (b) the spheroids model for $a/b = 1.4$, and (c) the axial-train (or stacked-coins) model. Assume that $\beta = 2$ and $\lambda = 0.6$. Compare the results with those of Poiseuille's law.

2.7 Blood flows in an aorta of diameter $d = 1.78$ cm and length $l = 30$ cm. Treating blood as a Newtonian fluid with the absolute viscosity μ of 3.33 centipoise and ignoring the entrance effects, determine (a) the average pressure drop in the aorta, (b) resistance [pressure drop/volumetric flow rate, in mmHg/(ml s)], (c) Reynolds number, and (d) dynamic pressure (difference between the two levels of mercury in a U-tube manometer) in mmHg.

Note: 100 centipoise = 1 g/(cm·s) = 1 dyn·s/cm^2 = 0.1 Pa s.

2.8 Consider the steady flow of urine (assumed to be an inviscid, incompressible fluid) of density ρ from a bladder (reservoir) at pressure P_0 through a urethra (assumed to be a straight and uniform distensible tube). Both the fluid velocity v and the fluid pressure P are assumed uniform over the cross section of area A. Let Q be the volumetric flow rate through the tube. Utilizing the Bernoulli equation, derive the expression for a maximum Q (i.e., Q_{max}) as functions of A, P, and ρ. Determine the corresponding value of the fluid velocity v_0, which is the sonic velocity in urine. Notice that $P = P(A)$ and

$$3\frac{dP}{dA} + A\frac{d^2P}{dA^2} > 0$$

2.9 Derive the governing equation for $C(t = \tau)$ in the system shown in Fig. 2.27e.

2.10 Determine the outlet concentration $C(t)$ in response to a step change in the inlet concentration $C_i(t)$, for (a) two parallel compartments, (b) two series compartments, and (c) two interacting compartments.

2.11 Construct the equivalent electrical circuits for (a) two parallel compartments, (b) two series compartments, and (c) two interacting compartments.

2.12 Construct a block diagram for the transport process of the injected radioisotope in the body including the systemic and cerebral circulations.

2.13 Derive the dynamic equation for a spherical bubble of instantaneous radius $R(t)$ in Newtonian fluids through the integration of the continuity and radial-component momentum equations in spherical coordinates. The viscosity effects on the bubble surface due to the gas and liquid phases are generally small and may be neglected. Notice that the balance of forces at the bubble surface

$$P_l(R) + \frac{2\sigma}{R} = P_g(R)$$

where $P_l(R)$ and $P_g(R)$ denote the gas and liquid pressures on the bubble wall, respectively. Specify the appropriate initial conditions.

2.14 In order to determine the overall dialysis coefficient B of a membrane, it is stretched to a surface area of A between two compartments. The fluid in each compartment is well mixed by a mechanical stirrer. Write a mass balance on the solute for each compartment. Determine the time history of the solute concentration in each chamber. The initial concentrations are C_{10} and C_{20}. If the membrane is semipermeable to solute 1 (i.e., allows only species 1 to go through), will the history of the solute concentration differ? See Fig. 2.27d with $Q_1 = 0$.

2.15 In Problem 2.8, the fluid pressure $P(A)$ in the urethra varies linearly with the cross-sectional area A_c as

$$P[\text{Pa}] = 4 \times 10^3[\text{Pa}] + 10^8[\text{Pa/m}^2]A_c[\text{m}^2]$$

The urine pressure in the bladder P_0 is 5.5×10^3 Pa. The urine has the same density as water. Determine (a) the sonic velocity in urine filling the urethra, and (b) the corresponding urine flow rate and urine pressure.

2.16 On average, the surface area A of human beings is related to weight W and height H by a formula of the type $A = aW^bH^c$, where a, b, and c are constants. Measurements on a number of individuals of height 180 cm and different weights in kilograms give values of A in square meters in the following table:

W	70	75	77	80	82	84	87	90	94	98
A	2.10	2.12	2.15	2.20	2.22	2.23	2.26	2.30	2.33	2.37

Show that a power law $A = aW^b$ fits these data reasonably well and evaluate b.

2.17 When a drug is injected into a blood vessel, it is carried along by the bloodstream (by convection) and simultaneously spreads out along the blood vessel by diffusion. The drug concentration at any point z along the blood vessel increases to a maximum and then falls off. Its time history can be expressed as

$$C = at^{-1/2} \exp \left[b \left(t + \frac{f}{t} \right) \right]$$

where a, b, and f are constants. Determine the time instant when C achieves its maximum.

2.18 When a dose m (in milligrams) of a certain drug is administered to a subject, his blood pressure drops by an amount P (in mmHg) given by

$$P = \frac{m^2(a - m)}{3}$$

where a is a constant. Determine the dosage that provides the largest drop in blood pressure.

REFERENCES

Abel, J. J., Roundtree, L. G., and Turner, B. B. (1914) Removal of diffusible substances by dialysis, *J. Pharm. Exp. Therapeut.*, 5:275.

Altman, P. L. (1959) *Handbook of Circulation*, W. B. Saunders, Philadelphia.

Ariman, T. (1969) On the analysis of blood flow, *Int. J. Eng. Sci.*

Aroesty, J., and Gross, J. F. (1970) Convection and diffusion in the microcirculation, *Microvascular Res.*, 2:247–267.

Bayliss, L. E. (1960) The Anomalous Viscosity of Blood, in *Flow Properties of Blood* (ed. A. L. Copley and G. Stainsby), Pergamon Press, London, pp. 29–62.

Behnke, A. R. (1945) Decompression sickness incident to deep sea diving and high altitude ascent, *Medicine*, 24:381.

Bergel, D. H. (1964) Arterial Viscoelasticity, in *Pulsatile Blood Flow* (ed. E. O. Attinger), McGraw-Hill, New York.

Bird, R. B., Stewart, W. E., and Lightfoot, E. N. (1960) *Transport Phenomena*, Wiley, New York.

Blackshear, P. L., Jr. (1972) Mechanical hemolysis in flowing blood, in *Biomechanics—Its Foundations and Objectives* (ed. Y. C. Fung, N. Perrone and M. Anliker), Prentice-Hall, Englewood Cliffs, N.J., pp. 501–528.

Bloor, M. I. G. (1968) The flow of blood in the capillaries, *Phys. Med. Biol.*, 13:443–50.

Boycott, A. E., and Damant, G. C. C. (1908) Experiments on the influence of fatness on susceptibility to caisson disease, *J. Hyg.*, no. 8, 445.

Branemark, P. I. (1965) Intracapillary Rheological Phenomena, in *Proceedings of the Fourth International Congress on Rheology*, Vol. 4, *Symposium on Biorheology* (ed. A. L. Copley), Wiley Interscience, New York, pp. 459–473.

Branemark, P. I. (1971) *Intravascular Anatomy of Blood Cells in Man*, S. Karger, Basel, Switzerland.

Branemark, P. I., and Lindstrom, J. (1963) Shape of circulating blood corpuscles, *Biorheology* 1: 139–142.

Brody, S. (1945) *Bioenergetics and Growth*, Reinhold, New York, 566 pp.

Bugliarello, G. (1969) Rheology of Blood, in *Biomaterials* (ed. L. Stark and G. Agarwal), Plenum, New York, pp. 93–102.

Bugliarello, G., and Hsiao, G. C. C. (1964) Phase separation phenomena at bifurcations: A simplified hydrodynamic model, *Science*, 143:469–71.

Bugliarello, G., and Hsiao, C. C. (1967) Numerical simulation of three-dimensional flow in the axial plasmatic gaps of capillaries, *7th International Congress on Medical and Biological Engineering*, Stockholm, Sweden.

Burton, A. C. (1965) *Physiology and Biophysics of the Circulation*, Year Book of Medical Publishers, Chicago.

Chan, K. S., and Yang, W.-J. (1969a) Survey of literature pertinent to the problems of gas embolism in human body, *J. Biomech.*, 2:299–312.

Chan, K. S., and Yang, W.-J. (1969b) Behavior of gas emboli subjected to pressure variation in biological systems, *J. Biomech.*, 2:151–156.

Chan, K. S., and Yang, W.-J. (1969c) Bubble dynamics in a non-Newtonian fluid subject to periodically varying pressures, *J. Acoustic Soc. Amer.*, 47:205–210.

Charm, S., and Kurland, G. S. (1965) Viscometry of human blood for shear rates of 0–100,000 sec^{-1}, *Nature (London)* 206:617–618.

Chen, T. C., and Skalak, R. (1970) Sheroidal particle flow in a cylindrical tube, *Appl. Sci. Res.*, 22:403–441.

Clark, A. J. (1927) *Comparative Physiology of the Heart*, Cambridge University Press, London, pp. 9 and 149–150.

Davey, T. B., Kaufman, B., and Smeloff, E. A. (1983) Design and Testing of Prosthetic Heart Valves, in *Advances in Cardiovascular Physics, 5, Cardiovascular Engineering, Part IV: Prostheses, Assistant and Artificial Organs* (ed. D. N. Ghista, E. Van Vollenhoven, W.-J. Yang, H. Reul, and W. Bleifeld), S. Karger, Basel, Switzerland, pp. 1–15.

Davson, H. (1970) *A Textbook of General Physiology*, 4th ed., vol. 1, Churchill, London, pp. 322–332.

Downey, V. M. (1963) The use of overcompression in the treatment of decompression sickness. *Aerospace Med.*, 34:28–34.

Downey, V. M., Worley, T. W., Hackworth, R., and Whitley, J. L. (1963) Studies on bubbles in human serum under increased and decreased atmospheric pressure, *Aerospace Med.*, 34: 116–23.

Esmond, W. G., and Clark, H. (1966) Mathematical Analysis and Mass Transfer Optimization of a Compact, Low Cost, Pumpless System for Hemodialysis (Dialung), in *Biomedical Fluid Mechanics Symposium*, ASME, New York, pp. 161–191.

Fich, S., Welkowitz, W., and Hilton, R. (1966) Pulsatile Blood Flow in the Aorta, in *Biomedical Fluid Mechanics Symposium*, ASME, New York, pp. 34–44.

Fitzgerald, J. M. (1969a) Mechanics of red-cell motion through very narrow capillaries, *Proc. Roy. Soc. London B*, 174:193–227.

Fitzgerald, J. M. (1969b) Implications of a theory of erythrocytes motion in narrow capillaries, *J. Appl. Physiol.*, 27:912–918.

Fox, J. A., and Hugh, A. E. (1966) Localization of atheroma: A theory based on boundary layer separation, *Br. Heart J.*, 28:388–399.

Fredrickson, A. G. (1964) *Principles and Applications of Rheology*, McGraw-Hill, New York.

Fry, D. L., Thomas, L. J., and Greenfield, J. C. (1980) Flow in Collapsible Tubes, in *Basic Hemodynamics and Its Role in Disease Processes* (ed. D. J. Patel and R. N. Vaishnav), University Park Press, Baltimore, Md.

Fung, Y. C. (1966) Theoretical considerations of the elasticity of red cells and small blood vessels, *Fed. Proc.*, 25:1761–72.

Fung, Y. C. (1984) *Biodynamics—Circulation*, Springer-Verlag, New York.

Gaehtgens, P., Meiselman, H. J., and Wayland, H. (1970) Erythrocyte flow velocities in mesenteric microvessels of the cat, *Microvascular Res.*, 2:151–162.

Goldsmith, H. L., and Mason, S. G. (1962) The movement of single large bubbles in closed vertical tubes, *J. Fluid Mech.*, 14:45–58.

Greisheimer, E. M. (1955) *Physiology and Anatomy*, 7th ed., J. B. Lippincott, Philadelphia.

Guest, M. M., Bond, T. P., Cooper, R. G., and Derrick, J. R. (1963) Red blood cells: Change in shape in capillaries, *Science*, 142:1319–1321.

Hamilton, W. F. (1962) Measurement of the Cardiac Output, in *Handbook of Physiology*, vol. 1, sec. 2. Williams & Wilkins, Baltimore, Md.

Harvey, E. N. (1945) Decompression sickness and bubble formation in blood and tissues, *Bull. N.Y. Acad. Med.*, 21:505.

Haymaker, W., Johnston, A. D., and Downey, V. M. (1956) Fatel decompression sickness during jet aircraft flight, *J. Aviat. Med.*, 27:2–17.

Haynes, R. H. (1960) Physical basis of the dependence of blood viscosity on tube radius, *Am. J. Physiol.*, 198:1193–1200.

Haynes, R. H. (1961) The rheology of blood, *Trans. Soc. Rheol.*, 5:85–101.

Hill, L. (1912). *Caisson Sickness and the Physiology of Work in Compressed Air*, Arnold, London.

Hochmuth, R. M., and Sutera, S. P. (1970) Spherical caps in low Reynolds number tube flow, *Chem. Eng. Sci.* 25:593–604.

Hochmuth, R. M., Marple, R. N., and Sutera, S. P. (1970) Capillary blood flow: I. Erthrocyte deformation in glass capillaries, *Microvascular Res.*, 2, (4), 409–419.

Holenstein, R., Niederer, P., and Anliker, M. (1980) A viscoelastic model for use in predicting arterial pulse waves, *J. Biomech. Eng.*, 102:318–325.

Hyman, W. A., and Skalak, A. (1970) Viscous flow of a suspension of deformable liquid drops in a cylindrical tube, *Tech. Rep. No. 5*, Project 062-393, Columbia University, New York.

Intaglietta, M., Pawula, R. F., and Tompkins, W. R. (1970) Pressure measurements in the mammalian microvasculature, *Microvascular Res.* 2:212–220.

Kandarpa, K., and Davids, N. (1976) Analysis of the fluid dynamic effects on atherogenesis at branching sites, *J. Biomech.*, 9:735–741.

Karino, T., and Motomiya, M. (1983) Flow visualization in isolated transparent natural blood vessels, *Biorheology* 20:119–127.

Karino, T., Kwong, H. H. M., and Goldsmith, H. L. (1979) Particle flow behaviour in models of branching vessels: I. Vortices in 90 T-junctions, *Biorheology*, 16:231–248.

Klein, S. J. (1965) *Similitude and Approximation Theory*, McGraw-Hill, New York.

Kuchar, N. R., and Ostrach, S. (1966) Flows in the Entrance Regions of Circular Elastic Tubes, in *Biomedical Fluid Mechanics Symposium*, ASME, New York, pp. 45–69.

Kuwahara, M., Hirakawa, A., and Saito, M. (1979) Compartmental Analyses of Transport Processes in the Circulatory and Renal Systems by Radiocardiogram, Radioencephalogram and Renogram, in *Advances in Cardiovascular Physics, vol. 4, Foundations of Noninvasive Cardiovascular Diagnostic Processes* (ed. D. N. Ghista, E. van Vollenhoven, W. J. Yang, and H. Reul), S. Karger, Basel, Switzerland, pp. 41–105.

Lambert, J. W. (1958). On the nonlinearities of fluid flow in nonrigid tubes, *J. Franklin Inst.*, 266: 83–102.

Langhaar, H. L. (1942) Steady flow in the transition length of a straight tube, *Lafayette Ind., Trans.*, ASME 64:A-55 to A-58.

Lew, H. S., and Fung, Y. C. (1969) The motion of the plasma between the red cells in the bolus flow, *Biorheology*, 6:109–119.

Lighthill, M. J. (1968) Pressure forcing of tightly fitting pellets along fluid-filled elastic tubes, *J. Fluid Mech.*, 34:113–143.

Lighthill, M. J. (1972) Physiological fluid mechanics: A survey, *J. Fluid Mech.*, 52 (Pt. 3): 475–497.

Lyon, C. K., Scott, J. B., and Wang, C. Y. (1980) Flow through collapsible tubes at low Reynolds numbers: Applicability of the waterfall model, *Circ. Res.*, 47:68–73.

McDonald, D. A. (1964) Frequency Dependence of Vascular Impedance, in *Pulsatile Blood Flow* (ed. E. O. Attinger), McGraw-Hill, New York.

Monro, P. A. G. (1964) The deformation of red cells and groups of cells in blood flowing in small blood vessels, *Bibl. Anat.*, 7:376–382.

Moravec, S., and Liepsch, D. (1983) Flow investigation in a model of a three-dimensional human artery with Newtonian and non-Newtonian fluids, Part I, *Biorheology*, 20:745–759.

Morgan, G. W. (1952) On the steady laminar flow of a viscous imcompressible fluid in an elastic tube, *Bull Math. Biophys.*, 14:19–26.

Morgan, G. W., and Ferrante, W. (1955) Wave Propagation in elastic tubes filled with steaming liquid, *J. Acoust. Soc. Amer.*, 27:715–1055.

Morgan, G. W., and Kiely, J. P. (1954) Wave Propagation in a viscous liquid contained in a flexible tube, *J. Acoust. Soc. Amer.*, 26:323–328.

Myers, G. H., and Parsonnet, V. (1969) *Engineering in the Heart and Blood Vessels*, Wiley Interscience, New York.

Nerem, R. M. (ed.) (1981) Hemodynamics in the arterial wall. *J. Biomech. Eng.*, 103:171–212.

Netter, F. H. (1969) *The CIBA Collection of Medical Illustrations, 5, Heart*, CIBA Pharmaceutical Co.

O'Brien, V., Ehrlich, L. W., and Friedman, M. H. (1976) Unsteady flow in a branch, *J. Fluid Mech.*, 75:315–336.

Palmer, A. A. (1959) A study of blood flow in minute vessels of pancreatic region of the rat with reference to intermittent corpuscular flow in individual arteries, *Quart. J. Exp. Physiol.*, 44: 149–159.

Patel, D. J., and Vaishnav, R. N. (eds.) (1980) *Basic Hemodynamics and Its Role in Disease Processes*, University Park Press, Baltimore, Md.

Pedley, T. J. (1980) *The Fluid Mechanics of Large Blood Vessels*, Cambridge University Press, London.

Peirce, E. C., II (1969) *Extracorporeal Circulation for Open-Heart Surgery*. Charles Thomas, Springfield, Ill.

Phillips, W. M. (1983) Design Criteria and Evaluation of a Sac Arteficial Heart, in *Advances in Cardiovascular Physics*, vol. 5 *Cardiovascular Engineering*, Part IV: *Protheses, Assistant and Artificial Organs* (ed. D. N. Ghista, E. Van Vollenhoven, W.-J. Yang, H. Reul, and W. Bleifeld), S. Karger, Basel, Switzerland, pp. 184–215.

Prandtl, L., and Tietjens, O. (1957) *Applied Hydro and Aeromechanics*, Dover, New York.

Reul, H. (1983a) In-vitro Evaluation of Artificial Heart Valves, in *Advances in Cardiovascular Physics*, vol. 5, *Cardiovascular Engineering*, Part IV: *Protheses, Assistant and Artificial Organs* (ed. D. N. Ghista, E. Van Vollenhoven, W.-J. Yang, H. Reul, and W. Bleifeld), S. Karger, Basel, Switzerland, pp. 16–30.

Reul, H. (1983b) Blood Pumps—General Design Considerations, in *Advances in Cardiovascular Physics*, vol. 5, *Cardiovascular Engineering*, Part IV: *Protheses, Assistant and Artificial Organs* (ed. D. N. Ghista, E. Van Vollenhoven, W.-J. Yang, H. Reul, and W. Bleifeld), S. Karger, Basel, Switzerland, pp. 55–71.

Rudinger, G. (1966) Review of Current Mathematical Methods for the Analysis of Blood Flow, in *Biomedical Fluid Mechanics Symposium*, ASME, New York, pp. 1–33.

Rushmer, R. F. (1961) *Cardiovascular Dynamics*, W. B. Saunders, Philadelphia.

Sabbah, H. N., Hawkins, E. T., and Stein, P. D. (1984) Flow separation in renal arteries, *Arteriosclerosis* 4:28–33.

Seed, A. (1972) Hot Film Anemometry in Human Aorta, in *Blood Flow Measurements* (ed. Roberts), Sector Publishing, London.

Sevilla-Larrea, J. F. (1968) Detailed characteristics of pulsatile blood flow in small glass capillaries, Ph.D. thesis, Carnegie-Mellon University, Pittsburgh.

Sheshadri, V., Hochmuth, R. M., Croce, P. A., and Sutera, S. P. (1970) Capillary blood flow: III. Deformable model cells compared to erythrocytes in vitro, *Microvascular Res.*, 2(4):434–442.

Siouffi, M., Pelissier, R., Farahifar, D., and Rieu, R. (1984) The effect of unsteadiness on the flow through stenoses and bifurcations, *J. Biomech.*, 17:299–315.

Skalak, R. (1972) Mechanics of the microcirculation, in *Biomechanics—Its Foundations and Objectives* (ed. Y. C. Fung, N. Perrone, and M. Anliker), Prentice-Hall, Englewood Cliffs, N.J.

Skalak, R., and Branemark, P. I. (1969) Deformation of red blood cells in capillaries, *Science* 164: 711–712.

Stahl, W. R. (1963) The analysis of biological similarity, *Adv. Biol. Med. Phys.* 9:355–464.

Stalker, A. L., Engeset, J., and Matheson, N. A. (1966). Leucocyte impaction and erthrocytes flow in arterioles, Proc. Fourth European Conf. on Microcirculation, Cambridge Abstract F. 4.

Stehbens, W. E. (1967) Aspects of blood flow and platelet agglutination, *Biorheology* 4:85.

Stettler, J. C., Niederer, P., and Anliker, M. (1980) Theoretical analysis of arterial hemodynamics including the influence of bifurcations, *Ann. Biomed. Eng.*, 9:145–175.

Strasberg, M. (1956) Undissolved Air Cavities as Cavitation Nuclei. *Symp. Cavitation in Hydrodynamics*. Her Majesty's Stationary Office, London.

Streeter, V. L., and Wylie, E. B. (1985) *Fluid Mechanics*, 8th ed. McGraw-Hill, New York.

Streeter, V. L., Keitzer, W. F., and Bohr, D. F. (1963) Pulsatile pressure and flow through distensible vessels, *Circ. Res.*, 13:3–20.

Streeter, V. L., Keitzer, W. F., and Bohr, D. F. (1964) Energy Dissipation in Pulsatile Flow Through Distensible Tapered Vessels, in *Pulsatile Blood Flow* (ed. E. O. Attinger), Blakiston Div. McGraw-Hill, New York, pp. 149–177.

Sugawara, M., Matsuo, H. Kajiya, F., and Kitabatake, A. (1985) *Blood Flow* (in Japanese), Kodansha Scientific, Tokyo.

Tanasawa, I., and Yang, W.-J. (1969) Dynamic behavior of a gas bubble in viscoelastic liquids, *J. Appl. Phys.*, 41:4526–4531.

Tanasawa, I., Wotton, D. R., Yang, W.-J., and Clar, D. W. (1970) Experimental study of air bubbles in a simulated cardiopulmonary bypass system with flow constraint, *J. Biomech.*, 3: 417–424.

Tomonaga, G., Kajiya, F., Ogasawara, Y., Mito, K., Tsujioka, K., Hironaga, K., Kano, H., and Nishihara, H. (1983) Analysis of coronary flow dynamics by laser Doppler, in *The Application of Laser Doppler Velocimetry* (ed. T. Takahei), Power, Tokyo.

Trowell, O. A. (1943) Liver vacuoles and anoxemia, *Nature (London)*, 150:730.

Vorhauer, B. W., and Byars, E. F. (1966) The engineering design of heart valve protheses, *Biomedical Fluid Mechanics Symposium*, ASME, *Soc. Mech.*, New York, pp. 122–126.

Walburn, F. J., Sabbah, H. N. and Stein, P. D. (1981) Flow visualization in a mold of an atherosclerotic human abdominal aorta, *J. Biomech. Eng.*, 103:168–170.

Wang, H., and Skalak, R. (1969) Vicous flow in a cylindrical tube containing a line of spherical particles, *J. Fluid Mech.* 38:75–96.

Whitmore, R. L. (1967) A theory of blood flow in small vessels, *J. Appl. Physiol.*, 22:767–771.

Whitmore, R. L. (1968) *Rheology of the Circulation*, Pergamon, Oxford.

Womersley, J. R. (1955) Method for the calculation of velocity, rate of flow and viscous drag in arteries when the pressure gradient is known, *J. Physiol.*, 127:553–563.

Womersley, J. R. (1957) An Elastic Tube Theory of Pulse Transmission and Oscillating Flow in Mammalian Arteries, Wright Air Dev. Center Tech. Report TR. 56-614, Dayton, Ohio.

Womersley, J. R. (1958) Oscillatory flow in arteries: The constrained elastic tube as a model of arterial flow and pulse transmission, *Phys. Med. Biol.*, 2:178–187.

Yang, W.-J. (1971) Dynamics of gas bubbles in whole blood and plasma, *J. Biomech.*, 4:119–125.

Yang, W.-J. (1974) A Major Cause of Blood Trauma in Extracorporeal Circulation, in *Advances in Bioengineering*, ASME, New York, pp. 167–168.

Yang, W.-J., and Liang, C. Y. (1972) Behavior of gas bubbles in viscoelastic materials in a creep process, *J. Appl. Phys.*, 43:3060–3064.

Yang, W.-J., Echigo, R., Wotton, D. R., and Hwang, J. B. (1971a) Experimental studies of the dissolution of gas bubbles in whole blood and plasma, Part I: Stationary bubbles, *J. Biomech.*, 4:275–281.

Yang, W.-J., Echigo, R., Wotton, D. R., and Hwang, J. B. (1971b) Experimental studies of the dissolution of gas bubbles in whole blood and plasma, Part II: Moving bubbles or liquids, *J. Biomech.*, 4:283–288.

Yang, Wen-Jei, Echigo, R., Wotton, D. R., Ou, J. W., and Hwang, J. B. (1972) Mass transfer from gas bubbles to impinging flow of biological fluids with chemical reaction, Biophys. J., 12:1391–1404.

Yen, R. T., and Foppiano, L. (1981) Elasticity of small pulmonary veins in the cat, *J. Biomech. Eng.*, 103:38–42.

Yen, R. T., Fung, Y. C., and Bingham, N. (1980) Elasticity of small pulmonary arteries in the cat, *J. Biomech. Eng.*, 102:170–177.

Young, D. F., and Cholvin, N. R. (1966) Application of the Concept of Similutude to Pulsatile Blood Flow Studies, *Biomedical Fluid Mechanics Symposium,* ASME, New York, pp. 78–88.

CHAPTER

THREE

THERMODYNAMICS

3.1 SOME FUNDAMENTAL CONCEPTS

Thermodynamics deals with energy and entropy. Entropy is a thermodynamic property used to determine whether or not an event is to occur. Once the event does occur, energy associated with it must be conserved.

3.1.1 System and Control Volume

The first step in a thermodynamic analysis is to define a system and the surroundings. One can define a quantity of substance or a volume of empty space as a system. Then everything external to the system is called the surroundings. The system is separated from its surroundings by a boundary. The system may be classified into two categories—open and closed—depending on the involvement of mass flow across the boundary.

A closed system is not associated with mass flow. A nonvasculated tissue, such as fat, Fig. 3.1, is an example. The energy produced by metabolism, $\dot{Q}$ is dissipated across the boundary into the environment or the surrounding tissues as heat, Q.

A system that involves a mass flow across the boundary is called an open system or control volume. Its boundary is commonly referred to as the control surface. When a human being is defined as the system, it is an open system. During breathing, there is a flow of substances, such as air, water vapor, and carbon dioxide, into and out of the body. In addition to energy dissipation through breathing, a vast amount of waste energy is released through the skin

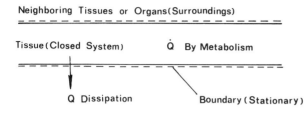

Figure 3.1 Closed system with stationary boundary.

into the environment in order to maintain the body temperature. A layer of vasculated tissue is another example of an open system. Figure 3.2 depicts a blood vessel branching inside the tissue. In addition to heat flow, there are mass flows into and out of the control volume.

One may define the cardiac space enclosed by the cardiac muscles as an open system. Figure 3.3a shows that during systole, blood flows across the control surface into the aorta. Heat transfer from the cardiac muscles into the blood within the control volume takes place simultaneously. During diastole, shown in Fig. 3.3b, blood enters into the control volume accompanied by heat transfer from the cardiac muscles into the blood. In this example, the boundary moves, while the systems in the previous two examples possess stationary boundaries.

3.1.2 Properties and State of a Substance

Properties determine the state of a substance. This is true in thermodynamics, rheology, physiology, as well as in finance (a person's financial state is determined by his or her financial properties, for example). Thermodynamic properties are divided into intensive and extensive properties. An intensive property is independent of the mass, for example, pressure and temperature. The value of an extensive property varies with the mass, such as mass and total volume. A division of the system results in a change in the value of extensive properties. An intensive property can be determined by dividing an extensive property by the mass, for example, specific volume. Two independent intensive properties may determine the state of a pure substance.

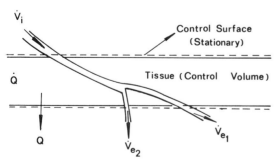

Figure 3.2 Open system with stationary boundary.

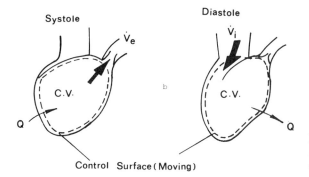

Figure 3.3 Open system with moving boundary. (*a*) Systole; (*b*) diastole. C.V., control volume.

Control Surface (Moving)

3.1.3 Path of a Change in State

A system changes in state when one or more of its properties change. The path of a change in state is called a process. If the system returns to the initial state after a change in state, the path is called a cycle. For example, a systolic process and a diastolic process form a cycle of the heartbeat. Figure 3.4 illustrates a cycle that consists of two processes, A and B.

3.2 HEAT, WORK, AND ENERGY

Both heat and work are a form of energy. They are interconvertible. There are three similarities between heat Q and work W:

1. Both are path functions that determine their magnitude. $+Q$ denotes heat added to the system, while $-W$ represents work done on the system. Both cases result in the addition of energy to the system.
2. Both are boundary phenomena signifying energy crossing the system boundary.
3. Both are transient phenomena during a change of states.

There are three modes of heat transfer: conduction, convection, and radiation, which are presented later in detail. Similarly, there are various work modes: work done at a moving boundary $P\,dV$, flow work $d(PV)$, pumping work $V\,dP$, stretching and contraction work in muscles, electrical work, etc.

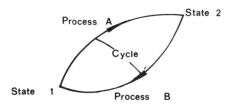

Figure 3.4 Cycle and processes.

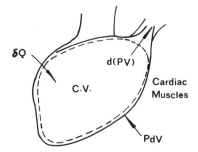

Figure 3.5 Work and heat in the heart.

Take the heart, as an example. Let us define the space enclosed by the cardiac muscles as the control volume. Figure 3.5 shows that during systolic process, the system is subject to work done by the cardiac muscles $(-P\,dV)$, convection from the cardiac muscles $(+\delta Q)$, and work of the effluent blood $[-d(PV)]$.

In sharp contrast to heat and work being *path* functions, the energy of the system or control volume, E, is a property, namely, a *point* function (this will be proven in the next section). E consists of the internal and external energy. The internal energy U includes all forms of energy that are associated with the thermodynamic state of the system, such as atomic, chemical, and electrical energy. Energy released by metabolism is a form of internal energy. Many physiologists treat metabolic energy as a stored chemical energy. External energy refers to kinetic and potential energy, KE and PE, respectively, whose magnitude depends on the coordinate frame selected. By definition,

$$E = U + KE + PE \tag{3.1}$$

where $U = mu$; $KE = \frac{1}{2}mv^2$; $PE = mgZ$; and m, g, u, v, and Z denote the mass, gravitational acceleration, specific internal energy, velocity, and elevation of the system, respectively. They are all relative quantities in reference to a certain state.

3.3 FIRST LAW OF THERMODYNAMICS

Heat, work, and energy must be conserved during the course of changing states in the system. This natural law, which describes the principle of conservation of energy, is called the first law of thermodynamics.

3.3.1 Cycle

Observations conclude that the amount of work and heat are always proportional during a cycle:

$$\oint \delta Q = \oint \delta W \tag{3.2}$$

in which $\oint$ is the symbol for cyclic integration.

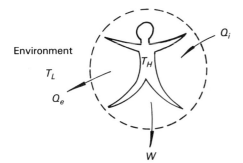

Environment

T_L

Q_e

Q_i

T_H

W

Figure 3.6 Work and heat in a man.

Example 3.1 Human body undergoing an energy cycle Let a man be the system and consider the conservation of energy in his body over a day, Fig. 3.6. The one-day cycle is divided into two periods: awake and asleep. Then, Eq. (3.2) can be written as

$$\int_{\text{Awake}} \delta Q + \int_{\text{Asleep}} \delta Q = \int_{\text{Awake}} \delta W + \int_{\text{Asleep}} \delta W \qquad (3.3)$$

During the awake time, foods (stored chemical energy) are consumed and, through metabolism, release an appropriate amount of thermal energy Q_i. External work W is performed in various forms such as exercise, labor, and study. Throughout the 24-h period, heat dissipation into the environment, Q_e, continues to take place through breathing and evaporation, convection, and radiation from the skin surface. Therefore, a cyclic balance of energy reads

$$(Q_i - Q_e)_{\text{awake}} - (Q_e)_{\text{asleep}} = (W)_{\text{awake}} \qquad (3.4)$$

Here, the heat produced by metabolism in an average-sized adult of 70 kg is

$$Q_i = (70 \text{ kcal/h})(24 \text{ h}) = 1680 \text{ kcal/day}$$

3.3.2 Process

Consider a system that undergoes a cycle changing from state 1 to state 2 following path A and returning from state 2 to state 1 by process B as shown in Fig. 3.7. Then, one can rewrite Eq. (3.2) as

$$\int_{1A}^{2A} \delta Q + \int_{2B}^{1B} \delta Q = \int_{1A}^{2A} \delta W + \int_{2B}^{1B} \delta W \qquad (3.5)$$

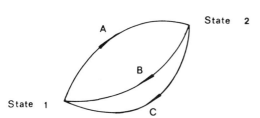

A

B

State 2

State 1

C

Figure 3.7 Cycles AB and AC.

Similarly, for another cycle which takes the same process A from state 1 to state 2 but returns to state 1 by process C, one obtains

$$\int_{1A}^{2A} \delta Q + \int_{2C}^{1C} \delta Q = \int_{1A}^{2A} \delta W + \int_{2C}^{1C} \delta W \tag{3.6}$$

When Eq. (3.6) is subtracted from Eq. (3.5), and the resulting equation is rearranged, it yields

$$\int_{2B}^{1B} (\delta Q - \delta W) = \int_{2C}^{1C} (\delta Q - \delta W) \tag{3.7}$$

Equation (3.7) suggests that $(\delta Q - \delta W)$ depends only on the initial and final states; in other words, it is independent of the path. It is, therfore, a point function and thus is a physical property, which is the energy of the system E:

$$\delta Q - \delta W \equiv dE$$
$$= dU + d(KE) + d(PE)$$
$$= d(mu) + d(\tfrac{1}{2}mv^2) + d(mgZ) \tag{3.8}$$

This is the first law of thermodynamics for a closed system expressed in differential form. For convenience in thermodynamic analyses, the first laws often appears in other forms such as the integral (or finite-difference) form:

$$\int_1^2 \delta Q - \int_1^2 \delta W = \Delta U + \Delta(KE) + \Delta(PE)$$

or

$$_1Q_2 - {}_1W_2 = m(u_2 - u_1) + \tfrac{1}{2}m(v_2^2 - v_1^2) + mg(Z_2 - Z_1) \tag{3.9}$$

where

$$_1Q_2 = \int_1^2 \delta Q$$

$$_1W_2 = \int_1^2 \delta W$$

Another form of the first law is the rate form:

$$\frac{\delta Q}{dt} - \frac{\delta W}{dt} = \frac{dU}{dt} + \frac{d(KE)}{dt} + \frac{d(PE)}{dt} \tag{3.10}$$

Equation (3.10) is used in the field of heat transfer.

For many thermodynamic processes in closed systems, the internal energy change dU is the only significant energy change. Then,

$$\delta Q - \delta W = dU \tag{3.11}$$

In the absence of friction, the only significant work done by the system is the work at a moving boundary:

$$\delta W = P\,dV \tag{3.12}$$

$$\delta Q = dU + P \, dV \tag{3.13}$$

with the definition of

$$H \equiv U + PV \tag{3.14}$$

and its differential form

$$dH = dU + P \, dV + V \, dP \tag{3.15}$$

The combination of Eqs. (3.14) and (3.15) produces

$$\delta Q = dH - V \, dP \tag{3.16}$$

The $(-V \, dP)$ term is the work required to pump a fluid against the pressure difference dP. Equations (3.13) and (3.16) can be simplified for an isobaric process, $dP = 0$:

$$\delta Q = dH = m \, dh \tag{3.17}$$

and for an isochoric process, $dV = 0$:

$$\delta Q = dU = m \, du \tag{3.18}$$

With the aid of the definition of specific heats, Eqs. (3.17) and (3.18) may be expressed in terms of the temperature change dT for an isobaric process:

$$\delta Q = mC_p \, dT \tag{3.19}$$

For an isochoric process:

$$\delta Q = mC_v \, dT \tag{3.20}$$

Here C_p and C_v are the specific heats under constant pressure and volume, respectively. In living bodies, reactions generally occur at constant pressure.

Example 3.2 Process in resting humans No significant mechanical work is done by resting humans. Therfore, $_1W_2 = 0$. The first law is reduced to

$$\Delta U = {_1}Q_2 \quad \text{or} \quad \frac{dU}{dt} = \dot{Q} = \text{Basal metabolism}$$

in which $\dot{Q} \cong 70$ kcal/h for an average-sized adult male.

Example 3.3 Body temperature regulation The average specific heat C_p of the human body is approximately 0.86 kcal/kg °C. If the body temperature is 1°C below normal, determine the amount of pure glucose that must be completely oxidized to provide the energy required to restore a body to its normal temperature.

$$_1Q_2 = \Delta H = mC_p \, \Delta T$$

$$= (70 \text{ kg})(0.86 \text{ kcal/kg °C})(1°C)$$

$$= 60 \text{ kcal}$$

In most biological systems, chemical reactions, such as metabolisms, occur essentially at isothermal conditions. The heat of reaction for glucose at $P = 1$ atm and $T = 298$ K is $\Delta H_R^\circ = -673$ kcal/mol (exothermic):

$$C_6H_{12}O_6 + 6O_2 \longrightarrow 6H_2O + 6O_2 - 673 \text{ kcal}$$

Therefore, the amount of glucose required is

$$m_g = \frac{{}_1Q_2 \, M_g}{\Delta H_R^\circ} = \frac{(60 \text{ kcal})(180 \text{ g/mol})}{(673 \text{ kcal/mol})}$$

$$= 16 \text{ g}$$

Here, the molecular weight of glucose M_g is 180 g.

Example 3.4 Cardiac muscles Figure 3.8 shows the cardiac muscles pumping blood at a rate of 5 L/min. A fraction of the metabolic energy generated in the muscles is dissipated into the blood. The mechanical work, assuming P is constant, is

$$\dot{W} = \int P \frac{dv}{dt} = P\dot{V}$$

$$= (100 \text{ mmHg})(5 \text{ L/min})$$

$$= 22.9 \text{ kcal/day}$$

The metabolic energy is

$$\dot{U} = (\dot{m}_{O_2})(\text{Calorific } O_2 \text{ equivalent})$$

$$= (5 \text{ ml/min})(4.825 \text{ kcal/L } O_2)$$

$$= -34.7 \text{ kcal/day}$$

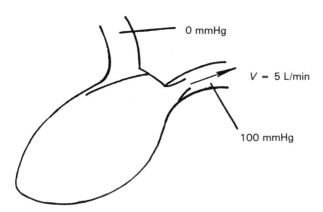

O mmHg

V = 5 L/min

100 mmHg

Figure 3.8 Action of cardiac muscles.

The first law for the cardiac muscles as a closed system reads

$$\dot{Q} = \dot{W} + \dot{U}$$

$$= -11.8 \text{ kcal/day}$$

The efficiency of the heart is 22.9/34.7, or about 66% of the available energy.

Each organ is a net consumer of stored chemical energy, either to do mechanical work, as in the heart and muscles, to perform synthesis of high-energy compounds, as in the liver, to conduct concentration processes, as in the stomach acid glands, or to transport species against gradients, as in the tubules of the nephron. The efficiency and modes of energy conversion of the physicochemical processes to accomplish these jobs vary from organ to organ.

3.3.3 Control Volume

In applying the first law of thermodynamics to a control volume, the principle of conservation of mass, or the continuity equation for short, must be taken into consideration. It reads

$$\frac{dm_{cv}}{dt} + \Sigma \dot{m}_e - \Sigma \dot{m}_i = 0 \tag{3.21}$$

Here m and $\dot{m}$ denote the mass and mass flow rate. The subscripts cv, e, and i signify within, exit, and enter (the control volume), respectively.

Now, the energy balance on the control volume must include all kinds of energy and work that are carried across the control surface by the crossing masses $\dot{m}_e$ and $\dot{m}_i$. Hence, each of the flow terms carries e and Pv^* (flow work), which can be written as

$$e + Pv^* = u + Pv^* + \tfrac{1}{2}V^2 + gZ$$

$$= h + \tfrac{1}{2}V^2 + gZ \tag{3.22}$$

where $h = u + Pv^*$ is the specific enthalpy and V is the flow velocity.

The rate equation of the first law for a control volume reads

$$\frac{\delta Q_{cv}}{dt} + \Sigma \frac{dm_i}{dt} (h_i + \tfrac{1}{2}V_i^2 + gZ_i)$$

$$= \frac{dE_{cv}}{dt} + \Sigma \frac{dm_e}{dt} (h_e + \tfrac{1}{2}V_e^2 + gZ_e) + \frac{\delta W_{cv}}{dt} \tag{3.23}$$

Two processes are of special interest: steady state, steady flow (SSSF) and uniform state, uniform flow (USUF). The former considers the control volume as being fixed relative to the coordinate frame and the mass at each point in the control volume remaining constant with time. Steady flows through kidneys, lungs, and vasculated tissues are typical examples. The USUF process assumes a uniform state throughout the entire control volume at any instant of time and the total mass crossing the control surface remaining constant with time (although the mass flow rates may be time varying). With the heart defined as a control volume, the filling of blood during the diastolic process and the discharge of blood during the systolic process are typical examples. The equations, in summary, are, for the SSSF process

Continuity:

$$\dot{m}_i = \dot{m}_e = \dot{m} = \frac{dm}{dt} \tag{3.24}$$

First law:

$$\frac{\delta Q_{cv}}{dt} + \dot{m}(h_i + \tfrac{1}{2}V_i^2 + gZ_i) = \dot{m}(h_e + \tfrac{1}{2}V_e^2 + gZ_e) + \frac{\delta W_{cv}}{dt} \tag{3.25}$$

For the USUF process, the equations are

Continuity:

$$\frac{dm_{cv}}{dt} = \Sigma \dot{m}_i - \Sigma \dot{m}_e \tag{3.26}$$

First law:

$$\frac{\delta Q_{cv}}{dt} + \Sigma \dot{m}_i(h_i + \tfrac{1}{2}V_i^2 + gZ_i)$$

$$= \Sigma \dot{m}_e(h_e + \tfrac{1}{2}V_e^2 + gZ_e) + \frac{d}{dt}[m(u + \tfrac{1}{2}V^2 + gZ)]_{cv} + \frac{\delta W_{cv}}{dt} \tag{3.27}$$

3.3.4 Environmental Control

A special case of the SSSF process bears an important physiological implication: the humidification process. It involves an air-water vapor mixture which is inhaled into the lungs, moisturized in the alveolar space, and exhaled as a higher humidity gas mixture at the lung temperature. Figure 3.9 illustrates the humidification process in the lungs. Neglecting changes in kinetic and potential energy, the first law reduces to

$$\dot{Q}_{cv} + \dot{m}_{a1}h_{a1} + \dot{m}_{v1}h_{v1} + \dot{m}_{l2}h_{l2} = \dot{m}_{a2}h_{a2} + \dot{m}_{v2}h_{v2}$$

The continuity equations for the air and water (vapor and liquid) are

$$\dot{m}_{a1} = \dot{m}_{a2} = \dot{m}_a$$

$$\dot{m}_{v1} = \dot{m}_{v2} + \dot{m}_{l2}$$

The first law can be rewritten in the form

$$\frac{\dot{Q}_{cv}}{\dot{m}_a} + h_{a1} + \omega_1 h_{v1} + (\omega_2 - \omega_1)h_{l2} = h_{a2} + \omega_2 h_{v2}$$

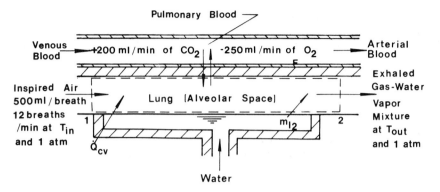

Figure 3.9 Adiabatic humidification process in normal lung.

or, with the assumption of an ideal gas behavior, as

$$\frac{\dot{Q}_{cv}}{\dot{m}_a} + \omega_1 h_{v1} + (\omega_2 - \omega_1)h_{l2} = C_{Pa}(T_2 - T_1) + \omega_2 h_{v2} \qquad (3.28)$$

where ω denotes the specific humidity,[1] defined as m_v/m_a; h_l is the enthalpy of saturated liquid at the existing temperature, and the subscripts a, v, and l signify air, water vapor, and liquid water. The specific humidity can be determined by

$$\omega = 0.622 \frac{P_v}{P_a} \qquad (3.29)$$

which is derived under the assumption of ideal gases (air and water vapor). P_v and P_a are the partial pressures of air and water vapor, respectively. Their sum is equal to the total pressure of the air-water vapor mixture P_t:

$$P_t = P_a + P_v \qquad (3.30)$$

3.4 SECOND LAW OF THERMODYNAMICS

For a cycle or a process to occur, the principle of conservation of energy must hold. An important question preceding the problem of energy conservation is whether or not the cycle or process under consideration will occur at all. What is the criterion for a spontaneous transformation of a substance?

---

[1] The relative humidity ϕ is defined as $(P_v$ at $T_1)/(P_{v\ sat}$ at same $T_1)$. ϕ is related to ω by $\phi = (P_a\omega)/(0.622\ P_{v\ sat}$ at $T_1)$; $P_a = P - P_v$.

3.4.1 Cycle

As for the case of the first law, observations have established the inequality of Clausius:

$$\oint \frac{\delta Q}{T} \leq 0 \tag{3.31}$$

This is the criterion for the occurrence of a cycle. An irreversible cycle takes place if the cyclic integral of $\delta Q/T$ is less than zero, while a reversible cycle occurs when the integral is zero. The cycle will never occur if the cyclic integral of $\delta Q/T$ is greater than zero. Equation (3.31) is a consequence of the second law of thermodynamics.

Example 3.5 Inequality of Clausius in humans Reconsider a human body undergoing an energy cycle in Example 3.1. Equation (3.4) indicates

$$\oint \delta Q = (Q_i - Q_e)_{\text{awake}} - (Q_e)_{\text{asleep}} > 0$$

As shown in Fig. 3.6, Q_i, the intake of foods, may be considered heat transferred into the system at body temperature T_H. The waste heat Q_e is dissipated into the environment at T_L through breathing and the skin. Since T_H is greater than T_L and $Q_i > Q_e$, one gets

$$\oint \frac{\delta Q}{T} = \frac{Q_i}{T_H} - \frac{Q_e}{T_L} < 0$$

Therefore, the cycle does occur, as all of us experience daily.

3.4.2 Process

The first law was stated in terms of a cycle, and then it led to the definition of a property, energy E. In a similar fashion, we have stated the second law for a cycle and are now proceeding to the introduction of another property, entropy S. Consider a reversible cycle consisting of a process from states 1 to 2 along path A and returning to the original state along path B as shown in Fig. 3.7. Equation (3.31) yields

$$\int_{1A}^{2A} \frac{\delta Q}{T} + \int_{2B}^{1B} \frac{\delta Q}{T} = 0 \tag{3.32}$$

For another reversible cycle following paths A and C, one obtains

$$\int_{1A}^{2A} \frac{\delta Q}{T} + \int_{2C}^{1C} \frac{\delta Q}{T} = 0 \tag{3.33}$$

The subtraction of Eq. (3.32) by Eq. (3.33) yields

$$\int_{2B}^{1B} \frac{\delta Q}{T} = \int_{2C}^{1C} \frac{\delta Q}{T} \tag{3.34}$$

It is concluded that for all reversible processes between states 2 and 1, the quantity $\int \delta Q/T$ is independent of the path, which means it is a point function, and is thus a property. One defines entropy as

$$dS \equiv \left(\frac{\delta Q}{T}\right)_{\text{rev}}$$

(3.35)

Suppose only process C is irreversible, while both processes A and B are reversible. Then, Eq. (3.32) still holds for a reversible cycle consisting of paths A and B. For an irreversible cycle consisting of processes A and C, the inequality of Clausius gives

$$\int_{1A}^{2A} \frac{\delta Q}{T} + \int_{2C}^{1C} \frac{\delta Q}{T} < 0$$

(3.36)

Subtraction of Eq. (3.33) by Eq. (3.36) yields

$$\int_{2B}^{1B} \frac{\delta Q}{T} > \int_{2C}^{1C} \frac{\delta Q}{T}$$

However, since process B is reversible and since entropy is independent of path, one gets

$$\int_{2B}^{1B} \frac{\delta Q}{T} = \int_{2B}^{1B} dS = \int_{2C}^{1C} dS$$

Therefore,

$$\int_{2C}^{1C} dS > \int_{2C}^{1C} \frac{\delta Q}{T}$$

It is thus concluded that

$$dS \geqslant \frac{\delta Q}{T}$$

(3.37)

in which the equality holds for a reversible process while the inequality holds for an irreversible case. The inequality leads to the concept of lost work LW by friction or other irreversibilities. It is manifested by a temperature increase, a velocity decrease, a pressure drop, or some combination thereof. Equation (3.37) can be written as

$$(dS)_{\text{system}} = \frac{\delta Q + \delta LW}{T}$$

(3.38)

The heat transferred to the surroundings at T_0 is $-\delta Q$ which causes a change in entropy by

$$(dS)_{\text{surroundings}} = \frac{-\delta Q}{T_0}$$

(3.39)

The total change of entropy in the universe is

$$(dS)_{universe} = (dS)_{system} + (dS)_{surroundings} \qquad (3.40)$$

$$= \frac{\delta Q + \delta LW}{T} - \frac{\delta Q}{T_0} \geq 0$$

The criterion for a process to occur is inequality for irreversible processes and equality for reversible processes. This is called the principle of the increase of entropy.

Example 3.6 Criterion for occurrence of metabolism As a result of metabolism, -673 kcal/mole of glucose (a negative quantity due to exothermal reaction) of internal energy is produced in biological systems at 37°C. This energy is dissipated into the surroundings at 20°C in order to maintain the tissue temperature. The process causes entropy changes in both the tissue and the surroundings. Letting the tissue be the system,

$$_1Q_2 = {}_1W_2 + \Delta U = -673 \text{ kcal/mol}$$

$$\Delta S_{system} = \frac{_1Q_2}{T} = \frac{-673}{310} = -2.171 \text{ kcal/K mol}$$

In reality, metabolism is an irreversible process and thus ΔS_{system} should be greater than -2.171 kcal/K. The entropy change of the surroundings is

$$\Delta S_{surroundings} = \frac{-{}_1Q_2}{T_0} = \frac{673}{293} = 2.297 \text{ kcal/K mol}$$

Hence, the total entropy change is

$$\Delta S_{system} + \Delta S_{surroundings} = 0.126 \text{ kcal/k mol} > 0$$

For a control volume, the second law reads, for the SSSF process,

$$\Sigma \dot{m}_e s_e - \Sigma \dot{m}_i s_i \geq \int_A \frac{\dot{Q}_{cv}/A}{T} \, dA \qquad (3.41)$$

where s is the specific entropy and A is the control surface area. The equation for the USUF process is

$$\Sigma \dot{m}_e s_e - \Sigma \dot{m}_i s_i + \frac{d}{dt}(ms)_{cv} \geq \int_A \frac{\dot{Q}_{cv}/A}{T} \, dA \qquad (3.42)$$

The principle of the increase of entropy reads

$$\frac{d}{dt} \int_V \rho s \, dV + \left(\frac{dS}{dt}\right)_{surroundings} \geq 0 \qquad (3.43)$$

3.5 FREE ENERGY

The second law of thermodynamics employs the properties of the entropy function to describe the conditions of spontaneity of material transformation. However, the changes in entropy involve both the transforming system as well as its surroundings. Therefore, the entropy function is not a convenient one to use in practice. It would be much more convenient to define the criteria of spontaneity based on a new function which is related to changes in properties of only the changing substance. This new function, called the free energy function, the thermodynamic potential, or Gibbs function, was created by fusing the first and the second laws into a single mathematical statement.

3.5.1 Criterion of Material Transformation

Gibbs free energy G is defined as

$$G = U - TS + PV = H - TS \qquad (3.44)$$

where U is the internal energy; P, pressure; V, volume; and H, enthalpy defined as

$$H = U + PV \qquad (3.45)$$

In biological systems, it is usually not necessary to consider the changes in kinetic and potential energies, that is, $U = E$. For most reactions that occur in solution, especially those of biological interest, the $P\Delta V$ term is negligible because the volume change ΔV is very small during the course of the transformation. Hence for most practical purposes, ΔH and ΔE are indistinguishable. Furthermore, conditions normally characteristic of biological reactions are at constant pressure (usually that of the atmosphere) and constant temperature. Under conditions of constant pressure and temperature, Eq. (3.44) yields

$$\Delta G = \Delta H - T\Delta S$$
$$\cong \Delta E - T\Delta S \qquad (3.46)$$

or

and Eq. (3.12) gives the expression for expansion work as

$$W = P\Delta V \qquad (3.47)$$

The first law requires that

$$Q_{surroundings} = -Q_{system} = -(\Delta E + P\Delta V)_{system}$$

or

$$\Delta S_{surroundings} = -(\Delta E + P\Delta V)_{system} \qquad (3.48)$$

Equation (3.48) is incorporated into the second law. It results in

$$\Delta E_{system} + P\Delta V_{system} - T\Delta S_{system} \leq 0 \qquad (3.49)$$

Since the left-hand side of Eq. (3.49) is identical with the right-hand side of Eq. (3.46), one obtains

$$\Delta G \leq 0 \qquad (3.50)$$

Now each thermodynamic quantity in the criterion of feasibility of material transformation refers to the system itself, while the properties of the surroundings have been eliminated. Thus, it can be stated that for a specified reaction to proceed spontaneously, the free energy change for the reaction must be negative. On the other hand, if the free energy change is zero, the system remains at equilibrium. When ΔG is a positive number, the reverse reaction must be the spontaneous one. In short, ΔG is a convenient fusion of the first and second laws which provides a quantitative indication of the potential ability of a substance to undergo a chemical or physical transformation.

As indicated in Eq. (3.46), the changes in both internal energy and entropy during the transformation affect the sign and magnitude of the free energy change ΔG. ΔS is positive for most practical transformations. Therefore, if ΔE is negative, then ΔG should be negative, i.e., useful work can be obtained from the free energy drop. Even if ΔE is zero, ΔG could still be negative.

3.5.2 Chemical Equilibrium Constant

Consider a simple chemical reaction in which conversion of A into B takes place. When the reaction reaches equilibrium at which no further net chemical changes occur, the rate of conversion of A into B is exactly counterbalanced by the rate of conversion of B into A:

$$A \rightleftharpoons B \qquad (3.51)$$

The thermodynamic equilibrium constant K is defined as

$$K = \frac{[B]}{[A]} \qquad (3.52)$$

K is employed to express the chemical equilibrium reached by this system. Here, the brackets represent the thermodynamically active masses of the reactant and product at equilibrium state, with the units of gram-moles or gram-ions per liter. The constant K depends on the temperature and pressure of the system and on the initial compositions of each component.

When the reaction has more than two components, the equilibrium constant is defined as the product of the active masses of the reaction products divided by the product of the active masses of reactants at equilibrium. For the reaction

$$A + B \rightleftharpoons C + D \qquad (3.53)$$

the equilibrium constant is

$$K = \frac{[C][D]}{[A][B]} \qquad (3.54)$$

The equilibrium constant is a mathematical function of the standard free energy change of the reaction. The relationship between the standard free energy change of the reaction ΔG_0 and the equilibrium constant K is known to be

$$\Delta G° = -RT \ln K \tag{3.55}$$

in which R is the universal gas constant and is equal to 1.987 cal mol^{-1} deg^{-1}. When $K = 1$, there is no change in the standard free energy change and the reaction is at equilibrium. For the value of K greater than unity, $\Delta G°$ is negative, that is, the reaction proceeds with a decline of free energy. However, if the value of K is less than unity, the standard free energy change is positive. Therefore, the reaction does not proceed toward completion. This implies that energy must be supplied to the system in order to transform 1 mole of reactant to 1 mole of product when each is present at 1 molal concentration. Since an enzyme or catalyst cannot change the equilibrium point, the value of $\Delta G°$ is the same in the presence or absence of enzyme.

3.5.3 Evaluation of Standard Free Energy Change

Depending on the data available, there are a number of general procedures for evaluating $\Delta G°$.

3.5.3.1 From Equilibrium Constants. If equilibrium data are available, $\Delta G°$ for the reaction can be calculated from Eq. (3.55).

3.5.3.2 From Enthalpy and Entropy Changes. Under the standard conditions ($T = 298$ K), Eq. (3.46) can be written as

$$\Delta G° = \Delta H° - T \Delta S° \tag{3.56}$$

The standard free energy change $\Delta G°$ can be obtained if the changes in both the enthalpy and entropy for the reaction, $\Delta H°$ and $\Delta S°$, respectively, can be calculated.

For any reaction, one can write

$$\Delta H° = (\Sigma \Delta H_f°)_{\text{products}} - (\Sigma \Delta H_f°)_{\text{reactants}} \tag{3.57}$$

$$\Delta S° = (\Sigma \Delta S_f°)_{\text{products}} - (\Sigma \Delta S_f°)_{\text{reactants}} \tag{3.58}$$

in which $\Delta H_f°$ is the standard enthalpy of formation of a substance from its elements at the same temperature, and $\Delta S_f°$ is the absolute entropy of a substance at temperature T. Both $\Delta H_f°$ and $\Delta S_f°$ are listed in several sources. A list of $\Delta H_f°$'s and $\Delta S_f°$'s is presented in Table 3.1.

3.5.3.3 From Free Energies of Formation. For any reaction, the standard free energy change can be expressed as

$$\Delta G° = (\Sigma \Delta G_f°)_{\text{products}} - (\Sigma \Delta G_f°)_{\text{reactants}} \tag{3.59}$$

Table 3.1 Heats of formation ΔH_f°, absolute entropies ΔS_f°, and standard free energies of formation ΔG_f° at 25°C for some substances in solid state

| | | | ΔG_f°, kcal/mol | |
Substance	ΔH_f°, kcal/mol	ΔS_f°, cal/mol/°C	Pure solid	In aqueous solution
L-Alanine	−134.60	31.6	−88.40	−88.75
DL-Alanine	−135.19	31.6	−	−89.11
L-Aspartic acid	−233.33	41.5	−174.76	−172.31
L-Cysteine	−127.88	40.6	−82.08	−81.21
L-Cystine	−251.92	68.5	−163.55	−159.00
Glycine	−126.66	26.1	−88.61	−89.14
L-Leucine	−153.39	49.5	−82.63	−81.68
DL-Leucine	−154.46	49.5	−	−81.76
DL-Leucylglycine	−207.10	67.2	−	−110.90
CO_2 (gas)	−94.05	51.1	−94.26	−92.31
Water (liquid)	−68.32	16.72	−56.69	−

ΔG_f°, the standard free energy of formation of a substance from its elements, is available for many substances involved in metabolic and biosynthetic processes. Table 3.1 presents a list of ΔG_f°'s for several substances.

3.5.3.4 From Oxidation-Reduction Potentials. The standard free energy change ΔG° is related to the standard oxidation-reduction potentials ϵ° as

$$G^\circ = -n\mathscr{F}\epsilon^\circ \tag{3.60}$$

where n is the number of moles of electrons transferred and $\mathscr{F}$ is the Faraday constant, 96,487 coulombs/equivalent (C/equiv) or 23,061 cal/volt/equiv. Tables of ϵ° are available for many substances of biochemical interest.

3.5.4 Computation of Free Energy Change ΔG

One may now proceed to calculate the change in free energy ΔG from the known change at standard state ΔG°. The free energy G is related to the activity a as

$$G = RT \ln a + G^\circ \tag{3.61}$$

For the process of changing the state of a substance from one condition to another, ΔG for the reaction can be expressed as

$$\Delta G = G_{\text{products}} - G_{\text{reactants}} = \Delta G^\circ + RT \ln K \tag{3.62}$$

It should be noted that at equilibrium $\Delta G = 0$, Eq. (3.62) reduces to Eq. (3.55). The activity is a generalized concentration and can be reduced to experimentally measured concentrations. These relationships between a and measured concentrations are summarized in Table 3.2.

In the case of a gas, a can be replaced by P, the pressure or partial pressure in a mixture. The gas pressure is usually less than 1 atm in most biological processes. At such low pressures, ideal behavior may be postulated for most gases. However, for real gases or at higher pressures, the fugacity f, defined as γP, should be used in free energy calculations. The corrected pressure is f and the correction factor is γ. The activity takes the value of unity for a pure solid or a pure liquid.

In the case of substances in solution, the activity of the solvent a_1 is always defined as $N_1\gamma_1$, where N_1 is the mole fraction of the solvent and γ_1 is the correction factor. Both N_1 and γ_1 are often close to unity in many solutions except those containing high concentrations of solute. The molality m is the most common or practical unit of concentration being employed for the solute of both the electrolytes and nonelectrolytes. The activity of the solute a_2 is defined as $m\gamma_2$ for a nonelectrolyte and as $(m\gamma_2)^2$ for a univalent electrolyte. a_2 can be approximated as m or m^2 for solutes in a very dilute solution in which γ_2 approaches unity. a_2 is defined as $N_2\gamma_2$ when the unitary or rational standard state is used.

3.5.5 The van't Hoff Isobar

An interesting and important relation between the equilibrium constant and enthalpy change of reaction, which is known as the van't Hoff isobar, may be derived from Eq. (3.55) and the definition of the Gibbs function. One can write

$$\ln K = \frac{-\Delta G^\circ}{RT} = -\frac{\Delta H^\circ}{RT} + \frac{\Delta S^\circ}{R} \qquad (3.63)$$

Table 3.2 Activity relations

State of substance	Activity a
Gas	
Ideal	$a = P$ (pressure, in atm.)
Real	$a = f$ (fugacity) $= P$
Pure solid	$a = 1$
Pure liquid	$a = 1$
Nonelectrolytic Solution	
Solvent	$a_1 = N_1\gamma_1$
Solute (practical standard state)	$a_2 = m\gamma_2$
Solute (unitary or rational standard state)	$a_2 = N_2\gamma_2$
Electrolytic solution	
Solvent	$a_1 = N_1\gamma_1$
Solute (e.g., NaCl)	$a_2 = (m\gamma_2)^2$

Since the equilibrium constant is a function of temperature only and is independent of pressure, the differentiation of this relation with respect to temperature yields

$$\frac{d \ln K}{dT} = \frac{\Delta H^\circ}{RT^2}$$

or

$$\frac{d \ln K}{d(1/T)} = -\frac{\Delta H^\circ}{R} \tag{3.64}$$

Thus, when $\ln K$ is plotted against the reciprocal of the absolute temperature, the slope of the line is equal to $-\Delta H^\circ/R$. An endothermic chemical reaction gives a negative slope, while an exothermic reaction has a postive slope.

Example 3.7 Free energy change in digestive glands Glands in the digestive system undergo a hydrogen concentration process to produce approximately 2.5 L/day of 0.16 N hydrochloric acid using fluids with plasma concentration as raw material. Assuming that there is adequate chloride in plasma to produce this level of acid concentration, determine the energy required to accomplish this concentration.

The concentration of H^+ in the plasma fluid is at pH $= 7.4$ or $[H^+]_{reactant} = 4 \times 10^{-8}$ equiv/L, while the concentration of H^+ in acid is $[H^+]_{product} = 0.16$ equiv/L. Therefore, the equilibrium constant K is 4×10^6 and thus

$$\begin{aligned} \Delta G &= RT \ln (4 \times 10^6) \quad \text{(cal/mol acid)} \\ &= 9.5 \text{ kcal/mol acid} \\ &= (9.5)(0.16)(2.5)^. = 3.8 \text{ kcal/day} \end{aligned}$$

In view of the lost energy in this process, these glands are likely to expend at least twice the 3.8 kcal/day calculated to operate.

Example 3.8 Energy flow in biological systems The energy flow from the sun to cellular conversion can be divided into three stages: (1) release of solar energy, (2) energy conversion in the plant world, and (3) energy conversion in the animal world. Sunlight is the origin of biological energy. In the immensely high temperature of the sun, estimated to be 6000 K, a part of the energy locked in the nucleus of hydrogen atoms is released as the atoms are converted into helium atoms and electrons by nuclear fusion:

$$4_1H^1 \longrightarrow He^4 + 2_1e^0 + h\nu$$

Here, $h\nu$ represents a quantum or photon which is released in the form of gamma radiation. h is Planck's constant (6.6256×10^{-34} J s) and ν is the wavelength of the gamma radiation.

The radiant energy flows from the sun to the plant world where it is absorbed by chlorophyll for photosynthesis:

$$6CO_2 + 6H_2O + nh\nu \rightarrow C_6H_{12}O_6 + 6O_2$$

This is an endergonic process resulting in a large increase in free energy and a decrease in entropy:

At 25°C:
$$\Delta G° = +686 \text{ kcal/mol}$$
$$\Delta H = +673 \text{ kcal/mol}$$
$$\Delta S = -43.6 \text{ cal/mol K}$$

It should be noted that at isothermal conditions,

$$\Delta G° = \Delta H - T\Delta S$$

Now the solar energy is converted into chemical energy that is stored by green plants.

The animal world ultimately derives its foodstuffs from plants. The next stage in the flow of biological energy takes place in respiration, the process of biological combustion of foodstuff molecules by cells. For example, the chemical reaction for oxidation of glucose is an exergonic process:

$$\text{Glucose} + 6O_2 \longrightarrow 6O_2 + 6H_2O$$

At 25°C,
$$\Delta G° = -686 \text{ kcal/mol}$$
$$\Delta H = -673 \text{ kcal/mol}$$
$$\Delta S = +43.6 \text{ cal/mol K}$$

$\Delta G°$ represents the maximum energy available. However, the actual cellular machinery carries out this process only at about 40–50% efficiency. It is this last stage of energy flow in the biological world that is the subject of our interest here.

3.6 TRANSFER POTENTIALS

Some chemical transformations involve the transfer of a proton, an electron, or a substituent group in a compound from one molecule to another. Such a transformation requires an input of energy called the transfer potential. The transfer potential may be defined in terms of the change in chemical potential $\Delta G°$. A number of transfer potentials are encountered in biochemical energetics. For biochemical applications, the proton-transfer potential, electron-transfer potential, and group-transfer potential are presented here. Each of these chemical transformations is usually expressed in a very concise form but only as a shorthand notation. The actual reaction must be described by an equation with some acceptor species.

3.6.1 Proton-Transfer Potential in Dissociation of Weak Acids

The process of removing a proton from the weak acid AH, called the dissociation of weak acids, can be expressed in a concise form as

$$AH \longrightarrow A^- + H^+ \tag{3.65}$$

The dissociation constant K_a is defined as $[H^+][A^-]/[HA]$ and $-\log K_a$ is
called pK_a. Actually the proton is transferred from AH to some acceptor species
such as water:

$$AH + H_2O \longrightarrow A^- + H_3O^+ \tag{3.66}$$

For this transfer reaction, the energy change, the proton-transfer potential or
acidity, can be measured in terms of pK_a, following from Eq. (3.55), as

$$pK_a = \frac{\Delta G^\circ}{2.303RT} \tag{3.67}$$

Here, pK_a is proportional to ΔG° for the transfer of a mole of protons H^+ from
the acid to a reference receptor molecule. A higher value of pK_a, or a larger
positive ΔG°, implies a weaker acid in which the dissociation of a proton would
not be favored. On the other hand, a lower value of pK_a implies a stronger
acid.

3.6.2 Electron-Transfer Potential in Ionization of Weak Acids

The concise form of an ionization process of a weak electrolyte A

$$A \longrightarrow A^+ + e \tag{3.68}$$

is used to describe the actual reaction

$$A + H^+ \longrightarrow A^+ + \tfrac{1}{2}H_2 \tag{3.69}$$

in the presence of an acceptor. The energy change in the electron transfer
process, called the electron-transfer potential, oxidation-reduction potential or
redox potential, can be measured in terms of ϵ°. It follows from Eq. (3.60) that

$$\epsilon^\circ = -\frac{\Delta G^\circ}{n\mathcal{F}} \tag{3.70}$$

It is important to note that, in practice, ϵ° is used for the reverse action of Eq.
(3.68). Equation (3.70) indicates that ϵ° is proportional to ΔG° for the transfer
of a mole of electrons from an acceptor to the substance, namely the reverse
of Eq. (3.69). A positive ϵ° implies a negative ΔG° indicating spontaneity in
the uptake of an electron by A^+. A higher positive ϵ° indicates a stronger
oxidizing potential for A^+.

3.6.3 Group-Transfer Potential

The group-transfer potential is sometimes referred to as the high-energy bond, which is the energy input into a compound such as adenosine triphosphate (ATP) to remove one of its terminal groups with high-energy bonding. The expression for the chemical transformation of a high-energy phosphate, such as various foodstuffs, reads, in concise form,

$$A \sim PO_4 \longrightarrow A + PO_4 \tag{3.71}$$

or in the presence of some acceptor molecule, usually H_2O,

$$A \sim PO_4 + H_2O \longrightarrow A{-}OH + HPO_4 \tag{3.72}$$

The change in chemical energy ΔG_g°, when the terminal phosphate group PO_4 is transferred to H_2O, is ΔG°. That is,

$$\Delta G_g^\circ = \Delta G^\circ \tag{3.73}$$

per mole of PO_4 transferred to H_2O. Since a negative ΔG° signifies the spontaneity of a reaction, a compound with high ΔG_g° would have a greater phosphate-transfer potential.

3.7 SOME LAWS OF PHYSICOCHEMICAL BEHAVIOR

There are relationships between the free energy change ΔG and experimentally measurable quantities such as concentration differences, electric potentials, pressure differences, centrifugal fields, temperature, surface properties, tension, and magnetic properties. Through the applications of the two fundamental laws of thermodynamics and the concept of free energy, one can derive a host of principles governing the behavior of matter. In this section, laws of physicochemical behavior for electrochemical cells, osmosis, and sedimentation are to be derived.

3.7.1 Electrochemical Potentials

Consider a simple cell containing two solutions of a salt, one at a concentration C_1 on one side and the other at a concentration C_2 on the other side, which are separated by an appropriate permeable membrane. Two identical electrodes of metal A connected to a potentiometer are inserted into the cell, one on each side. Let the electrons of current flow through the circuit from left to right through the potentiometer. Then, electrons are generated at the left electrode from the metal in the reaction

$$A \longrightarrow A^+(C_1) + e \tag{3.74}$$

and are absorbed at the right electrode in the reaction

$$e + A^+(C_2) \longrightarrow A \qquad (3.75)$$

The net reaction in the cell is the sum of the above equations:

$$A^+(C_2) \longrightarrow A^+(C_1) \qquad (3.76)$$

The free energy change associated with the reaction ΔG can be related to the concentrations as

$$\Delta G = RT \ln \frac{C_1}{C_2} \qquad (3.77)$$

The maximum useful work from the reaction W_{max} is

$$W_{max} = n\mathscr{F}\epsilon \qquad (3.78)$$

wherein n is the number of moles of electrons, $\mathscr{F}$ is the charge carried by one mole of electrons, that is, 96,500 coulombs, and ϵ is the electrochemical potential for the cell. Since the free energy change is related to the maximum useful work as

$$\Delta G = -W_{max} \qquad (3.79)$$

one obtains

$$\epsilon = \frac{RT}{n\mathscr{F}} \ln \frac{C_2}{C_1} \qquad (3.80)$$

The latter is the well-known Nernst equation for electrochemical potentials for the cell. For more complicated electrochemical systems, the above procedure may be followed in deriving equations for the electromotive force.

3.7.2 Osmosis

The process of net diffusion of water across a semipermeable cell membrane caused by a concentration difference is called osmosis. Osmosis of water molecules can be opposed by applying a pressure across the membrane in the direction opposite to that of the osmosis. The amount of pressure required to stop osmosis completely is called the osmotic pressure.

Consider a semipermeable membrane separating two columns of fluid in the two arms of a U tube. At first, both columns are filled with water (pure solvent) to equal height. The system is at equilibrium, that is, $\Delta G = 0$, or the chemical potentials of the solvent in both columns are equal.

When some solute is added to one side, say the left-hand side, the chemical potential of the solvent is lowered due to the presence of dissolved solute, while the solvent on the right retains its initial chemical potential. As a result, osmosis of water from the right column into the left column occurs, causing the rise of the level of the left column. This level difference results in a pressure difference between the right and left columns. Equilibrium is reestablished in the system only when a pressure difference is developed that is great enough to oppose the osmotic effect. The pressure difference across the membrane at this time is called the osmotic pressure of the solution.

The lowering of the chemical potential of the solvent in the left column is proportional to the number of moles of added solute n_2, that is, n_2RT. However, due to the osmotic pressure π, the chemical potential of the solvent on the left is increased by πV, where V is the volume of the solvent in the left column. When equilibrium is reestablished in the system, ΔG has to be zero, resulting in $n_2RT = \pi V$ or

$$\pi = RT \frac{n_2}{V} = RTC_2 \qquad (3.81)$$

in which $C_2 = n_2/V$ is the number of moles per liter of dissolved solute. Equation (3.81) signifies that if π can be measured, then C_2, in moles per liter, or the molecular weight of a solute can be calculated.

3.7.3 Sedimentation

The molecular weight of a solute can also be determined from sedimentation equilibrium using an ultracentrifuge. A solution is placed in a small cell which is set in a rotor of an ultracentrifuge. When the rotor is rotating at a high speed, say 15,000 rpm, a strong centrifugal force is produced to move the molecules toward the outer end of the cell. However, as the molecules congregate at the outer end of the cell, a concentration difference is created between the outer and inner ends of the cell. As a result, a diffusion force is set up to move molecules in the direction opposite to the centrifugal force. Equilibrium is reestablished in the system when the centrifugal and diffusion forces are eventually balanced. Under this condition, the concentrations C_1 and C_2 are measured, usually by optical means, at two points in the cell at location r_1 and r_2 from the center of the rotor.

The free energy difference ΔG_{co} induced by the concentration difference $(C_1 - C_2)$ is

$$\Delta G_{co} = RT \ln \frac{C_1}{C_2} \qquad (3.82)$$

The amount of centrifugal work required to move an object of mass M from r_1 to r_2 is

$$W = \frac{M\Omega^2(r_1^2 - r_2^2)}{2}$$

where Ω is the angular velocity of the rotor. If the object is in a solvent, its effective mass is $M(1 - v\rho)$ from Archimedes' principle. Here, v is the specific volume of the solute, while ρ is the density of the solution. From the relationship between free energy and maximum work, one can write the free energy change due to the centrifugal force as

$$\Delta G_{ce} = \frac{M(1 - \upsilon\rho)\Omega^2(r_1^2 - r_2^2)}{2} \qquad (3.83)$$

At equilibrium, $\Delta G_{co} = -\Delta G_{ce}$ results in

$$M = \frac{-2RT \ln (C_1/C_2)}{(1 - \upsilon\rho)(r_1^2 - r_2^2)\Omega^2} \qquad (3.84)$$

which is the expression for the molecular weight of the solute.

3.8 ENERGETICS OF CELL AND BODY METABOLISM

To facilitate an understanding of how, where, and how much heat is produced in a cell during its functions, the structure and functions of cells are briefly introduced. The mechanics of heat production in cells is then described from the aspects of biophysics/biochemistry or bioenergetics or the combination of the two. Besides the heat produced during cell function to sustain its life, heat is generated during cell growth and division. Also present is heat production resulting from the firing of action potentials in nerve and muscular cells (Yang, 1980).

3.8.1 Structure and Functions of Cells

Each of the 100 trillion cells in the human being is a living structure that can survive indefinitely and in most cases can reproduce itself if its surrounding fluids remain constant. A typical cell is depicted in Fig. 3.10. In consists of two major parts, the nucleus and the cytoplasm. The nucleus, which is filled with nucleoplasm, is separated from the cytoplasm by a nuclear membrane, while the cytoplasm is separated from the surrounding fluids by a cell membrane. The cell membrane is approximately 75–100 Å and elastic. It is composed almost entirely of proteins and lipids and is semipermeable, allowing some substances to pass through it but excluding others. The nuclear membrane is double with a wide space between the two unit membranes. It is quite permeable.

The nucleus is the control center of the cell. It regulates both the chemical reactions that take place in the cell and the reproduction of the cell.

Mitochondria are present in the cytoplasm of all cells (Fig. 3.11). The shape, size, and number vary greatly in different cells. They may be only a few hundred millimicrons in diameter and spherical or 1 μm in diameter, 7 μm in length, and elongated in shape, depending on their metabolic activity. There may be only a few to many thousands per cell, depending on the rate of energy required by each cell to perform its functions. Each mitochondrion is constructed from

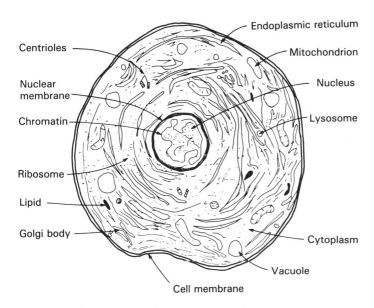

Centrioles

Nuclear membrane

Chromatin

Ribosome

Lipid

Golgi body

Endoplasmic reticulum

Mitochondrion

Nucleus

Lysosome

Cytoplasm

Vacuole

Cell membrane

Figure 3.10 Structure of a cell.

two membranes similar in structure to the nuclear membrane. The inner membrane is convoluted to form shelves or cristae which project into a central aqueous cavity, while the outer membrane forms a skin-like limiting layer to the mitochondrion. A second mitochondrial cavity exists between the two membranes. While the inner membrane is permeable to only relatively small molecules, the outer membrane is permeable to molecules of up to about 10,000 molecular weight (MW). Stacked on the shelves inside the mitochondrion are packets of enzymes in granules. When nutrients and oxygen come into contact with these enzymes, they combine to form carbon dioxide and water, and the liberated energy is used to synthesize ATP, a high-energy phosphate compound. The ATP then diffuses throughout the cell and releases its stored energy wherever it is needed for performing cellular functions. The mitochondria are thus the power-generating units of the cell. They probably self-replicate whenever there is a need in the cell for increased amounts of ATP. The enzymes on the outside surface of the mitochondria are mainly concerned with biological ox-

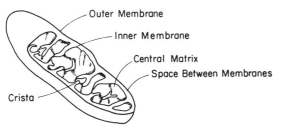

Outer Membrane

Inner Membrane

Central Matrix

Space Between Membranes

Crista

Figure 3.11 Structure of a mitochondrion.

idations, providing raw materials for the citric acid cycle and respiratory chain systems inside the mitochondrion.

3.8.2 Mechanics of Heat Production in Cells

Biological reactions in cells may be considered as occurring at constant pressure and at constant volume (changes in volume being negligible). More importantly, cells can function when they are at a uniform temperature. Although temperature gradients may be temporarily produced in cells when heat is generated, they are incidental. In other words, uniform temperature is essential to cell functions.

Consider the chemical reaction

$$A + B \rightleftharpoons C + D \qquad (3.85)$$

which takes place in a cell under the conditions of constant pressure, constant volume, and uniform temperature. The first law of thermodynamics for an isolated cell reads

$$\Delta H = \Delta E \qquad (3.86)$$

Here, ΔH denotes the heat of reaction (enthalpy of the products minus the enthalpy of the reactants) and ΔE signifies the total energy change. A process is exothermic for a negative ΔH and endothermic for a positive ΔH. Some of these values are given in Table 3.3. The fundamental equation relating changes in entropy S and free energy to changes in enthalpy is

$$\Delta H = \Delta G + T\Delta S \qquad (3.87)$$

in which ΔG is the free energy change (free energy of products minus free energy of reactants) and T is the absolute temperature. The magnitude of the entropy change measures the degree of irreversibility. ΔG represents the utilizable energy change while $T\Delta S$ is the nonutilizable energy change. A reaction is exergonic for a negative ΔG (oxidation of cell nutritions) and endergonic for

Table 3.3 Heat of combustion of various organic compounds

Organic compound	Heat of combustion, cal/mol
Stearic acid	2711×10^3
Sucrose	1349
Glucose	673
Glycerol	397
Alanine	387
Ethyl acid	327
Lactic acid	326
Acetaldehyde	279
Acetic acid	209
Pyruvic acid	279

Table 3.4 Standard free energy changes at pH 7 and 25°C of some chemical reactions

Reaction	$\Delta G° \times 10^{-3}$, cal/mol
Oxidation	
Glucose + $6O_2 \rightarrow 6CO_2 + 6H_2O$	-686
Lactic acid + $3O_2 \rightarrow 3CO_2 + 3H_2O$	-326
Palmitic acid + $23O_2 \rightarrow 16CO_2 + 16H_2O$	-2338
Hydrolysis	
Sucrose + $H_2O \rightarrow$ glucose + fructose	-5500
Glucose 6-phosphate + $H_2O \rightarrow$ glucose + H_3PO_4	-3300
Glycylglycine + $H_2O \rightarrow$ 2 glycine	-4600
Rearrangement	
Glucose 1-phosphate $\rightarrow$ glucose 6-phosphate	-1745
Fructose 6-phosphate $\rightarrow$ glucose 6-phosphate	-400
Ionization	
$CH_3COOH + H_2O \rightarrow H_3O^+ + CH_3COO^-$	$+6310$
Elimination	
Malate $\rightarrow$ fumarate + H_2O	$+750$

a positive ΔG (photosynthesis). The thermodynamic equilibrium constant for the reaction in Eq. (3.85) is

$$K = \frac{[C][D]}{[A][B]} \tag{3.88}$$

It is related to the free energy change by

$$\Delta G° = -RT \ln K \tag{3.89}$$

where R is the gas constant (1.987 cal/mol °C). The symbol $\Delta G°$ designates the standard free energy change at pH 7.0 (the pH level at which most biological reactions occur) and 25°C. Table 3.4 gives $\Delta G°$ values for some chemical reactions. When K exceeds unity, a forward reaction occurs with $\Delta G < 0$; when K is less than unity, a backward reaction takes place with $\Delta G > 0$; and there is no reaction when K is at unity.

In biochemical reactions in the cell, the free energy is not truly released but is conserved in high energy bonds. Energy from glucose (or other cell nutrients) made available during its anaerobic (without utilization of oxygen) and aerobic decomposition in the cells is transferred to an inorganic phosphate which is then linked by a resonant, high energy phosphate bond to adenosine diphosphate (ADP) making it into ATP. The ATP may transfer its energy to high energy phosphate to creatine (C) forming creatine phosphate (CP) as a storage depot of high energy bonds; or it may release its energy, by splitting away a phosphoric acid radical to form ADP, which is used to perform all kinds of cell work. In the latter case, once again, energy derived from the cellular nutrients causes the ADP and phosphoric acid to recombine to form new ADP.

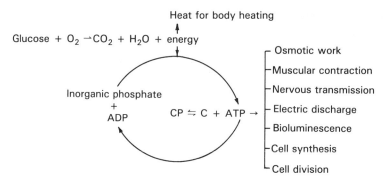

Figure 3.12 Transfer of energy in cells by the ATP-ADP system.

Thus, the entire process of spending and remaking ADP can continue over and over again (Fig. 3.12).

The cells split nutrients up step by step so that at no one step is there a massive release of energy. For example, there are more than 70 separate, sequential enzymatic steps in the oxidation of glucose in heterotropic cells. However, the process of converting the energy of oxidation to ATP energy is not completely efficient and thus some of the energy turns up in the form of heat. This process is at least 40–50% efficient. Since the free energy of oxidation of glucose is 686,000 cal/mol and 38 molecules of ATP are formed, each of which require a minimum input of 7000 cal/mol, the approximate efficiency of energy conversion is 42%. However, making some adjustment in the $\Delta G°$ value for the concentrations of ADP, ATP, and phosphate actually occurring in the cell, the efficiency of the process in the intact cell should be considerably higher, 60–70% (Krebs and Kornberg, 1957). The balance of the energy is lost as heat. In addition to the inefficiency of energy generating processes, most heat is produced in the cells due to inefficiency of energy utilizing processes.

3.8.3 Energetics of Cell Growth and Division

Temperature rise is necessary during growth and cell replication. Consider a simplified model of a sphere which grows from a volume of V to $2V$ followed by splitting into two spheres. The average cell volume during this process is $3V/2$ and the average radius is $(9V/8\pi)^{1/3}$. If the cell density is ρ, the nonaqueous mass increase is $\rho V/4$. During the division time of τ, the cell may be considered as a uniform spherical heat source of strength $\dot{Q}$ cal/cm³ s equal to

$$\dot{Q} = \frac{(\rho V/4)\Delta H_m}{(3V/2)\tau} = \frac{\rho \Delta H_m}{6\tau} \qquad (3.90)$$

Here, ΔH_m is the heat generated per unit mass of biomass synthesized and equals $\Delta H_c + f\Delta H_w$. ΔH_c represents the enthalpy per unit mass for cell synthesis, while ΔH_w is the enthalpy per unit mass of waste produced. f denotes the ratio of the mass of waste generated to that of which the cell produced.

The second law of thermodynamics states that the net entropy change is greater or equal to zero for the synthesis of a living cell. Within the limitation of the second law, the entropy changes accompanying cell division are summed. It results in the inequality which defines the minimum amount of waste material accompanying cellular synthesis (Morowitz, 1960):

$$f \geq -\frac{1}{\Delta F_w} \left(\frac{RT_0}{M} \log \frac{C_c}{C_m} + RT_i \ln 2 + \Delta H_c \right) \qquad (3.91)$$

Here, ΔF_w is the free energy change per unit mass of waste produced; M, the average molecular weight of the nutrient; and C_c and C_m, the concentrations of nutrient in the cell and waste, respectively.

3.8.4 Energetics of Body Metabolism

The natural thermoregulatory mechanisms of shivering, sweating, and vascular adjustment are related to the rate of body metabolism. Data for an average-

Table 3.5 Average man

Parameter	Measurement
Age	30 yr
Height	173 cm
Weight	68.2 kg
Surface area	1.80 m²
Normal body core temperature	37.0°C
Normal mean skin temperature	34.2°C
Heat capacity	0.86 kcal/kg °C
Percent body fat	15%
Basal metabolism	40 kcal/m²h, 72 kcal/h, 1730 kcal/day
Oxygen consumption	250 ml/min (at STP)
CO_2 production	200 ml/min
Respiratory quotient	0.80
Blood volume	5 L
Resting cardiac output	5 L/min
Systemic blood pressure	120/80 mmHg
Heart rate	65 beats per min
Total lung capacity	6000 ml
Vital capacity	4200 ml
Ventilation rate	6000 ml/min
Alveolar ventilation rate	4200 ml/min
Tidal volume	500 ml
Dead space	150 ml
Breathing frequency	12 breaths per min
Pulmonary capillary blood volume	75 ml
Arterial O_2 content	0.195 ml O_2/ml blood
Arterial CO_2 content	0.480 ml CO_2/ml blood
Venous O_2 content	0.145 ml O_2/ml blood
Venous CO_2 content	0.520 ml CO_2/ml blood

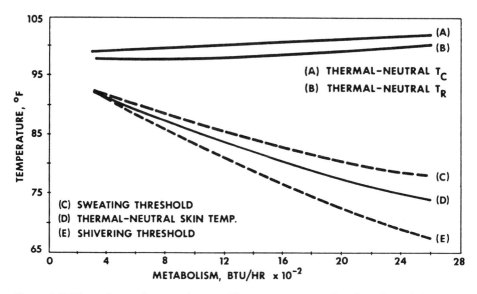

Figure 3.13 Thermal-neutral core and mean skin temperature as a function of metabolism. (A) Thermal-neutral T_C; (B) thermal-neutral T_R; (C) sweating threshold; (D) thermal-neutral skin temperature; (E) shivering threshold. (Buchberg and Harrah, 1968.)

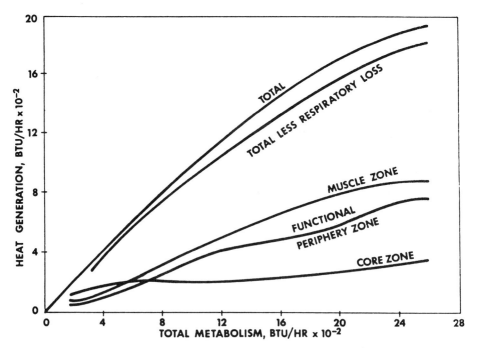

Figure 3.14 Steady-state heat generation as a function of total metabolism (Buchberg and Harrah, 1968).

Table 3.6 Physical properties of body zones

Zone	Area		Volume			Thickness		Mass		Density	
	ft²	m²	ft³	× 10³m³		ft	mm	lb	kg	lb/ft³	kg/m³
Skip	15.4	1.43	0.125	3.54		0.006	2.47	8.25	3.74	66.1	1059
Functional periphery	15.4	1.43	1.000	28.3		0.0650	19.81	64.4	29.21	64.3	1030
Musculature	15.4	1.43	0.721	20.4		0.0468	14.26	47.4	21.50	66.1	1059
Core	15.4	1.43	0.514	14.5		0.0334	10.18	34.6	15.69	66.1	1059

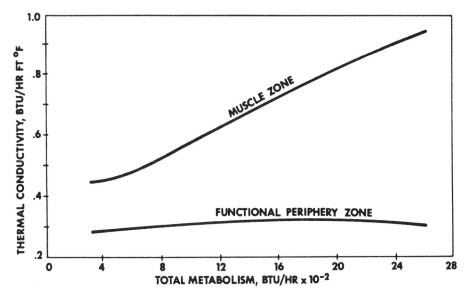

Figure 3.15 Thermal conductivity of the muscle zone and the functional periphery zone as a function of total metabolism (from Buchberg and Harrah, 1968).

sized human are presented in Table 3.5. Figure 3.13 shows the variation of the thermoneutral core and the mean skin temperature as a function of metabolism for a lean well-acclimatized man. All curves were measured except B, which was obtained by numerical analysis. Curves A and B correspond to the measured and calculated results of the body core temperature.

The total metabolism indicates the amount of energy that is released by oxidation-reduction reactions in a living organism. Only a fraction of the total metabolism is utilized to perform the work required in the organism. In order to maintain thermoneutrality, the remaining portion of the total metabolism becomes useless, called the waste energy, and must be dissipated into the environment. The waste energy is referred to as the total metabolic heat generation. Figure 3.14 illustrates the total metabolic heat generation in the various zones of the human body as a function of total body metabolism.

In Fig. 3.14, the core zone comprises the visceral organs of the body and inert tissue having low metabolic rates. Thermal gradients in the core zone are very small. The muscle zone consists of the skeletal muscle, heart, and blood, while the functional periphery zone includes the peripheral vascular system and a small portion of the musculature. The physical properties of each zone are listed in Table 3.6. Figure 3.15 depicts the thermal conductivity of the muscle zone and the functional peripheral zone as a function of total metabolism. The skin zone is characterized by a constant thermal conductivity, 1.374 W/m^2K (Buchberg and Harrah, 1968).

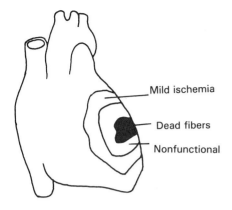

Figure 3.16 Large area of coronary ischemia.

3.9 CORONARY ISCHEMIA

When some cardiac muscular fibers fail to function and some are too weak to contract (Fig. 3.16), the pumping ability of the affected ventricle is proportionately reduced. When the normal portions of the ventricular muscle contract, the ischemic (dead or nonfunctional) muscle is forced outward by the pressure developing inside the ventricle. As a result, much of the pumping force of the ventricle is dissipated by bulging of the area of dead and nonfunctional cardiac muscle (called the systolic stretch). The heart thus becomes incapable of contracting with sufficient force to pump enough blood into the arteries, causing cardiac failure and death of the peripheral tissues. When the dead muscle fibers begin to degenerate and the dead cardiac musculature grows thinner, the extent of systolic strength becomes greater and greater until finally the heart ruptures.

To study the effect of ischemic cardiac muscle on the cardiac output, a

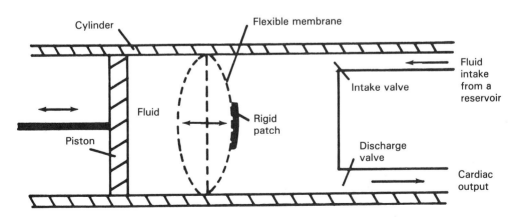

Figure 3.17 Simulation of cardiac ischemia.

model is developed consisting of a flexible membrane (simulating the healthy ventricular muscle) attached to a rigid patch (simulated ischemic cardiac area). The size, mass, and location of the rigid patch can vary. The membrane is operated by a fluid enclosed in a cylinder between a piston and the membrane as shown in Fig 3.17. The liquid to be pumped out from the other side of the membrane simulates the cardiac output.

PROBLEMS

3.1 Define the space enclosed by the cardiac muscles as the control volume and determine changes in the internal energy during systole and diastole. Estimate changes in the kinetic and potential energy during both strokes. The blood velocity is 63 cm/s in the aorta during systole and is 15 cm/s in the vena cava during diastole, as seen in Table 3.7 (p. 167). The size of different blood vessels is given in Fig. 3.18 (p. 168).

3.2 Dry air at 20°C is inhaled and the body in return exhales a water-saturated gas mixture at 37°C. The flow rate is 6000 ml/min and the average specific heat is 0.25 kcal/kg °C for the gases. Determine the energy expended in normal breathing, ignoring the flow of O_2 and CO_2.

3.3 A 70-kg adult male performs some external work $\dot{W}_b$ which requires an additional production of metabolic heat $\dot{W}_{me}$ beyond the basal rate $\dot{W}_{mo}$ (in kcal/h) required at resting conditions. Only a certain fraction of $\dot{W}_{me}$ is actually usefully employed in performing the work. In other words, the difference $(\dot{W}_{me} - \dot{W}_b)$ is either dissipated to the atmosphere as heat or used to accomplish internal heating of the body. Assume that the rate of heat loss from the body to the ambient air remains the same at rest and with work. The body temperature rises at a rate of 3°C/h above its normal value of 37°C while performing external work at 45 kcal/h. Determine the efficiency of the muscles. The specific heat of the body is 0.86 kcal/(kg °C) and units of $\dot{W}$ are given in kilocalories per hour.

3.4 Consider a human body (with mass m and specific heat C) as the system. Normally, the effects of both kinetic and potential energies are negligible. Then the first law gives

$$\frac{\delta Q}{dt} - \frac{\delta W}{dt} = \frac{dU}{dt}$$

Chemical energy conversion (metabolism) is considered as external work (like an electrical current crossing the boundary) done on the system, $-W_m$. One writes

$$W = W_b - W_m$$

where W_b = work done by man. In the rate forms

$$\dot{W} = \dot{W}_b - \dot{W}_m$$

while

$$\dot{W}_m = \dot{W}_{mo} + \dot{W}_{me}$$

where $\dot{W}_{mo}$ denotes the basal rate of metabolic heat generation and $\dot{W}_{me}$ is an increase in the metabolic heat generation rate beyond the basal rate required to perform the work $\dot{W}_b$. The efficiency of the muscles in performing external work W_b has been estimated to be about 20% of the energy conversion required. Derive the expression for the rate of change in the body temperature during the work.

3.5 Determine the amount of pure glucose (enthalpy of formation = 673 kcal/mol) consumed per hour in the human body. If heat dissipation to the atmosphere were completely prevented, how much would the body temperature rise per hour?

Table 3.7 Systemic circulation*

Structure	Diameter, cm	Number	Total cross-sectional area, cm²	Length, cm	Total volume, cm³	Blood velocity, cm/s	Tube Reynolds number
Dog							
Aorta	1.0	1	0.8	40	30	50	1670
Large arteries	0.3	40	3.0	20	60	13	130
Main arterial branches	0.1	600	5.0	10	50	8	27
Terminal branches	0.06	1800	5.0	1	5	6	12
Arterioles	0.002	40×10^6	125	0.2	25	0.3	0.02
Capillaries	0.0008	12×10^9	600	0.1	60	0.07	0.002
Venules	0.003	80×10^6	570	0.2	110	0.07	0.007
Terminal veins	0.15	1800	30	1	30	1.3	6.5
Main venous branches	0.24	600	27	10	270	1.5	12
Large veins	0.6	40	11	20	220	3.6	72
Vena cava	1.25	1	1.2	40	50	33.0	1375
Human being							
Ascending aorta	2.0–3.2					63†	3600–5800
Descending aorta	1.6–2.0					27†	1200–1500
Large arteries	0.2–0.6					20–50†	110–850
Capillaries	0.0005–0.001					0.05–0.1‡	0.0007–0.003
Large veins	0.5–1.0					15–20‡	210–570
Venae cavae	2.0					11–16‡	630–900

*Assuming viscosity of blood is 0.03 poise for dogs and 0.035 poise for human being.
† Mean peak value.
‡ Mean velocity over infinite period of time.
Source: Whitmore (1968).

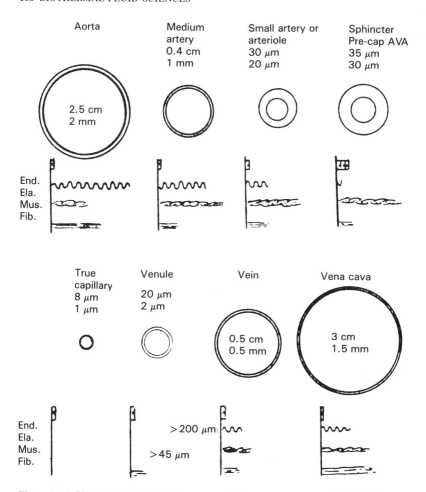

Figure 3.18 Size, thickness of wall, and four basic tissues in the wall of different blood vessels. The figures directly under the name of the vessel represent the diameter of the lumen; next, the thickness of the wall. End., endothelial lining cells; Ela., elastin fibers; Mus., smooth muscle; Fib., collagenous fibers (Burton, 1944).

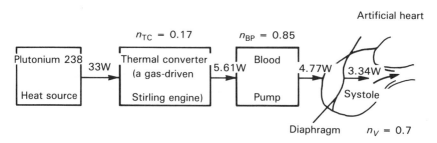

Figure 3.19 Bench model of artificial heart driven by atomic energy.

3.6 An implantable nuclear-powered heart consists of two main subsystems: a thermal converter and a blood pump, as shown in Fig. 3.19 (p. 168). The thermal converter is to change thermal energy into mechanical energy for driving the blood pump. It is a gas-driven Stirling cycle engine which draws 33 W of heat from a plutonium 238 heat source. At an efficiency of 17%, the converter delivers 5.61 W mechanical power to the blood pump which in turn transmits 4.77 W to the diaphragm-type ventricles at 85% efficiency. The pumping ventricles then deliver 3.34 W to the blood at an efficiency of 70%. Operated by the pump at 120 beats per minute, the ventricles have 12 L/min of cardiac output. Determine:

  (a) Heat transfer from the converter to the surroundings,
  (b) Heat transfer from the pump to the surroundings,
  (c) Average cardiac pressure (in mmHg) on blood during the systolic process. This is the diaphragm pressure applying on blood.

Note: 1 Pa = 1 N/m^2; 1 J = Nm; 760 mmHg = 1.01325 × 10^5 Pa; 1 W = 1 J/s.

REFERENCES

Buchberg, H., and Harrah, C. B. (1968) Conduction cooling of the human body—A biothermal analysis, *Thermal Problems in Biotechnology* (ed. J. C. Chato). ASME, New York, pp. 82–95.
Burton, A. C. (1944) Relation of structure to function of tissues of the wall of blood vessels, *Physiol. Rev.* 34:619–642.
Krebs, H. A., and Kornberg, H. L. (1957) *Energy Transformations in Living Matter,* Springer, Berlin.
Morowitz, H. J. (1960) Some consequences of the application of the second law of thermodynamics to cellular systems. *Biochim. Biophys. ACTA,* 40:340–345.
Whitmore, R. L. (1968) *Rheology of the Circulation,* Pergamon, Oxford.
Yang, W.-J. (1980) Human Micro and Macro Heat Transfer—In Vivo and Clinical Applications, *Perspective in Biomechanics,* vol. 1 (ed. D. N. Ghista), Harwood Academic, New York, pp. 227–311.

FOUR

HEAT TRANSFER

Heat transfer is an extension of classical thermodynamics. Its major task is to employ the rate form of the first law to determine temperature distribution in the system from which the rate of heat transfer can be evaluated by utilizing one of three natural laws for conduction, convection, and radiation. The principal difference between thermodynamics and heat transfer is that the former treats physical properties, such as temperature and pressure, to be uniform (but they may vary with time) throughout the entire system (called equilibrium of state) and deals with the relations among equilibrium states. On the other hand, heat transfer takes into account both the mechanism and rate of heat flow. Fluid mechanics and thermodynamics are both characterized by three primary dimensions. The dimensions for fluid mechanics are mass M, length L, and time t, while M, L, and temperature T are the dimensions for thermodynamics. Heat transfer deals with four dimensions: M, L, t, and T. Just as the velocity difference Δv is the driving force for fluid motion or momentum transfer in fluid mechanics, the temperature difference ΔT is the driving force for heat flow in the system.

4.1 MODES OF HEAT TRANSFER

Heat can be transferred by three distinct mechanisms: conduction, convection, and radiation. While the operation of conduction and radiation depends on only ΔT, convection is induced by ΔT but its operation is also associated with fluid flow or mass transport.

Although heat flow may be induced by several of these mechanisms acting simultaneously, in the interest of simplification all but the dominant mechanism is neglected. Like the first law of thermodynamics, the physical laws governing the three heat transfer mechanisms were obtained empirically and are described in mathematical form in the following sections.

4.1.1 Conduction

Conduction is a process in which heat transfer is induced by molecular diffusion. It takes place in solids and quiescent fluids. A molecule at a higher temperature possesses a higher kinetic energy and thus is in a more excited state. The molecule transmits part of its energy to its neighboring molecule at a lower energy level. The rate of heat flow by this mechanism can be determined by the Fourier conduction equation (1822) which, in finite difference form, reads

$$q_k = kA \frac{T_H - T_L}{\delta} \tag{4.1}$$

or, in differential form [see Eq. (1.25) in Table 1.1] as

$$q_k = -kA \frac{dT}{dx} \tag{4.2}$$

Here, q denotes the rate of heat flow in watts (W); A is the area perpendicular to the direction of heat flow in square meters (m^2); k is the thermal conductivity of the material (in W/m K); and δ, the distance in meters between two locations with temperature difference $(T_H - T_L)$. The subscripts k, H, and L represent conduction, higher temperature, and lower temperature, respectively. Since heat transfer occurs in the direction of decreasing temperature, the temperature gradient dT/dx is always negative. A minus sign is introduced in Eq. (4.2) for the sole purpose of making q_k a positive quantity in the x direction. The quantity q/A is called heat flux. Any substances having very low values of k are called insulating materials, an example being fat. On the other hand, substances having high values of k are good thermal as well as electrical conductors. Equation (4.1) can be rewritten in the form of Eq. (1.26).

4.1.2 Convection

Convection is a process in which heat is transferred by the simultaneous action of molecular diffusion and mixing motion. It takes place only in a fluid. The rate of convection between a solid surface at T_s and a fluid at T_f can be evaluated by Newton's law of cooling (1701) [Eq. (1.26)]:

$$q_c = h_c A(T_s - T_f) \tag{4.3}$$

Here, h_c is the convective heat transfer coefficient in W/m^2 K, A represents the wetted area in square meters, and the subscript c denotes convection. The quantity h_c serves as a parameter that collects miscellaneous variables such as the system geometry (including shape, orientation, surface roughness, etc.), fluid properties, and velocity and ΔT. In practice, h_c may vary from location to location over a surface. For such cases, an average h_c is employed in Eq. (4.3).

4.1.3 Radiation

Radiation is a process in which heat is transferred by electromagnetic (EM) wave phenomena. It occurs in vacuum space, in fluids as well as in solids. For example, thermal energy at the surface of the sun, estimated to be at 6000°C, is converted into EM waves which propagate in the huge space of a vacuum. When intercepted by an object on the earth, the EM waves reconvert into thermal energy to heat up the object. The rate of radiation emitted from a perfect radiator is determined by the expression derived empirically by Stefan (1879) and theoretically by Boltzmann (1884):

$$q_r = \sigma A T^4 \tag{4.4}$$

where A denotes the radiating surface area in square meters; T, the surface temperature in the absolute scale K; and σ, the Stefan-Boltzmann constant equal to 5.67×10^{-8} W/m² K⁴. The subscript r signifies radiation. Equation (4.4) is called the Stefan-Boltzmann equation, or the fourth-power law. For radiation between two surfaces at different temperatures, T_1 and T_2, the rate of net radiation exchange can be evaluated by Eq. (1.26):

$$q_r = F_{1-2} A_1 \sigma (T_1^4 - T_2^4) \tag{4.5}$$

Here, F_{1-2} is called the shape factor whose magnitude depends on the geometry, characteristics, and orientation of the radiating surfaces, if the medium enclosed between the surfaces does not interfere with the radiation exchange. The shape factor takes a value between zero and unity.

4.1.3.1 Electrical analogy. Both Eqs. (4.1) and (4.5) may be expressed in the form of Eq. (4.3) by defining

$$h_k = \frac{k}{\delta} \quad \text{and} \quad h_r = F_{1-2} \sigma (T_1^3 + T_1^2 T_2 + T_1 T_2^2 + T_2^3) \tag{4.6}$$

Then, the unified expression $q = hA\Delta T$ may be rewritten as

$$q = \frac{\Delta T}{1/hA} \tag{4.7}$$

It is similar in form with Ohm's law

$$i = \frac{\Delta E}{R} \tag{4.8}$$

where i is the current passing through an electrical resistance R over a voltage drop of ΔE. Therefore, the quantity $(1/hA)$ in Eq. (4.7) is equivalent to thermal resistance and some heat transfer problems may be solved using the concept of electrical network. The similarity between Eqs. (4.7) and (4.8) describes the analogy between the flows of heat and electricity.

4.1.4 Combined Heat Transfer Mechanisms

The skin surface at T_s dissipates heat into the environment at T_∞ by convection and radiation simultaneously. The total heat flow is

$$
\begin{aligned}
q &= q_c + q_r \\
&= h_c A(T_s - T_\infty) + h_r A(T_s - T_\infty) \\
&= (h_c + h_r)A(T_s - T_\infty) \\
&= \frac{T_s - T_\infty}{R_c} + \frac{T_s - T_\infty}{R_r} = \frac{T_s - T_\infty}{R_t}
\end{aligned}
\tag{4.9}
$$

where A is the area of the skin surface exposed to the environment. R_t is the combined thermal resistance consisting of two resistances R_c and R_r in parallel (as shown in Fig. 4.1); that is

$$
\frac{1}{R_t} = \frac{1}{R_c} + \frac{1}{R_r}
$$

4.2 CONDUCTION

Ways of studying conduction can be classified as shown in Table 4.1, together with the methods to study them. We will not discuss the graphic approach

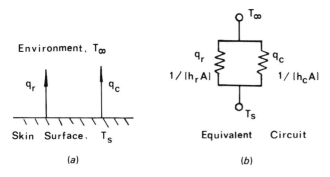

Figure 4.1 Heat dissipation from skin surface.

Table 4.1 Conduction studies in humans

| State | Phenomena | | | | Approaches | |
	Geometry	Structure	Physiological	Therapeutic	Heat transfer considerations	Experimental	Theoretical
1. Steady	1. One dimensional	1. Single	1. Without blood flow	1. Without radiant heating	1. Conduction only	1. In vivo	1. Analytic
2. Unsteady	2. Two dimensional	2. Composite	2. With blood flow	2. With radiant heating	2. With metabolism	2. In vitro	2. Numerical
	3. Three dimensional				3. Conduction-convection		3. Analogic
							4. Graphic

175

because of its limited application. The simplest case to start with is steady one-dimensional conduction.

4.2.1 Steady, One-dimensional Conduction

4.2.1.1 Effect of temperature dependence of k on q_k. If the value of k varies appreciably with temperature over the range of interest (T_H, T_L), what kind of average value should be used in Eq. (4.1)?

Equation (4.2) is rearranged as

$$\frac{q_k}{A} dx = -k(T) dT \tag{4.10}$$

Integration of Eq. (4.10) yields

$$\frac{q_k \delta}{A} = \int_{T_L}^{T_H} k(T) dT \tag{4.11}$$

where T_H and T_L are the temperatures at $x = 0$ and δ, respectively (Fig. 4.2a). A comparison of Eqs. (4.1) and (4.11) yields that k in Eq. (4.1) should be an integrated average value of $k(T)$ over the temperature range:

$$\bar{k}_{\text{int}} = \frac{1}{T_H - T_L} \int_{T_L}^{T_H} k(T) dT \tag{4.12}$$

$\bar{k}_{\text{int}}$ is also called the area average value of $k(T)$ between T_H and T_L on the k versus T plot. From now on, k refers to its integrated average value unless stated otherwise.

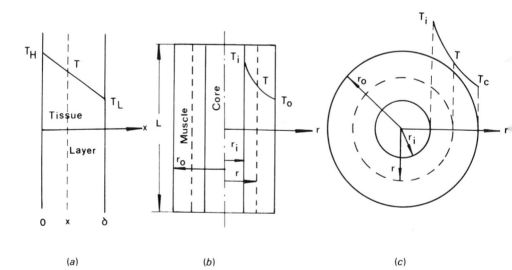

(a)   (b)   (c)

Figure 4.2 Basic system geometry. (a) Plain layer; (b) long hollow cylinder; (c) hollow sphere.

4.2.1.2 Effect of system geometry on q_k. A flat plate, long cylinder, and sphere are the three basic and simplest geometries treated in heat transfer.

Flat plates A layer of tissues with a large curvature can be considered a flat plate (Fig. 4.2a). Equation (4.10) is integrated twice: first from T_H to T_L for T and from 0 to δ for x, and the second from T_H to T and 0 to x. The results are combined to eliminate q_k, A, and k, yielding

$$T(x) = T_H - (T_H - T_L)\frac{x}{\delta} \tag{4.13}$$

which indicates that temperature varies linearly with location in flat plate.

Long hollow cylinders The body core may be treated as a solid cylinder, while the musculature zone over the core can be regarded as a hollow cylinder (Fig. 4.2b). For a long hollow cylinder of length l and inner and outer radii, r_i and r_o, respectively, A and x in Eq. (4.10) are replaced by $2\pi r l$ and r, respectively (see Fig. 4.2b). r is the radial distance from the center. Let the inner and outer surfaces be maintained at different values, T_i and T_o, respectively. By repeating the procedure of integrating twice on Eq. (4.2), one gets

$$q_k = 2\pi k l \frac{T_i - T_o}{\ln (r_o/r_i)} \tag{4.14}$$

for the heat transfer rate and

$$T(r) = T_i + (T_o - T_i)\frac{\ln (r/r_i)}{\ln (r_o/r_i)} \tag{4.15}$$

for the temperature distribution. Equation (4.15) reveals a logarithmic temperature profile in a hollow cylinder.

Hollow spheres Figure 4.2c illustrates a hollow sphere with T_i and T_o as the surface temperatures at the inner and outer radii, r_i and r_o, respectively. A and x in Eq. (4.10) are replaced by $4\pi r^2$ and r, respectively. Integrating Eq. (4.10) twice, one obtains, for the heat transfer rate,

$$q_k = \frac{4\pi r_i r_o k(T_i - T_o)}{r_o - r_i} \tag{4.16}$$

and for the temperature distribution,

$$T(r) = T_i - \frac{r - r_i}{r}\frac{r_o}{r_o - r_i}(T_i - T_o) \tag{4.17}$$

A hyperbolic temperature profile prevails in a hollow sphere as indicated in Eq. (4.17).

Now Eqs. (4.1), (4.14), and (4.16) are summarized in a unified form as

$$q_k = k\bar{A}\,\frac{\Delta T}{\delta} \tag{4.18}$$

where $\bar{A}$ denotes a mean area which is the cross-sectional area for flat plates

$$\bar{A} = \begin{cases} \bar{A}_{\log} \equiv \dfrac{A_o - A_i}{\ln\,(A_o/A_i)} = \dfrac{2\pi(r_o - r_i)l}{\ln\,(r_o/r_i)} \text{ for hollow cylinders} \\[2mm] \text{or} \quad \bar{A}_{\text{ari}} \equiv \dfrac{A_o + A_i}{2} \quad \text{if } \dfrac{r_o}{r_i} \leq 2 \\[2mm] \bar{A}_{\text{geo}} \equiv \sqrt{A_o A_i} = 4\pi r_i r_o \text{ for hollow spheres} \end{cases} \tag{4.19}$$

$\bar{A}_{\log}$, $\bar{A}_{\text{ari}}$, and $\bar{A}_{\text{geo}}$ stand for the logarithmic, arithmetic, and geometric mean areas, respectively. δ is the thickness.

4.2.1.3 Composite structures.

A composite structure consists of multiple layers of various substances with different thicknesses. Examples of composite structures are abundant in the human and animal bodies, composing various tissues. For theoretical analysis, the model consisting of multiple plain-layered tissues is employed when the layers are sufficiently thin compared with the curvature of the body (Yang and Wang, 1977; Baranski and Czerski, 1976; Lih, 1975). A composite layer of core, muscle, and skin in cylindrical configuration (Fig. 4.3a), has been used as the model of heat transfer in the body (torso) (Seagrave, 1971; Shitzer, 1975). As the model for the human head (Fig. 4.3b), three concentric spheres are used: the inner sphere as the cortex of the brain, the skull, the sinus, and the meninges, and the outer sphere as the scalp and external tissues (Nevins and Darwish, 1970). These multiple tissue layers are unique in that all tissues produce metabolic heat and that blood perfusion plays a very important role in dissipating the internal energy into the environment for maintaining homeostasis (i.e., maintaining a steady state in the internal environment). First, consider the in vitro case in which all tissues cease to generate metabolic heat. This corresponds to a situation where the tissues are excised from the body for an in vitro determination of their thermal properties.

Flat plates In Fig. 4.3c, the left surface is heated to T_c and there is a continuous heat flow q through the layers to the environment at T_a with the heat transfer coefficient h. Since the heat flow through a given area A is the same for any section, one writes

$$q = \frac{k_1 A}{\delta_1}\,(T_c - T_2) = \frac{k_2 A}{\delta_2}\,(T_2 - T_1) = \frac{k_3 A}{\delta_3}\,(T_1 - T_s)$$

$$= hA(T_s - T_a) \tag{4.20}$$

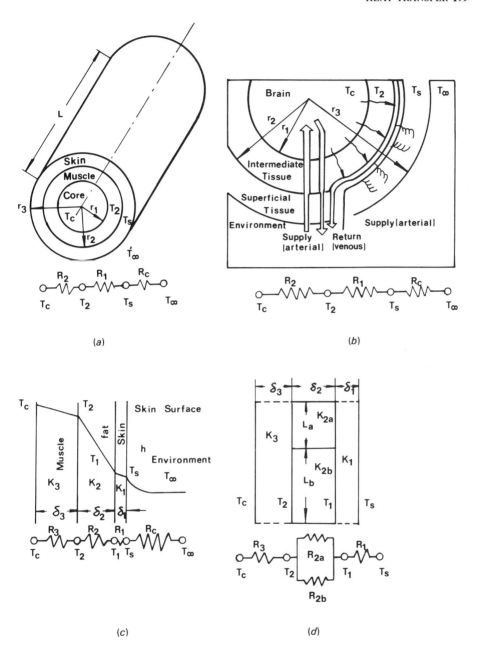

Figure 4.3 Composite structures (thermal circuits for no metabolism only). (*a*) Concentric cylinders; (*b*) concentric spheres; (*c*) plain tissues in series; (*d*) plain tissues in series and parallel arrangement.

The situation is analogous to a current flowing through a series of four resistances over the voltage drop equivalent to $(T_s - T_a)$. Equation (4.20) can be rewritten as

$$q = \frac{\Delta T_{ov}}{R_t} \qquad (4.21)$$

in which ΔT_{ov} is the overall temperature drop across the composite structure, R_t is the total thermal resistance of the composite structure $= \Sigma R_i$, and R_i is the thermal resistance of individual layers.

In numerous complex tissue structures, it is more realistic to consider combinations of series- and parallel-connected heat-flow paths. An example of such a case is shown in Fig. 4.3d. For simplicity, one-dimensional heat flow is assumed. The intermediate layer consists of two separate thermal paths in parallel and its thermal conductance is the sum of the individual conductances. Equation (4.21) determines the heat transfer rate across the composite structure.

Concentric cylinders The cylindrical model treats radial heat flow through long concentric cylinders of different thermal conductivity. For an in vitro case in the absence of metabolism, the rate of steady heat flow through each section is the same:

$$q = \frac{2\pi k_2 l}{\ln (r_2/r_1)} (T_c - T_2) = \frac{2\pi k_3 l}{\ln (r_3/r_2)} (T_2 - T_s)$$
$$= 2\pi r_3 lh(T_s - T_a) \qquad (4.22)$$

Again, Eq. (4.21) determines the rate of heat flow into the environment.

4.2.2 Conduction Involving Metabolism, Blood Perfusion, and Thermal Radiation Therapy

4.2.2.1 Conduction in tissues with metabolism. Metabolism takes place in living tissues and is equivalent to internal heat generation. Many systems in engineering have internal heat sources, such as electric coils, nuclear reactors, resistance heaters, and viscous heating.

Flat plates A differential element in Fig. 4.4a is defined as the system. The first law is applied which yields

$$q_{kx} - q_{k(x+dx)} + \dot{q}(A \, dx) = 0 \qquad (4.23)$$

Here, q_{kx} and $q_{k(x+dx)}$ are the heat flow rates across the left and right surfaces of the system, respectively, while q''' denotes the volumetric rate of internal heat generation. Now the term $q_{k(x+dx)}$ is expanded into a Taylor's series

$$q_{k(x+dx)} = q_{kx} + \frac{dq_{kx}}{dx} dx + \text{Higher-order terms} \qquad (4.24)$$

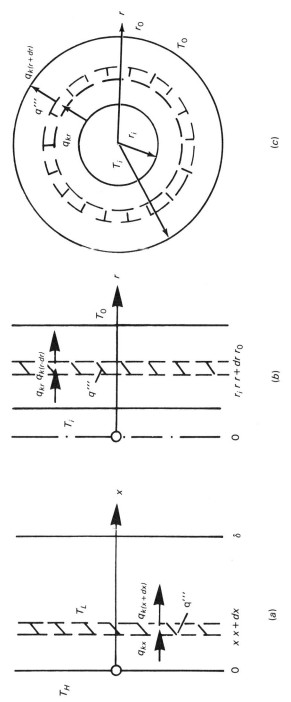

Figure 4.4 Conduction in systems with internal heat generation. (*a*) Flat plate; (*b*) hollow cylinder; (*c*) hollow sphere.

The first two terms of Eq. (4.24) are substituted into Eq. (4.23) followed by the replacement of q_{kx} by the Fourier equation, Eq. (4.2). For isotropic materials, we obtain

$$k \frac{d^2T}{dx^2} + q''' = 0 \tag{4.25}$$

Conduction   Internal
heat
generation

The appropriate boundary conditions for Fig. 4.4a are

$$T(0) = T_H \qquad T(\delta) = T_L \tag{4.26}$$

Equation (4.25) is integrated twice, followed by the substitution of Eq. (4.26) to determine the integration constants. This yields, for a system with uniform internal heat generation (metabolism in resting humans),

$$T(x) = -\frac{q'''}{2k} x^2 + \left(\frac{q''' \delta}{2k} - \frac{T_H - T_L}{\delta}\right) x + T_H \tag{4.27}$$

Equation (4.27) exhibits a parabolic temperature profile with its peak (at $dT/dx = 0$) shifted from the midplane toward the surface of higher temperature. It is important to note that no heat flows across the plane of $dT/dx = 0$ (i.e., maximum temperature) and that

$$q_1 = -kA \frac{dT(\delta)}{dx} \tag{4.28a}$$

and

$$q_{II} = -kA \frac{dT(0)}{dx} \tag{4.28b}$$

represent the rates of heat flow out of the flat plate across the right and left surfaces, respectively. The sum of q_1 and q_{II} is equal to $q''' \delta A$.

Hollow cylinders The procedure is repeated for a long hollow cylinder of length l in Fig. 4.4b. The differential element is a hollow cylinder with the inner and outer radii of r and $r + dr$, respectively. The application of the first law yields

$$\frac{1}{r} \frac{d}{dr} \left(r \frac{dT}{dr}\right) + \frac{q'''}{k} = 0 \tag{4.29}$$

The appropriate boundary conditions are

$$T(r_i) = T_i \qquad T(r_o) = T_o \tag{4.30}$$

The solution of Eqs. (4.29) and (4.30) is

$$T(r) = -\frac{q'''(r^2 - r_i^2)}{4k} + \frac{T_o - T_i + (q'''/4k)(r_o^2 - r_i^2)}{\ln(r_o/r_i)} \ln\left(\frac{r}{r_i}\right) + T_i \tag{4.31}$$

which is also a parabolic temperature profile.

Hollow spheres The governing equation for hollow spheres in Fig. 4.4c is

$$\frac{1}{r^2}\frac{d}{dr}\left(r^2\frac{dT}{dr}\right) + \frac{q'''}{k} = 0 \tag{4.32}$$

subject to the boundary conditions, Eq. (4.30). The temperature distribution is found to be

$$T(r) = -\frac{q'''(r^2 - r_i^2)}{6k} + \frac{(r - r_i)[q'''(r_o^2 - r_i^2)/(6k) + T_o - T_i]r_o}{(r_o - r_i)r} + T_i \tag{4.33}$$

which is a parabolic profile.

4.2.2.2 Conduction in tissues with blood perfusion. Consider a tissue layer perfused by blood entering at a lower temperature T_A. In the absence of metabolic heat generation, the heat transfer mechanism in the system is a combination of conduction and convection. For a control volume in a tissue layer (Fig. 4.5), the first law gives

$$q_{kx} - q_{x(x+dx)} - (\dot{w}C_p)_b A dx(T - T_A) = 0$$

With the Taylor's series expansion of $q_{k(x+dx)}$, the equation is reduced to

$$k\frac{d^2T}{dx^2} - (\dot{w}C_p)_b(T - T_A) = 0$$

$$\underset{\text{Conduction}}{\qquad\qquad} \underset{\text{Convection}}{\qquad\qquad}$$

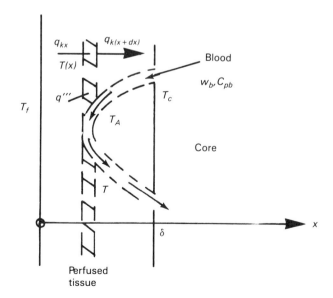

Figure 4.5 Perfused tissue layer with metabolic heat production.

Here, $\dot{w}_b$ denotes the perfusion rate of blood per unit volume of the tissue while T_A is the temperature of arterial blood. The equation signifies a balance of conduction with convection through blood perfusion in the tissue. It can be rewritten as

$$\frac{d^2T}{dx^2} - m^2(T - T_A) = 0 \tag{4.34a}$$

where $m^2 = (\dot{w}c_p)_b/k$. The appropriate boundary conditions are

at $x = 0$ (fat-tissue interface): $T(0) = T_f$

at $x = \delta$ (core-tissue interface): $T(\delta) = T_c$

The tissue temperature distribution is obtained as

$$\frac{T - T_A}{T_f - T_A} = \frac{T_c - T_A}{T_f - T_A}\frac{\sinh mx}{\sinh m\delta} + \cosh m\delta - \coth m\delta \sinh mx$$

4.2.2.3 Fins. Newton's convection equation, Eq. (4.3), suggests that with operating conditions (h and $T_s - T_f$) fixed, the rate of heat flow will increase in direct proportion to an increase in the area A. Additional areas being attached to the original (or primary) surface for such a purpose are called extended surfaces or fins. Such extended surfaces have wide industrial applications. In human beings, the ears, the nose, and fingers do, to a certain extent, function like extended surfaces in cold weather. Fig. 4.6 shows a one-dimensional fin of uniform cross-sectional area A and perimeter P. It is attached to a primary surface at T_s and is cooled along by a fluid at T_f with the heat transfer coefficient h. It is obvious from our experience that the fin temperature $T(x)$ varies along the fin axis between T_s and T_f. In other words, the driving force $(T - T_f)$ for heat flow into the fluid is less than $(T_s - T_f)$. The answer to how much less leads to the definition of fin efficiency.

In order to determine the temperature distribution in the fin, a differential element is defined in Fig. 4.6. There are three heat flows: conduction com-

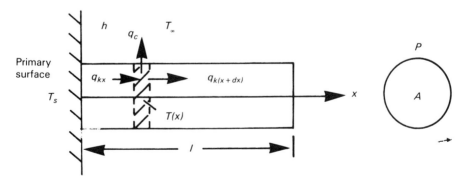

Figure 4.6 One-dimensional extended surface.

ponents q_{kx} and $q_{k(x+dx)}$ into and out of the left and right faces and convection q_c out from the surface of the element. The application of the first law followed by the Taylor's series expansion of $q_{k(x+dx)}$ and the substitution of the Fourier and Newton's equations produces

$$kA \frac{d^2T}{dx^2} - hP(T - T_f) = 0 \qquad (4.34b)$$

Conduction   Convection

Let $m^2 = hP/kA$. The solution of Eq. (4.34b) may take either of two different forms:

$$T(x) = c_1 e^{mx} + c_2 e^{-mx} + T_f \qquad (4.35)$$

or

$$T(x) = c_3 \sinh mx + c_4 \cosh mx + T_f \qquad (4.36)$$

which are appropriate for a very long fin and a finite fin, respectively. The boundary conditions are at the fin root:

$$T(0) = T_s \qquad (4.37)$$

At the fin tip:

$$T(\infty) = \text{Finite, for a very long fin} \qquad (4.38a)$$

In the case of a finite fin of length l, the heat balance at the tip surface (i.e., $q_k'' = q_c''$) leads to

$$-k \frac{dT(l)}{dx} = h[T(l) - T_f] \qquad (4.38b)$$

The integration constants c_1 and c_2 are determined by Eqs. (4.37) and (4.38a), and c_3 and c_4 are determined by Eqs. (4.37) and (4.38b). For very long fins, we get:

$$T(x) = (T_s - T_f)e^{-mx} + T_f \qquad (4.39a)$$

for finite fins,

$$T(x) = \frac{\cosh m(l - x) + \psi \sinh m(l - x)}{\cosh ml + \psi \sinh ml} (T_s - T_f) + T_f \qquad (4.39b)$$

where $\psi = h/mk$. Equations (4.39a) and (4.39b) indicate that the temperature profile in the fin is an exponential or hyperbolic function of x. When the tip surface is insulated, i.e., $h = 0$ and consequently $\psi = 0$, Eq. (4.39b) is reduced to

$$T(x) = \frac{\cosh m(l - x)}{\cosh ml} (T_s - T_f) + T_f \qquad (4.39c)$$

The heat flow rate from the fin, q_{fin}, is equal to that across the fin root. This can be determined by

$$q_{fin} = -kA \frac{dT(0)}{dx} \qquad (4.40)$$

The substitution of Eqs. (4.39a), (4.39b), and (4.39c) into Eq. (4.40) yields, respectively,

for very long fins: $\qquad q_{fin} = \sqrt{hPkA}(T_s - T_f) \qquad (4.41a)$

for finite fins:

$$q_{fin} = \sqrt{hPkA}(T_s - T_f) \frac{\sinh ml + \psi \cosh ml}{\cosh ml + \psi \sinh ml} \qquad (4.41b)$$

$$= \sqrt{hPkA}(T_s - T_f) \tanh ml \text{ with insulated tip} \qquad (4.41c)$$

4.2.2.4 Conduction in perfused tissues with metabolic heat generation. The body tissues are very poor thermal conductors with approximately the same thermal conductivity as cork, or somewhat less than water (see Table D.9 in Appendix D). Therefore, if the metabolic heat production has to be dissipated by conduction alone, it will require very high temperature gradients to fulfill its duty. It is blood perfusion in the body tissues that plays the vitally important role in the regulation of body temperature. When heat must be conserved in a cold environment, vasoconstriction occurs. In hot environments or during exercise, vasodilation aids in transmitting heat to the skin layer for dissipation. The rate of blood perfusion in tissue is regulated automatically and nonvoluntarily.

Plain tissues Consider a perfused tissue layer with metabolic heat generation, as shown in Fig. 4.5. The rate of blood perfusion per unit volume of the tissue bed is $\dot{w}_b$ and the specific heat of blood is C_{pb}. The arterial blood enters the differential element at T_A and leaves it at the local tissue temperature $T(x)$ (Pennes, 1948). The net rate of heat brought into the differential element by blood perfusion is $\dot{w}_b C_{pb}(T_A - T)$. Therefore, the first law leads to

$$\underbrace{\frac{d^2T}{dx^2}}_{\text{Conduction}} + \underbrace{\dot{w}_b C_{pb}(T_A - T)}_{\substack{\text{Convection} \\ \text{by blood} \\ \text{perfusion}}} + \underbrace{q'''}_{\substack{\text{Metabolic} \\ \text{heat} \\ \text{generation}}} = 0 \qquad (4.42)$$

which is equivalent to the superposition of Eqs. (4.25) and (4.34). It is subject to the appropriate boundary conditions

$$T(0) = T_f \qquad T(\delta) = T_c \qquad (4.43)$$

The solution is obtained and rearranged in dimensionless form as

$$\frac{T_c - T(x)}{T_c - T_f} = \left(1 + \frac{H_m}{H_p^2}\right)\frac{\sinh H_p(1 - \xi)}{\sinh H_p} + \frac{H_m}{H_p^2}\left(\frac{\sinh H_p \xi}{\sinh H_p} - 1\right) \quad (4.44)$$

where $H_p \equiv \sqrt{\dot{w}_b C_{pb}\delta^2/k}$ is the perfusion-conduction ratio and
$H_m \equiv \sqrt{q'''\delta^2/[k(T_c - T_f)]}$ is the metabolism-conduction ratio.

The heat flux q_0'' across the fat–muscle interface at $x = 0$ can be found as

$$q_o'' = k\frac{dT(0)}{dx} = \frac{k(T_c - T_f)}{\delta}\left(H_p \coth H_p + \frac{H_m}{H_p}\frac{\cosh H_p - 1}{\sinh H_p}\right) \quad (4.45)$$

Hollow cylinders Heat transfer in extremities was treated in Wissler (1961) by considering the entire arm or leg, instead of the layers (such as skin and fat) near the surface with the curvature neglected. Now, consider each limb to consist of only two tissue layers: deep and superficial tissues (Fig. 4.7), which form two concentric tissue cylinders of finite length. The two-dimensional analysis is treated later in Section 4.2.4. For simplicity, we consider the one-dimensional case here.

The governing equations read

$$\frac{k_i d}{r\, dr}\left(r\frac{dT_i}{dr}\right) + \dot{w}_{bi}C_{pbi}(T_A - T_V) + q_i''' = 0 \quad (4.46)$$

where $i = 1$ and 2 for the deep and superficial tissues, respectively. Note that the venous blood temperature T_V is used to replace the local tissue temperature T_i in the blood perfusion term. The boundary conditions are at center:

$$\frac{dT_1(0)}{dr} = 0 \quad (4.47)$$

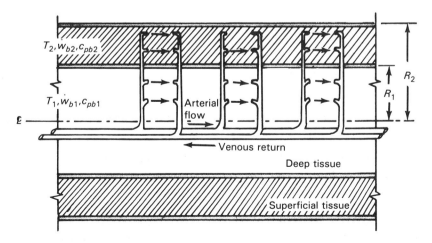

Figure 4.7 Two concentric tissue cylinders of the limb.

at tissue interface:

$$T_1(R_1) = T_2(R_1) \qquad \text{(continuity of temperature)} \qquad (4.48a)$$

$$k_1 \frac{dT_1(R_1)}{dr} = k_2 \frac{dT_2(R_2)}{dr} \qquad \text{(continuity of heat flux)} \qquad (4.48b)$$

at surface:

$$-k_2 \frac{dT_2(R_2)}{dr} = h[T_2(R_2) - T_a] \qquad (4.49)$$

The temperature profiles in the limb are obtained as

$$T_1 = -\frac{B_1 r^2}{4} + c_2 \qquad (0 \leqslant r \leqslant R_1) \qquad (4.50)$$

$$T_2 = -\frac{B_2 r^2}{4} + c_3 \ln r + c_4 \qquad (R_1 \leqslant r \leqslant R_2) \qquad (4.51)$$

where,

$$B_1 = \frac{1}{k_1}[\dot{w}_{b1} C_{pb1}(T_A - T_V) + q_1''']$$

$$B_2 = \frac{1}{k_2}[\dot{w}_{b2} C_{pb2}(T_A - T_V) + q_2''']$$

$$c_2 = \frac{R_1^2}{2}\left(B_2 - B_1\frac{k_1}{k_2}\right)$$

$$c_3 = \frac{1}{\ln(R_2/R_1)}\left[\frac{B_1 R_1^2}{4} + \frac{B_2 R_2^2}{4\mathrm{Bi}_2} + \frac{B_2}{4}(R_2^2 - R_1^2)\right.$$

$$\left. + T_a - c_2\left(1 + \frac{1}{2\mathrm{Bi}_2}\right)\right]$$

$$c_4 = \frac{R_1^2}{4}(B_2 - B_1) + c_2 - c_3 \ln R_1$$

$$\mathrm{Bi}_i = \frac{hR_i}{2k_i} = \text{Biot number} = \text{convection-conductance ratio}$$

4.2.2.5 Conduction in perfused tissues with metabolic heat generation subject to radiation diathermy. Since the past century, it has repeatedly been claimed that heat may exert an inhibitory effect on a malignant tumor or even cure it. Local heating of the tumor and its surroundings can be achieved by external heat applications, such as a water bath (convection), microwave, ultrasound, radio frequency, localized electric current fields, and infrared wave. Many studies (see Section 4.9.2) involving more than two decades, lead to the conclusion that electromagnetic (EM) and ultrasonic (US) diathermy are the most effective in localized deep tissue heating.

When a perfused tissue layer, such as the one in Fig. 4.6, is irradiated by a radiant heat flux, q''_{ro}, the local absorption of the radiant energy varies according to $q''_{rx} = q''_{ro}e^{-k'x}$, where k' is the absorption coefficient. The heat equation for the differential element then reads

$$k\frac{d^2T}{dx^2} + \dot{w}_b C_{pb}(T_A - T) + q''' + \frac{q''_{ro}}{\delta}e^{-k'x} = 0 \qquad (4.52)$$

| Conduction | Convection by blood flow | Internal heat generation | Radiation heating |

subject to the same boundary conditions, Eq. (4.43). The steady temperature profile is obtained as

$$T(x) = C_1 \sinh mx + C_2 \cosh mx + A(1 - \cosh mx)$$

$$+ \frac{B}{2m}\left[\frac{e^{(m-k')x} - 1}{m - k'}(\cosh mx - \sinh mx)\right.$$

$$\left.+ \frac{e^{-(m+k')x} - 1}{m + k'}(\cosh mx + \sinh mx)\right] \qquad (4.53)$$

where $m = \dfrac{\dot{w}_b C_{pb}}{k}$

$A = \dfrac{\dot{w}_b C_{pb} T_a}{k} + \dfrac{q'''}{k}$

$B = \dfrac{q_r}{k\delta}$

4.2.3 Formulation of Bioheat Transfer Problems

In many situations, heat flows in two or three directions and temperature becomes a function of two or three coordinates. The heat conduction through a rectangular slab or a finite cylinder is a typical example.

A differential element is defined as the system for derivation of the conduction equations. In Cartesian coordinates, the element has the shape of a rectangular parallelepiped with its edges dx, dy, and dz parallel to the x, y, and z axes, respectively (Fig. 4.8). Let (q_{kx}, q_{ky}, q_{kz}) and $(q_{k(x+dx)}, q_{k(y+dy)}, q_{k(z+dz)})$ be the heat flow components into and out of the element, respectively. With the aid of the Fourier conduction equation, the heat balance in each direction produces a conduction term: $\partial^2 T/\partial x^2$, $\partial^2 T/\partial y^2$, and $\partial^2 T/\partial z^2$ in the x, y, and z directions, respectively. Taking metabolic heat generation, radiation absorp-

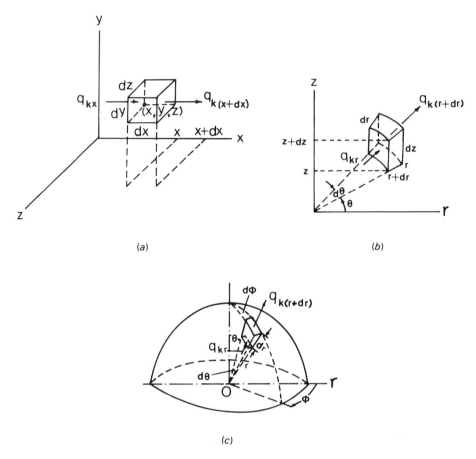

Figure 4.8 Differential elements in (a) Cartesian, (b) cylindrical, and (c) spherical coordinates.

tion, blood cooling, and timewise change of the system energy into account, the first law yields the following bioheat transfer equation:

$$\rho c_p \frac{\partial T}{\partial t} + \dot{w}_b c_{pb}(T - T_A) = k\left(\frac{\partial^2 T}{\partial x^2} + \frac{\partial^2 T}{\partial y^2} + \frac{\partial^2 T}{\partial z^2}\right) + q''' + q_r'''$$

| Unsteady state | Cooling by blood perfusion | Conduction components | Metabolic heat generation | Radiation absorption |

(4.54)

The last term on the right-hand side is needed only when the system is subject to radiation diathermy. At steady state, the first term on the left-hand side becomes zero. The first term on right-hand side consists of the conduction components in the x, y, and z direction. In the case of one-dimensional heat flow in the x direction, as in Section 4.2.2, the other two terms $\partial^2 T/\partial y^2$ and $\partial^2 T/\partial z^2$ diminish.

The bioheat equation (4.54) is most commonly used for regional variations in temperature and heat flux in living tissues. The model is valid at the macrostructural level for the purposes of studying thermal regulation, comfort, hyperthermia, hypothermia, and other phenomena. In this model, $\dot{w}_b C_{pb}(T_A - T_s)$ is the perfusion heating term using some arterial blood supply temperature T_A. Recently, several investigators such as Chen and Holmes (1980), Chato (1980), and Weinbaum and Jiji (1985) examined the effects of the microvascular organization on the mechanism for blood-tissue energy exchange. The primary thermal equilibration occurred in microvessels larger than 40–50 μm diameter between the terminal arterial branches and the precapillary arterioles, since the thermal relaxation length of these small microvessels (characteristic distance for the blood to equilibrate to the local tissue temperature) is much shorter than their length. In thermally significant microvessels, Chen and Holmes (1980) suggested the inclusion of two additional terms in the bioheat equation: a term proportional to local blood perfusion velocity and a term describing the perfusion thermal conductivity. Weinbaum and Jiji (1985) derived a new "simplified bioheat equation" to describe the effect of blood flow on blood-tissue heat transfer. The new theory claims that vascularization causes the tissue to behave as an anisotropic heat transfer medium. Hence, an expression is derived in the new bioheat equation for the tensor conductivity of the tissue as a function of the local vascular geometry and flow velocity in the thermally significant counterflow vessels.

In the case of a cylindrical or spherical system, it is more convenient to use the bioheat transfer equation in cylindrical or spherical coordinates. Due to the curvature of the cross section (i.e., the change in area) in the direction of the heat flow, the expression for conduction in the corresponding direction will differ. Other terms remain unchanged. Equation (4.54) in the cylindrical coordinate system (r, θ, z) reads

$$\rho C_p \frac{\partial T}{\partial t} + \dot{w}_b C_{pb}(T - T_A)$$
$$= k \left[\frac{1}{r} \frac{\partial}{\partial r} \left(r \frac{\partial T}{\partial r} \right) + \frac{1}{r^2} \frac{\partial^2 T}{\partial \theta^2} + \frac{\partial^2 T}{\partial z^2} \right] + q''' + q_r''' = 0 \qquad (4.55)$$

In the spherical coordinate system (r, θ, ψ), the general bioheat transfer equation becomes

$$\rho C_p \frac{\partial T}{\partial t} + \dot{w}_b C_{pb}(T - T_A)$$
$$= k \left[\frac{1}{r^2} \frac{\partial}{\partial r} \left(r^2 \frac{\partial T}{\partial r} \right) + \frac{1}{r^2 \sin \theta} \frac{\partial}{\partial \theta} \left(\sin \theta \frac{\partial T}{\partial \theta} \right) + \frac{1}{r^2 \sin^2 \theta} \frac{\partial^2 T}{\partial \psi^2} \right]$$
$$+ q''' + q_r''' = 0 \qquad (4.56)$$

The bioheat transfer equation is subject to appropriate initial and boundary conditions. The initial condition describes the state of the system at zero time. For example, the system is at a uniform temperature at $t = 0$:

$$T(x, 0) = T_0 \tag{4.57}$$

Generally, one-, two- and three-dimensional systems have two, four, and six boundaries, respectively, at which thermal conditions are to be described.

Typical boundary conditions include:

1. *Prescribed temperature.* For example, a thermocouple measures the temperature of a surface at $x = b$ as $T_b(t)$:

$$T(b, t) = T_b(t) \tag{4.58}$$

2. *Prescribed heat flux.* For example, a surface at $x = b$ is subject to a radiation heat flux of q_r'':

$$-\frac{\partial T}{\partial x}(b, t) = q_r'' \tag{4.59}$$

An insulated surface corresponds to the case of zero heat flux:

$$\frac{\partial T}{\partial x}(b, t) = 0 \tag{4.60}$$

3. *Heat transfer to the environment by convection.* When a surface is in contact with the environment at T_a, heat transfer takes place by convection, radiation, or a combination of the two. One describes it mathematically as

$$-k\frac{\partial T(b, t)}{\partial x} = h[T(b, t) - T_a] \tag{4.61}$$

in which h is the heat transfer coefficient by convection, radiation, or a combination of the two.

4. *Interface between two solids.* When two solids of different thermal conductivities, k_1 and k_2, are in tight contact (with no contact resistance), then the continuity of the temperature and heat flux at the interface is:

$$T_i(b, t) = T_2(b, t) \tag{4.62}$$

$$k_1\frac{\partial T_1}{\partial x}(b, t) = k_2\frac{\partial T_2}{\partial x}(b, t) \tag{4.63}$$

Now, the formulation of a conduction problem is completed. It is then followed by a procedure to obtain the theoretical solution which can be compared with empirical results by in vivo or in vitro tests. Theoretical solutions can be obtained by four distinct approaches: analytic, numerical, analogical, and graphic methods. Results presented in the preceding Section 4.2.2 are obtained by the analytic method.

4.2.4 Steady Two- and Three-dimensional Conduction

A few relatively simple conduction problems for rectangular slabs and finite cylinders are solved by the analytic method.

4.2.4.1 Slabs. Consider a rectangular slab with one surface heated to T_H and the other surfaces maintained at a reference temperature, 0. The coordinate axes are fixed as shown in Fig. 4.9. At steady state, the heat equation

$$\frac{\partial^2 T}{\partial x^2} + \frac{\partial^2 T}{\partial y^2} = 0 \qquad (4.64)$$

is subject to the boundary conditions

$$T(x, 0) = 0 \qquad T(a, y) = 0 \qquad \frac{\partial T}{\partial x}(0, y) = 0 \qquad (4.65a)$$

$$T(x, b) = T_H \qquad (4.65b)$$

The classical separation-of-variables technique is employed. A product solution of the form

$$T(x, y) = X(x) \cdot Y(y) \qquad (4.66)$$

is substituted into Eq. (4.64) followed by separation of variables. It yields

$$-\frac{X''}{X} = \frac{Y''}{Y} = \lambda^2 \qquad (4.67)$$

where the double prime denotes a second derivative. λ^2 is a constant to be determined and λ is called the eigenvalue. Now the partial differential equation is reduced to two ordinary differential equations (4.67) whose general solutions are readily available as

$$X = C_1 \sin \lambda x + C_2 \cos \lambda x \qquad (4.68a)$$

$$Y = C_3 \sinh \lambda y + C_4 \cosh \lambda y \qquad (4.68b)$$

When Eq. (4.66) is substituted into Eq. (4.65a), we obtain

$$Y(0) = 0 \qquad (4.69a)$$

$$X(a) = 0 \qquad (4.69b)$$

$$X'(0) = 0 \qquad (4.69c)$$

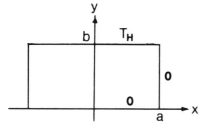

Figure 4.9 Rectangular plate at steady state.

The substitution of Eq. (4.69c) into Eq. (4.68a) results in $C_1 = 0$. When Eq. (4.69b) is substituted into Eq. (4.68a), one gets $C_2 \cos \lambda a = 0$. To satisfy this condition, $\cos \lambda a$ must be zero or $\lambda = (2n + 1)\pi/a$, where n is a positive integer including zero. Therefore, there exists a solution for each integer n. The substitution of Eq. (4.69a) into Eq. (4.68b) yields $C_4 = 0$. Summarizing all these results into Eq. (4.66), we obtain

$$T(x, y) = \sum_{n=0}^{\infty} a_n \sinh \lambda_n y \cos \lambda_n x \qquad (4.70)$$

When Eq. (4.65b) is substituted into Eq. (4.70),

$$T_H = \sum_{n=0}^{\infty} a_n \sinh \lambda_n b \cos \lambda_n x$$

One now utilizes the orthogonality of the cosine function to determine a_n:

$$a_n = \frac{\int_0^a T_H \cos \lambda_n x \, dx}{\int_0^a \sinh \lambda_n b (\cos \lambda_n x)^2 \, dx} = \frac{2(-1)^n T_H}{(\lambda_n a) \sinh \lambda_n b}$$

The solution therefore becomes

$$\frac{T(x, y)}{T_H} = 2 \sum_{n=0}^{\infty} \frac{(-1)^n}{\lambda_n a} \frac{\sinh \lambda_n y}{\sinh \lambda_n b} \cos \lambda_n x \qquad (4.71)$$

Here, $\cos \lambda_n x$ is called the eigenfunction, and the problem is referred to as an eigenvalue problem.

When metabolic heat generation is included, the bioheat transfer equation becomes

$$\frac{\partial^2 T}{\partial x^2} + \frac{\partial^2 T}{\partial y^2} + \frac{q'''}{k} = 0 \qquad (4.72)$$

which can be solved by

$$T(x, y) = \phi(x, y) + \psi(x) \qquad (4.73a)$$

or

$$T(x, y) = \phi(x, y) + \psi(y) \qquad (4.73b)$$

The resulting homogeneous equation of $\phi(x, y)$ is treated by the separation-of-variables method as with the preceding equations, while the ordinary differential equation, $\psi(x)$, carrying the nonhomogeneous term q'''/k can be integrated directly.

Adding the blood perfusion term,

$$\frac{\partial^2 T}{\partial x^2} + \frac{\partial^2 T}{\partial y^2} - \frac{\dot{w}_b C_{pb}}{k} (T - T_A) + \frac{q'''}{k} = 0 \qquad (4.74)$$

one can define

$$\theta(x, y) = T(x, y) - T_A - \frac{q'''}{\dot{w}_b C_{pb}} \qquad (4.75)$$

Equation (4.74) can be reduced to a homogeneous form and solved by the separation-of-variables technique.

4.2.4.2 Finite cylinders. Nevins and Darwish (1970) treated heat flow in the limb as two dimensional. The bioheat transfer equations for the two issue layers are

$$k_i \left[\frac{1}{r} \frac{\partial}{\partial r} \left(r \frac{\partial T_i}{\partial r} \right) + \frac{\partial^2 T_i}{\partial z^2} \right] + \dot{w}_{bi} C_{pbi}(T_A - T_V) + q_i''' = 0 \qquad (4.76)$$

where $i = 1$ and 2 for deep and superficial tissues, respectively. The following boundary conditions govern, in addition to those for the radial direction. Since the limb is connected to the torso, the temperature is the same as that of the arterial blood supply:

$$T_1(r, 0) = T_2(r, 0) = T_A \qquad (4.77)$$

Heat loss from the end of the limb, such as the fingers or the sole, is negligible:

$$\frac{\partial T_1(r, l)}{\partial z} = \frac{\partial T_2(r, l)}{\partial z} = 0 \qquad (4.78)$$

where l is the length of the limb. This assumption is justified because an internal heat exchange or heat shunt mechanism in the hand or foot tends to neutralize the end temperature to zero.

The procedure to obtain the analytic solution to this problem is quite complicated. Temperature profiles in the arm at various environmental temperatures are shown in Fig. 4.10.

4.2.5 Unsteady Conduction

Consider a flat plate of thickness 2δ and thermal conductivity k (Fig. 4.11). Both surfaces are exposed to the environment at T_a with the heat transfer coefficient h. Conduction from the midplane at T_c to the surface at T_s is balanced with convection between the surface and the environment:

$$\frac{kA(T_c - T_s)}{\delta} = hA(T_s - T_a) \qquad (4.79)$$

This can be rewritten as

$$\frac{T_c - T_s}{T_s - T_a} = \frac{\delta/kA}{1/hA} = \frac{h\delta}{k} \qquad (4.80)$$

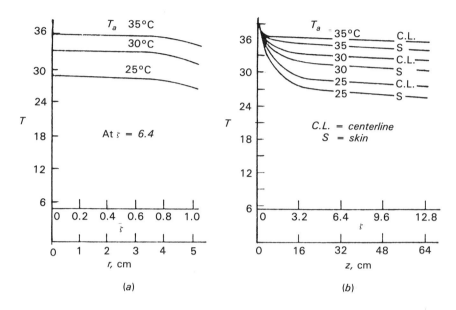

Figure 4.10 Temperature profiles in the arm at various environmental temperatures. (*a*) Radial; (*b*) longitudinal.

The middle term is the ratio of internal (or conductive) resistance to surface (or convective) resistance, while the right-hand term is the definition of the Biot number. Equation (4.79) implies that a uniform temperature can be achieved in the plate if the system is characterized by a very small Biot number. $Bi \leq 0.1$ has been accepted as the criterion for a uniform temperature at any time instant in a system of any geometry. Such a system is called a lumped system or a compartment.

4.2.5.1 Lumped system analysis. A system of any geometry is in thermal equilibrium with the environment at T_0. Its volume, surface area for heat transfer, density, and specific heat are denoted by V, A, ρ, and C_p, respectively. At zero time, the environment undergoes a step change in temperature to T_a. We are interested in determining the subsequent change in temperature of the system

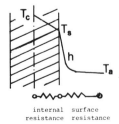

internal surface
resistance resistance   **Figure 4.11** Internal and surface resistances.

$T(t)$ (Fig. 4.12). Let the heat transfer coefficient between the system and the environment be h. The first law of thermodynamics leads to the expression

$$hA(T_a - T) = \rho V C_p \frac{\partial T}{\partial t} \tag{4.81}$$

subject to the initial condition

$$T(0) = T_0 \tag{4.82}$$

The solution is

$$\frac{T(t) - T_a}{T_0 - T_a} = \exp\left(-\frac{hAt}{\rho V C_p}\right) = \exp\left(-\mathrm{BiF_0}\right) = \exp\left(-\frac{t}{\tau}\right) \tag{4.83}$$

where $\mathrm{F_0} \equiv \dfrac{\alpha t}{2} = $ Fourier number

$\tau \equiv \dfrac{\rho V C_p}{hA} = $ time constant of the system

$L \equiv \dfrac{V}{A} = $ characteristic length of the system

$\alpha \equiv \dfrac{k}{\rho C_p} = $ thermal diffusivity

$(\alpha t)^{1/2}$ is a measure of the depth of heat penetration in the system during time t. α is the ratio of conduction speed to heat capacity of the system. τ is indicative of the rate of response of a single capacity (or first-order) system to a sudden change in the environmental temperature. It corresponds to the time when the temperature difference $(T - T_a)$, drops to 36.8% of the initial potential difference $(T_0 - T_a)$. The smaller the time constant, the faster is the response.

The instantaneous rate of heat transfer between the system and the environment q can be readily obtained as

$$\frac{q}{hA(T_0 - T_a)} = \exp\left(-\mathrm{BiF_0}\right) \tag{4.84}$$

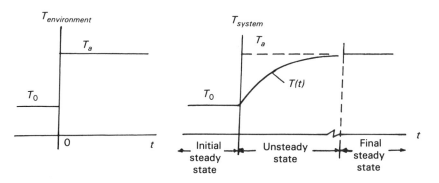

Figure 4.12 Response of system temperature to step change in environmental temperature.

Here the denominator signifies the maximum rate of heat transfer which exists only at the instant of the step change. The amount of heat transfer between zero time and any time instant Q can be obtained by integrating q as

$$Q = \int_0^t q \, dt$$

It yields

$$\frac{Q}{\rho V C_p (T_0 - T_a)} = 1 - \exp(-BiF_0) \qquad (4.85)$$

The denominator represents the maximum energy that the system possesses at zero time.

4.2.5.2 Distributed parameter analysis. When a system has a Biot number greater than 0.1, its temperature distribution must be treated as functions of both time and location.

Flat plates A flat plate of thickness 2δ is at a uniform temperature T_0. Both surfaces are suddenly exposed to an environment at zero temperature (a reference value). The heat transfer coefficient is very large. With the origin of the x axis fixed at the midplane, the heat equation

$$\frac{\partial T}{\partial t} = \alpha \frac{\partial^2 T}{\partial x^2} \qquad (4.86)$$

is subject to the initial and boundary conditions

$$T(x, 0) = T_0 \qquad (4.87a)$$

$$\frac{\partial T(0, t)}{\partial x} = 0 \qquad T(\delta, t) = 0 \qquad (4.87b)$$

Assume a product solution for the temperature as

$$T(x, t) = X(x) \, \tau(t) \qquad (4.88)$$

The substitution of Eq. (4.88) into Eq. (4.86) followed by separation of the variables yields

$$\frac{X''}{X} = \frac{1}{\alpha} \frac{\tau'}{\tau} = \lambda^2 \qquad (4.89)$$

or

$$X'' + \lambda^2 x = 0 \qquad \tau' + \alpha\lambda^2\tau = 0$$

Their solutions are, respectively,

$$X = c_1 \sin \lambda x + c_2 \cos \lambda x \qquad (4.90a)$$

$$\tau = c_3 e^{-\alpha\lambda^2 t} \qquad (4.90b)$$

With the substitution of Eq. (4.88) into Eq. (4.87b) one obtains

$$X'(0) = 0 \qquad X(\delta) = 0$$

which are used in Eq. (4.90a). It is found that $c_1 = 0$ and the eigenvalues are $\lambda_n = (2\eta + 1)\pi/(2\delta)$, where η is a positive integer including zero. Now, Eqs. (4.88) and (4.90) are combined to give

$$T(x, t) = \sum_{n=0}^{\infty} a_n e^{-\alpha \lambda_n^2 t} \cos \lambda_n x \qquad (4.91)$$

The initial condition (4.87a) is utilized to determine the coefficient a_n in Eq. (4.91) as

$$a_n = (-1)^n \frac{2T_0}{\lambda_n \delta}$$

The temperature-time history becomes

$$\frac{T(x, t)}{T_0} = 2 \sum_{n=0}^{\infty} \frac{(-1)^n}{(\lambda_n \delta)} e^{-\alpha \lambda_n^2 t} \cos \lambda_n x \qquad (4.92)$$

When metabolic heat generation is taken into account, the bioheat transfer equation reads

$$\frac{\partial T}{\partial t} = \alpha \frac{\partial^2 T}{\partial x^2} + \frac{q'''}{k}$$

The nonhomogeneous equation can be separated using

$$T(x, t) = \phi(x, t) + \psi(x)$$

into a homogeneous partial differential equation of $\phi(x, t)$ and a nonhomogeneous ordinary differential equation of $\psi(x)$.

Cylinders An example of unsteady heat conduction in a long cylinder with metabolic heat generation is observed in nerve cells. The solution of the problem is presented later in this chapter.

4.3 CONVECTION

The mechanism of convective heat transfer is a combination of molecular diffusion and fluid mixing. Therefore, knowledge of fluid motion is essential in understanding the convective mode of energy transport.

Two processes induce flow: natural and forced convection. Natural convection results from the action of gravity, centrifugal force, surface tension, electrostatic force, or magnetic force. When the motion is caused by some external agency, such as a pump or a blower, the fluid motion is called forced convection.

In both natural and force convection, flow patterns can be grouped into three types: laminar, transition, and turbulent. In laminar or streamline flow, the fluid particles move in layers. Each fluid particle takes a smooth and continuous path. Heat transfer in the laminar flow region is by molecular diffusion alone. No energy stored in fluid particles is transported across streamlines. The motion of fluid particles in turbulent flow is random and chaotic. Hence, heat is transferred by not only the conduction mechanism but also by mixing of energy-carrying particles. However, in the immediate vicinity of the solid surface, quasi-laminar motion still exists in a thin layer. This is called the laminar sublayer, in which heat is transferred by molecular diffusion. The transitional region between the laminar sublayer and the fully turbulent core is called the buffer layer.

The rate of heat transfer from the surface at T_s to the bulk fluid at T_f is expressed by Eq. (1.26) as

$$q_c = \bar{h}_c A(T_s - T_f)$$

However, heat transfer at the interface is by the conduction mechanism. The rate of heat flow from the surface to the fluid is

$$q_k = -k_f A \frac{\partial T(x, 0)}{\partial y}$$

in which k_f denotes the thermal conductivity of the fluid. Since $q_c = q_k$, one gets

$$-k_f \frac{\partial T(x, 0)}{\partial y} = \bar{h}_c(T_s - T_f) \tag{4.93}$$

After rearranging, it becomes

$$\frac{\bar{h}_c L}{k_f} = \frac{-\partial T(x, 0)/\partial y}{(T_s - T_f)/L} \tag{4.94}$$

The dimensionless parameter $\bar{h}_c L/k_f$ is called the Nusselt number; it describes the ratio of the temperature gradient in the fluid at the interface to a reference temperature gradient. L is the characteristic length of the system for convection, the hydraulic diameter for internal flows. For external flows, L is different depending on the method of induction of fluid motion and the geometry and orientation of the system. The fluid temperature gradient may be approximated as

$$-\frac{\partial T(x, 0)}{\partial y} \cong \frac{T_s - T_f}{\delta_t} \tag{4.95}$$

in which δ_t is the thickness of the thermal boundary layer. The combination of Eqs. (4.93) and (4.95) produces

$$\bar{h}_c \cong \frac{k_f}{\delta_t} \tag{4.96}$$

Equation (4.96) reveals an important answer to the key question in convection research: How can one enhance the convective heat transfer? The answer is to reduce the thickness of the thermal boundary layer. This can be accomplished by numerous methods which are categorized by Bergles and his associates (see, for example, Bergles, 1978).

Dimensional analysis leads to the following functional relationships.

Natural convection: $\quad\quad\quad\quad$ Nu $= f$(Gr, Pr) $\quad\quad\quad\quad\quad\quad\quad$ (4.97)

Forced convection: $\quad\quad\quad\quad$ Nu $= f$(Re, Pr) $\quad\quad\quad\quad\quad\quad\quad$ (4.98)

where Gr $\equiv \dfrac{g\beta(T_s - T_f)L^3}{\nu^2} =$ Grashof number;

$\quad\quad$ Pr $\equiv \dfrac{\nu}{\alpha} =$ Prandtl number

β represents the coefficient of thermal expansion, and g is gravitational acceleration.

There are three general methods for determining the magnitude of h_c, the expression of Nu in terms of Gr or Re and Pr: (1) dimensional analysis combined with experiments; (2) mathematical analyses of the boundary layer phenomena; and (3) the analogy between heat, mass, and momentum transfer. Results for Nu in the form of Eqs. (4.97) and (4.98) are available in handbooks (e.g., Rohsenow and Hartnett, 1986) and textbooks (e.g., White, 1988). Only those which have biomedical applications are included here.

4.3.1 Natural Convection

A typical example of natural convection is a nude subject in still air. For vertical plates and cylinders:

$$\text{Nu} = 0.555\text{Ra}^{1/4} \quad\quad (\text{Gr} < 10^9) \quad\quad\quad\quad (4.99)$$

$$\text{Nu} = 0.13\text{Ra}^{1/3} \quad \text{or} \quad \text{Nu} = 0.021\text{Ra}^{2/5} \quad (\text{Gr} > 10^9) \quad (4.100)$$

where L is defined in Re, Gr, Ra, and Nu. The explanation of L is the most important thing in convection. Ra, the Rayleigh number, is equal to GrPr. L is the vertical length of the plate or cylinder.

For inclined plates and cylinders, Eqs. (4.99) and (4.100) may be used, provided the body force is modified by replacing g by $g \cos \theta$ where θ is the angle of inclination. For horizontal cylinders:

$$\text{Nu} = 0.53\text{Ra}^{1/4} \quad (10^9 \geq \text{Gr} \geq 10^3 \quad \text{and} \quad \text{Pr} > 0.5) \quad (4.101)$$

L is the cylinder diameter.

For spheres:

$$\text{Nu} = 2 + 0.43\text{Ra}^{1/4} \quad (1 < \text{Ra} < 10^5 \quad \text{Pr} \cong 1) \quad (4.102)$$

4.3.2 Forced Convection, External Flow

A typical example of forced convection, external flow is a nude subject in a room with ventilation.

The transition criterion is $Re = 5 \times 10^5$. For flat plates:

$$Nu = 0.664Re^{1/2}Pr^{1/3} \quad (Re < 5 \times 10^5) \quad (4.103)$$

$$Nu = (0.037Re^{4/5} - 871)Pr^{1/3} \quad (Re > 5 \times 10^5) \quad (4.104)$$

L corresponds to the plate length.

For cylinders:

$$Nu = CRe^nPr^{1/3} \quad (4.105)$$

C and n vary with Re as shown in Table 4.2. L is the outer diameter of the cylinder.

For spheres:

$$Nu = 0.37Re^{0.6} \quad (25 < Re < 100,000 \text{ in gases}) \quad (4.106)$$

$$\frac{Nu}{Pr^{1/3}} = 0.97 + 0.68Re^{1/2} \quad (1 < Re < 2,000 \text{ in liquids}) \quad (4.107)$$

L is the diameter.

4.3.3 Forced Convection, Internal Flow

An example of forced convection with internal flow is flow in blood vessels. Irrespective of the geometry of the flow cross section, the hydraulic radius D_H is used as L.

$$D_H = \frac{4 \times \text{flow cross-sectional area}}{\text{Wetted surface area}} \quad (4.108)$$

For long tubes or ducts:

$$Nu = 1.86Gz^{1/3}(\mu_b/\mu_s)^{0.14} \quad (Re \leq 2100 \text{ in laminar flow}) \quad (4.109)$$

$$Nu = 0.23Re^{0.8}Pr^{1/3} \quad (Re \geq 10,000 \text{ in turbulent flow}) \quad (4.110)$$

Table 4.2 C and n values for cylinders

Re	C	n
0.4–4	0.891	0.330
4–40	0.821	0.385
40–4,000	0.615	0.466
4,000–40,000	0.174	0.618
40,000–400,000	0.0239	0.805

Gz is the Graetz number, defined as Re Pr D_H/l, where l is the length. Subscripts b and s refer to the bulk and surface temperatures, respectively.

For short tubes:

$$\frac{h_c^*}{h_c} = 1 + \left(\frac{D_H}{l}\right)^{0.7} \qquad \left(\text{for } 20 \geq \frac{l}{D_H} \geq 2 \qquad \text{Re} \geq 10,000\right) \quad (4.111)$$

$$\frac{h_c^*}{h_c} = 1 + 6 \left(\frac{D_H}{l}\right) \qquad \text{for} \qquad 60 > \frac{l}{D_H} \geq 20 \qquad (4.112a)$$

When Re is between 2100 and 10,000, fluid flow is in the transition region. Since the flow may be unstable in this region, there exists a large uncertainty in the heat transfer performance. There is also uncertainty because of pulsatile nature of flow. The value of Nu or h_c can only be estimated.

4.3.4 Heat Transfer in the Microcirculatory System

About 80% of the pressure drop in the circulatory system occurs in the arterioles, capillaries, and venules which are normally 10–150 μm in diameter. It is commonly believed that a large portion of internal heat exchange within the body must also take place in these vessels because of their large surface area.

Knowledge of the heat transfer mechanism of the body is important to the advancement of technology in such areas as open heart surgery, design and development of pumps, valves, dialyzers, filters, heat exchangers and other blood handling equipment, cryotherapy, and cryosurgery.

For blood flow inside glass tubes with diameters of approximately 1070, 640, and 300 μm, heat transfer coefficients were correlated by the equation (Fazzio and Jacobs, 1972)

$$\text{Nu} = 3.15 \, \text{Gz}^{1/3} \left(\frac{\mu_b}{\mu_s}\right)^{0.14} \qquad (4.112b)$$

The viscosity terms account for the effect of temperature variation on the physical properties. This equation differs from the Seider-Tate equation (4.109) by a higher coefficient, possibly due to an alteration in the velocity profile in smaller diameter tubes. Due to the difficulty of flow in small tubes, the tendency of blood settling, and the problem of coagulation, it may be necessary in future studies to use living tissue to investigate flow through capillaries of about one red blood cell in diameter. Refer to Chen and Holmes (1980) for microvasculature contributions in tissue heat transfer.

4.3.5 Heat Exchangers

Various types of heat exchangers are employed in industry. One that is commonly used in warming or cooling blood extracorporeally is the counterflow type concentric-pipe heat exchanger, as shown in Fig. 4.13. It consists of two pipes of equal length but different diameters placed concentrically. Two liquids,

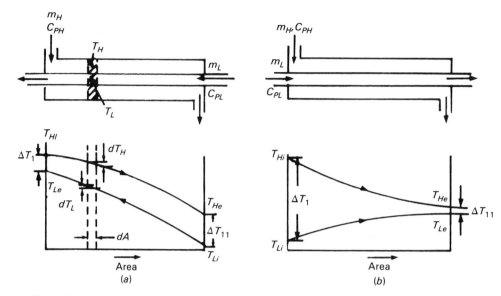

Figure 4.13 Temperature distribution in concentric-pipe heat exchangers. (*a*) Counter-flow type; (*b*) parallel-flow type.

at different temperatures T_H and T_L, flow in opposite directions, one within the inner pipe and the other in the annular space. Assuming no heat loss to the environment, a heat balance over a differential area dA yields

$$dq = -\dot{m}_H C_{pH}\, dT_H = \dot{m}_L C_{pL}\, dT_L = U\, dA(T_H - T_L) \qquad (4.113)$$

where U is the overall heat transfer coefficient. When the pipe surfaces are clean, three thermal resistances exist between the hot and cold liquids: $1/(hA)_H$, $\delta/(kA_{\log})$ and $1/(hA)_L$. δ, k, and $A_{\log}$ denote the thickness, thermal conductivity, and logarithmic-average surface area of the inner pipe, respectively. The total thermal resistance R_t is the sum of the three resistances. The reciprocal of R_t is defined as UA, where U may be based on any chosen surface area A:

$$\frac{1}{UA} = \frac{1}{(hA)_H} + \frac{\delta}{kA_{\log}} + \frac{1}{(hA)_L} \qquad (4.114)$$

Integration of Eq. (4.113) over the total length of exchanger l yields

$$\dot{m}_H C_{pH}(T_{Hi} - T_{He}) = \dot{m}_L C_{pL}(T_{Le} - T_{Li}) \qquad (4.115)$$

and

$$q = UA\, \overline{\Delta T}_{\log} \qquad (4.116)$$

where $\overline{\Delta T}_{\log}$ is the logarithmic mean temperature difference, defined as

$$\overline{\Delta T}_{\log} \equiv \frac{\Delta T_{\mathrm{I}} - \Delta T_{\mathrm{II}}}{\ln(\Delta T_{\mathrm{I}}/\Delta T_{\mathrm{II}})} \qquad (4.117)$$

in which

$$\Delta T_I = T_{Hi} - T_{Le} \qquad \Delta T_{II} = T_{He} - T_{Li} \qquad (4.118)$$

Under the same operating and design conditions, $(\dot{m}C_p)_{max}/(\dot{m}C_p)_{min}$ and $UA/(\dot{m}C_p)_{min}$, the counterflow type has the highest heat exchanger effectiveness while the parallel flow has the lowest of all heat exchangers. Here, $(\dot{m}C_p)_{max}$ is the larger of the $(\dot{m}_H C_p)_H$ and $(\dot{m}_L C_p)_L$ magnitudes. The heat exchanger effectiveness is defined as the ratio of the actual rate of heat transfer in a heat exchanger to its maximum possible rate of heat exchange limited by the second law of thermodynamics, $q_{max} = (\dot{m}C_p)_{max}(T_{Hi} - T_{Li})$. Equations (4.115)–(4.117) are all applicable to the parallel-flow type, provided that

$$\Delta T_I = T_{Hi} - T_{Li} \qquad \Delta T_{II} = T_{He} - T_{Le} \qquad (4.119)$$

Counterflow heat exchange between arterial and venous blood is the heat shunt or internal heat exchange mechanism in physiology. It functions according to Eq. (4.115) where the H and L sides correspond to the arterial and venous blood, respectively. As mentioned previously, the arterial blood is precooled through this mechanism before entering certain organs and tissues to remove the metabolic heat. It plays a vital role in the regulation of body temperature. The counterflow type heat exchanger is used to adjust the temperature of extracorporeally circulating blood before entering the body.

Because it has the lowest effectiveness in heat exchange, heat exchanger are not designed to operate on the parallel flow principle in industrial applications. However, numerous mass exchange (analogous to heat exchange) processes in biological systems take place on the parallel flow arrangement.

Within the human body, heat is transferred by conduction through tissues and by convection through blood-tissue interaction across the blood vessel. In the latter case, heat transfer to individual blood vessels may take place in three configurations (Chato, 1980): a single vessel, two vessels in counterflow, and a single vessel near the skin surface. All vessels function as heat exchangers consisting of either a single pipe or double pipes. Two blood vessels, an artery and a vein, in counterflow are also called a heat shunt in physiology. They exist in many mammals.

Chato (1980) studied heat transfer to individual blood vessel in the three configurations. For a single vessel, the controlling factor is the Graetz number, defined as Re Pr D/l, where D and l are the diameter and length of blood vessel, respectively, and Pr denotes Prandtl, number. The arterioles, capillaries, and venules have very low Graetz numbers (Gz < 0.4) and act as perfect heat exchangers in which the blood quickly reaches the tissue temperature. On the other hand, the large arteries and veins with Gz > 1000 have virtually no heat exchange with the tissue, and blood leaves them at near the entering temperature. Heat transfer between parallel vessels in counterflow is influenced most strongly by the relative distance of separation and by the mass transferred from the artery to the vein along the length. These two effects are of the same order of magnitude. In the case of a single vessel near the skin surface, the effect of a blood vessel on the heat dissipation to the environment and on the skin temperature distribution increases with decreasing depth-to-radius ratio and decreasing Biot number based on radius.

4.4. RADIATION

Radiation propagates as electromagnetic (EM) waves with the speed of light V_l in free space (the wave theory). The transfer of energy takes place in the form of quanta, the smallest unit of energy (the quantum theory). Thus, for convenience in explaining radiation phenomena, radiation exhibits dual characteristics. Its frequency ν and wavelength λ are related as

$$\lambda\nu = V_l = 3 \times 10^8 \text{ m/s}$$

All bodies at a temperature higher than absolute zero continuously emit radiation. Figure 4.14 illustrates the EM wave spectrum. EM waves are known by different names depending on the range of frequency or wavelength: cosmic rays, gamma rays, x-rays, ultraviolet rays, infrared rays, microwaves, radiowaves, etc. Only within the narrow band from 0.38 to 0.76 μm does radiation affect the optical nerves as light in the so-called visible range. We can only sense radiation in the spectrum region between 0.1 and 100 μm, the so-called thermal radiation, which is of a prime interest in engineering. However, therapeutic application of radiation for deep tissue heating is found to be most effective using microwaves with the frequency range between 20 and 1000 MHz (Yang and Wang, 1977).

The discussion of radiation is divided into two parts. Section 4.4.1 deals

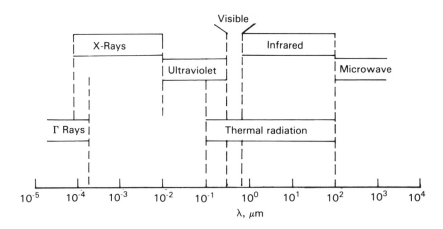

Figure 4.14 Spectrum of electromagnetic radiation.

with thermal radiation from an individual surface, while Section 4.4.2 treats radiation heat exchange between two or more surfaces.

4.4.1 Thermal Radiation

One primary interest in thermal radiation is the spectral distribution of monochromatic (single wavelength) emissive power (power means heat flux) for an ideal radiator (also called a black body) $E_{b\lambda}$, as shown in Fig. 4.15. The distribution curve is the starting point for an eventual calculation of the net gain or loss in radiation exchanges between surfaces. For each distribution curve corresponding to the radiating surface temperature, we are interested in the shape and peak of the curve and the area enclosed between the curve and the abscissa. Planck's law describes the shape of the $E_{b\lambda}$ distribution curve to be

$$E_{b\lambda} = \frac{C_1\lambda^{-5}}{e^{C_2/\lambda T} - 1} \qquad (4.120)$$

where $C_1 = 3.7420 \times 10^8$ Wμm^4/m^2
$\quad\;\; C_2 = 1.4388 \times 10^4$ μmK

The height determines the heat flux at a particular wavelength.

The wavelength corresponding to the peak, $E_{b\lambda}$, can be determined by Wien's displacement law

$$\lambda_{max} T = 2897.7 \ \mu\text{mK} \qquad (4.121)$$

The major portion of radiation is emitted within a relatively narrow band to both sides of the peak, as seen from these curves.

The area enclosed between the curve and the abscissa measures the heat

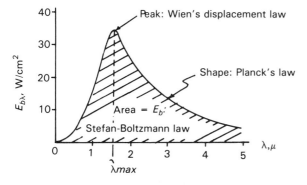

Figure 4.15 Spectral distribution of monochromatic emissive power for an ideal radiator at 1650°C.

flux emitted from the surface at a specified temperature. The Stefan-Boltzmann law predicts the area to be

$$E_b = \int_0^\infty E_{b\lambda} d\lambda = \sigma T^4 = \text{areas on } E_{b\lambda} - \lambda \text{ plot} \qquad (4.122)$$

where E_b = total emissive power of a black body (in W/m²)
σ = Stefan-Boltzmann constant = 5.670×10^{-8} W/m² K⁴

When radiation with heat flux G impinges on a surface, a fraction G_R is reflected, another fraction G_A is absorbed, and the remaining fraction G_T is transmitted. The heat fluxes R, A, and T are called reflective, absorptive, and transmissive power, respectively. Heat balance requires that

$$G = G_R + G_A + G_T$$

A division of both sides by G results in

$$1 = \frac{G_R}{G} + \frac{G_A}{G} + \frac{G_T}{G}$$

With the definition of $\rho = G_R/G$, $\alpha = G_A/G$, and $\tau = G_T/G$,

$$\rho + \alpha + \tau = 1 \qquad (4.123)$$

in which ρ, α, and τ are the reflectivity, absorptivity, and transmissivity, respectively, and are limited to values between zero and unity. For a black body, which has a surface that completely absorbs all radiation incident upon it, α takes the value of unity, since $\rho = \tau = 0$.

Two more laws in thermal radiation are important: Kirchhoff's law and Lambert cosine law. As Fig. 4.16 depicts, n small bodies of surface areas A_1, A_2, A_n are placed in a large vacuum enclosure. The inner surface of the enclosure emits radiation with the heat flux of G, while the outer surface is well insulated. Under thermal equilibrium these small surfaces of the absorptivities α_1, α_2, ..., α_n emit radiant energy E_1, E_2, ..., E_n in heat flux, respectively. A heat balance on each surface yields $\alpha_1 G A_1 = E_1 A_1$; $\alpha_2 G A_2 = E_2 A_2$; ...; $\alpha_n G A_n = E_n A_n$. Upon rearrangement, one obtains

$$G = \frac{E_1}{\alpha_1} = \frac{E_2}{\alpha_2} = \cdots = \frac{E_n}{\alpha_n} \qquad (4.124)$$

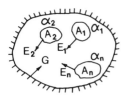

Figure 4.16 An insulated enclosure with n surfaces at thermal equilibrium.

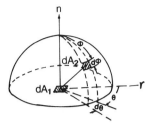

Figure 4.17 Emission from dA_1 to dA_2 on hypothetical hemisphere centered at point on dA_1.

This relation is called Kirchhoff's law. It implies that under thermal equilibrium, all enclosed surfaces have the same ratio of emissive power to absorptivity.

Since the enclosure surface forms a black body cavity regardless of the nature of its radiative properties, $G = E_b$. Consequently, a rearrangement of Eq. (4.124) produces

$$\alpha_1 = \frac{E_1}{E_b} \qquad \alpha_2 = \frac{E_2}{E_b} \qquad \cdots \qquad \alpha_n = \frac{E_n}{E_b} \qquad (4.125)$$

Since the maximum value of α is limited to unity, no real surface can have an emissive power greater than that of a black body. Introducing the definition of the emissivity of a surface as

$$\epsilon \equiv \frac{E}{E_b} \qquad (4.126)$$

Equation (4.125) gives an alternative form of Kirchhoff's law as

$$\frac{\epsilon_1}{\alpha_1} = \frac{\epsilon_2}{\alpha_2} = \cdots = \frac{\epsilon_n}{\alpha_n} = 1 \qquad (4.127)$$

It states that at thermal equilibrium, the absorptivity and the emissivity of a surface are equal. Similarly, $\epsilon_\lambda = \alpha_\lambda$ for monochromatic waves. It is obvious that the emissivity is limited to values between zero and unity and the maximum emissive power occurs at $\epsilon = \alpha = 1$. Thus, a black body is a perfect emitter. The combination of Eqs. (4.122) and (4.126) results in the expression for radiation from real surfaces as

$$E = \epsilon \sigma T^4 \qquad (4.128)$$

A real surface whose emissive power obeys Eq. (4.128) is called a gray surface.

A hemispheric surface of area A_2 is placed over a circular flat surface of area A_1. Two differential elements dA_1 and dA_2 are defined on A_1 and A_2, respectively, as shown in Fig. 4.17. Lambert's cosine law states that the rate of radiant energy emitted from dA_1 and being intercepted by dA_2 is proportional to dA_1, dA_2, and $\cos \phi$ but is inversely proportional to the square of the distance between the two elements R:

$$dq_{1\to 2} \propto \frac{\cos \phi \, dA_1 \, dA_2}{R^2}$$

where ϕ is the angle from the normal of dA_1 to the surface element dA_2. The observation may be described in mathematical form as

$$dq_{1 \to 2} = I \frac{\cos \phi \, dA_1 \, dA_2}{R^2} \qquad (4.129)$$

in which I is the proportionality constant known as radiation intensity. Both sides of Eq. (4.129) are divided by dA_1, followed by the surface integral of the right-hand side over the entire hemisphere. Remembering $dq_{1 \to 2}/dA = dE$, it yields

$$E = I \int_{A_2} \frac{\cos \phi \, dA_2}{R^2}$$

Upon integration, we get

$$E = \pi I \qquad (4.130)$$

Equation (4.130) relates the emissive power to its intensity. Thus, through measuring the intensity, the emissive power can be determined. Similarly, one can write

$$E_b = \pi I_b \qquad E_\lambda = \pi I_\lambda \qquad E_{b\lambda} = \pi I_{b\lambda} \qquad (4.131)$$

E and E_b are known as the total emissive powers for all wavelengths, while E_λ and $E_{b\lambda}$ are called the monochromatic emissive powers for wavelength λ.

4.4.2 Radiation Heat Transfer

So far the emission and absorption of radiation by a single surface have been considered. When two or more surfaces are present, the net exchange between them is of prime concern. The equivalent electrical network can be used to aid in understanding the process of radiation heat exchange. Only relatively simple cases are treated in which two radiating surfaces are separated by a medium that does not interfere with radiation heat exchange processes, for example, air.

4.4.2.1 Radiation heat transfer between two black surfaces. Consider a differential element dA_1 on the surface A_1 in radiation exchange with another differential element dA_2 on the surface A_2, as depicted in Fig. 4.18. R is the distance between them, ϕ_1 and ϕ_2 are the angles that line R makes with the normals n_1 and n_2, respectively. According to the cosine law, the radiation emitted from dA_1 that falls upon dA_2 is

$$dq_{1 \to 2} = I_{b1} \frac{dA_1 \, dA_2 \cos \phi_1 \cos \phi_2}{R^2}$$

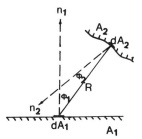

Figure 4.18 Notation for the geometric shape factor.

Similarly, the radiation from dA_2 that is received by dA_1 is

$$dq_{2\to 1} = I_{b2}\,\frac{dA_1\,dA_2\,\cos\phi_1\,\cos\phi_2}{R^2}$$

The net exchange between them, $dq_{1\rightleftharpoons 2}$, is their difference. Taking into account $E_{b1} = \pi I_{b1}$, one gets

$$dq_{1\rightleftharpoons 2} = (E_{b1} - E_{b2})\,\frac{\cos\phi_1\,\cos\phi_2\,dA_1\,dA_2}{\pi R^2}$$

The equation is then integrated over the entire surfaces A_1 and A_2. It yields

$$q_{1\rightleftharpoons 2} = (E_{b1} - E_{b2}) \int_{A_1}\int_{A_2}\frac{\cos\phi_1\,\cos\phi_2\,dA_1\,dA_2}{\pi R^2} \qquad (4.132)$$

The double surface integral on the right-hand side can be defined as either $F_{12}A_1$ or $F_{21}A_2$, that is,

$$\int_{A_1}\int_{A_2}\frac{\cos\phi_1\,\cos\phi_2\,dA_1\,dA_2}{R^2} = F_{12}A_1 = F_{21}A_2 \qquad (4.133)$$

where F is called the shape factor. The first and second subscripts denote the location of the observer and the object, respectively. The rate of net radiation exchange can be expressed as

$$q_{1\rightleftharpoons 2} = F_{12}A_1(E_{b_1} - E_{b_2}) = F_{21}A_2(E_{b_1} - E_{b_2}) \qquad (4.134)$$

This expression is similar to Newton's law of cooling: F is equivalent to h_c, while the temperature difference $(T_s - T_f)$ is replaced by the difference in the emissive powers of the two surfaces. In an equivalent circuit, ΔE_b is the driving potential, while the thermal resistance is $1/FA$.

There are three methods available for the evaluation of shape factors: (1) direct integration of the definition of shape factor, Eq. (4.133); (2) decomposition of the shape factor; and (3) use of a mechanical integrator. The first two methods produce an exact value of F, while the third technique yields an approximate one.

In the case of n black bodies forming a complete enclosure, the net radiation from a surface A_i is

$$q_i = \sum_{j=1}^{n} A_i F_{ij}(E_{bi} - E_{bj}) \qquad (4.135)$$

The shape factors based on the surface A_i must obey the summation rule

$$\sum_{j=1}^{n} F_{ij} = 1 \qquad (4.136)$$

4.4.2.2 Radiation heat transfer between gray surfaces. The radiosity J, a heat flux, is used in the analysis of radiation exchange between gray surfaces. It represents the sum of radiation reflected, emitted, and transmitted as

$$J = \rho G + \epsilon E_b + \tau G \qquad (4.137)$$

The net rate of radiation leaving a gray surface which is irradiated by G is

$$\frac{q_r}{A} = J - G \qquad (4.138)$$

The combination of Eqs. (4.137) and (4.138) gives

$$q_r = \frac{\epsilon}{\rho + \tau} A(E_b - J) = \frac{\epsilon}{1 - \epsilon} A(E_b - J) \qquad (4.139)$$

Here, $(1 - \epsilon)/(\epsilon A)$ is the thermal resistance between E_b and J. The effect of the system geometry on the net radiation between two gray surfaces, A_1 and A_2, emitting radiation at the rate J_1 and J_2, respectively, is the same as for similar black surfaces with the emissive powers E_{b1} and E_{b2}. The net rate is

$$q_r = F_{12}A_1(J_1 - J_2) = F_{21}A_2(J_1 - J_2) \qquad (4.140)$$

In an enclosure of n gray surfaces, the net radiation from a surface A_i is

$$q_i = \sum_{j=1}^{n} F_{ij}A_i(J_i - J_j) \qquad (4.141)$$

Equations (4.139) and (4.141) are equal, that is,

$$\frac{\epsilon_i}{1 - \epsilon_i} A_i(E_{bi} - J_i) = \sum_{j=1}^{n} F_{ij}A_i(J_i - J_j) \qquad (4.142)$$

Equation (4.141) is used to determine T_i when q_i for each surface is given. Equation (4.142) determines q_i when T_i of each surface is specified. The resulting set of n linear algebraic equations are then solved for the n unknowns J_1, J_2, . . . , J_n. After J_i's are determined, Eq. (4.139) is employed to evaluate T_i for a known q_i and vice versa. When the number of surfaces, n, is small, the problem can be readily solved by using a network. However, if many surfaces are involved, the use of the matrix inversion method is a common practice.

4.5 MICROHEAT TRANSFER

It is an extremely hard task to study heat production in microorganisms, typically a cell, and subsequent intracellular and intercellular transfer of heat because (1) cells are very small, (2) the rate of heat production in a cell is very low, (3) the resulting temperature rise and difference are difficult to measure, and (4) the constituents in cells are complex and changing. Thus, microheat transfer is a difficult area for both theoretical and experimental studies. Nevertheless, the information on microheat transfer in cells and macromolecules is important to the understanding of mechanisms of many fundamental biological phenomena in cells and macromolecules, such as metabolism, synthesis, and thermal denaturation of proteins.

Microheat transfer may be divided into three areas based on the temperature range of the environment that surrounds the microorganisms: (1) heat production in cells dealing with energy metabolism under normal conditions; (2) cell hypothermia concerning microorganisms in the environment below the normal temperature range; and (3) cell hyperthermia pertaining to cells held in above-normal temperature range. A fourth topic treats thermal denaturation of macromolecules by heating.

Although microheat transfer is a basic transport phenomenon vital to the life and reproduction of microorganisms, it is also a subject that has attracted less attention and is not yet organized. It is hoped that this chapter will serve as a bridge between physiology and thermal engineering and that some interest on the subject will be generated among the readers through this bold assault on the broad frontier of the microheat transfer field.

4.5.1 Microheat Transfer in Cell Growth and Division

For mathematical development, two simple models of biological cells are employed. Both predict practically identical results under the same boundary conditions. The first model treats the cell as if it were a sphere surrounded by a thin membrane with negligible thermal resistance immersed in a substance of infinite thermal conductivity. It is a promising model for unsteady-state studies due to simplicity in the boundary conditions. The second model also treats the cell as a sphere but with a membrane of finite thickness and thermal resistance surrounded by a material of finite thermal conductivity. In both models, heat transfer occurs by conduction.

4.5.1.1 Thin membrane model. The heat-balance equation reads, from Eq. (B.28),

$$\frac{1}{\alpha}\frac{\partial T}{\partial t} = \frac{1}{r^2}\frac{\partial}{\partial r}\left(r^2\frac{\partial T}{\partial r}\right) + \frac{q'''}{k} \tag{4.143}$$

where r is the radial distance from the cell center. It is subject to the initial and boundary conditions

$$T(r, o) = T_o \qquad \frac{\partial T(o, t)}{\partial r} = 0 \qquad T(a, t) = T_o \qquad (4.144)$$

in which T_o signifies the temperature of the environment and a is the cell radius. Through the transformation $u = rT + qr^3/6k$, one obtains the solution for constant q''' as

$$\frac{T(r, t) - T_o}{q'''a^2/6k} = 1 - \left(\frac{r}{a}\right)^2 + \frac{12a}{r} \sum_{n=1}^{\infty} \frac{(-1)^n}{n^3} \sin\frac{n\pi r}{a} \exp\left(-\frac{n^2\pi^2\alpha t}{a^2}\right) \qquad (4.145)$$

Figure 4.19 illustrates the temperature rise in the cell during the exothermic process of cell duplication. Let T_i be the temperature at the center of the sphere. The temperature difference, $\Delta T = T_i - T_o$, can be readily determined as

$$\frac{\Delta T}{q'''a^2/6k} = 1 + 12 \sum_{n=1}^{\infty} \frac{(-1)^n}{n^2\pi^2} \exp\left(-\frac{n^2\pi^2\alpha t}{a^2}\right) \qquad (4.146)$$

Let ΔT_c be the maximum tolerable temperature rise beyond which cell replication would be interfered. By combining Eqs. (3.90) and (4.146), one finds

$$36 \frac{C\Delta T_c}{\Delta H_m} \frac{\alpha\tau}{a^2} = 1 + 12 \sum_{n=1}^{\infty} \frac{(-1)^n}{n^2\pi^2} \exp\left(-\frac{n^2\pi^2\alpha\tau}{a^2}\right) \qquad (4.147)$$

where C is the specific heat. The equation relates the minimum division time of cells to the maximum tolerable temperature rise.

Since the second term on the right-hand side is negligible in comparison

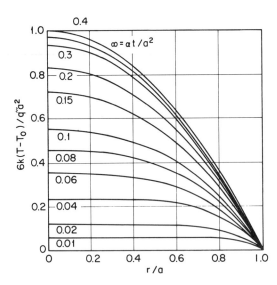

Figure 4.19 Temperature-time history in the cell during synthesis with heat production rate q''' in cal/cm³ s.

Table 4.3 Minimum division time of cells as a function of size and metabolic efficiency

	τ, s	
a, cm	$f = 0.1$	$f = 10$
10^{-4}	6.54×10^{-6}	3.91×10^{-4}
10^{-3}	6.54×10^{-4}	3.91×10^{-2}
10^{-2}	6.54×10^{-2}	3.91
10^{-1}	6.54	391 (6.52 min)
1	654 (10.9 min)	$39,100$ (10.86 h)
10	$65,400$ (18.17 h)	$3,910,000$ (45.25 days)

to unity, Eq. (4.147) may be approximated as

$$\frac{\alpha\tau}{a^2} = \frac{\Delta H_m}{36C\,\Delta T_c} \tag{4.148}$$

Table 4.3 presents the cell division time τ as a function of the cell radius a for two extreme values of the metabolic efficiency f, $\alpha = 0.0013$ cm^3/s, $C = 1.0$ cal/g °C, $\Delta H_c = -124$ cal/g, $\Delta H_w = -1812$ cal/g, and $\Delta T_c = 10$°C are imposed (f, ΔH_m, ΔH_c, and ΔH_w are defined in Section 3.8.3). In principle, Table 4.3 should hold for a tissue mass as well as a cell, provided that conduction is the only method available to dissipate heat. It is then obvious that as a cell or tissue gets larger than 1 mm or so, it must develop methods of getting rid of heat or it must divide at a rather slow rate. This limitation originates solely from the second law of thermodynamics and the auxiliary fact that living systems are subject to thermal inactivation.

4.5.1.2 Rigid shell model. The cell membrane is approximately 75–100 Å thick and is elastic. Its postulated molecular organization consists of a central double layer of lipids covered by protein layers and a thin mucopolysaccharide layer on the external surface of the cell membrane (Fig. 4.20a). The movement of molecules of different sizes (lipid-insoluble substances of very small sizes such as water and urea molecules) between the extracellular and intracellular fluids prompts one to believe in the existence of many minute pores that pass from one side to the other of the cell membrane. The cell membrane is composed of approximately 55% protein, 40% lipids, and perhaps 5% polysaccharides.

Since the temperature variation in the cell reaches a steady state rapidly in comparison to the change in cell size, the steady-state model of Fig. 4.20b is employed. Let R^* be the thermal resistance of the cell membrane with the inner and outer radii of a_1 and a_2, respectively. k_1, k_m, and k_2 are the average

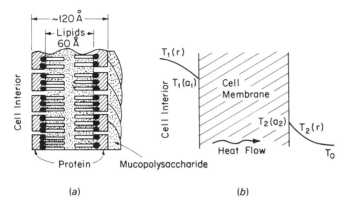

Figure 4.20 Postulated molecular organization of cell membrane and its thermal model in exothermic process during cell division. (*a*) Membrane structure; (*b*) thermal model.

thermal conductivities of the cell substance, cell membrane, and environmental material, respectively. The thermal resistance R^* is equal to $(a_2 - a_1)/4\pi k_m a_1 a_2$. Let T_1 and T_2 be the temperatures of the cell substance and the environmental material, respectively. They are governed by the heat equations

$$\left.\begin{array}{l} \dfrac{1}{r^2}\dfrac{d}{dr}\left(r^2\dfrac{dT_1}{dr}\right) + \dfrac{q'''}{k_1} = 0 \qquad a_1 > r \geq 0 \\[4mm] \dfrac{1}{r^2}\dfrac{d}{dr}\left(r^2\dfrac{dT_2}{dr}\right) = 0 \qquad r > a_2 \end{array}\right\} \qquad (4.149)$$

subject to the boundary conditions

$$\left.\begin{array}{c} \dfrac{dT_1(0)}{dr} = 0 \\[4mm] 4\pi a_1^2 k_1 \dfrac{dT_1(a_1)}{dr} = \dfrac{T_2(a_2) - T_1(a_1)}{R} = 4\pi a_2^2 k_2 \dfrac{dT_2(a_2)}{dr} \\[4mm] T_2(\infty) = T_0 \end{array}\right\} \qquad (4.150)$$

The solutions are readily obtained as

$$\frac{T_1 - T_0}{q'''a_1^2/6k_1} = 1 - \left(\frac{r}{a_1}\right)^2 + 2\frac{k_1}{k_m}\left(1 - \frac{a_1}{a_2}\right) + \frac{2a_1 k_1}{a_2 k_2} \qquad (4.151)$$

$$\frac{T_2 - T_0}{q'''a_1^2/6k_1} = \frac{2a_1 k_1}{r k_2} \qquad (4.152)$$

It is interesting to note that the maximum temperature rise is

$$\frac{\Delta T}{q'''a_1^2/6k_1} = 1 + 2\frac{k_1}{k_m}\left(1 - \frac{a_1}{a_2}\right) + 2\frac{a_1 k_1}{a_2 k_2} \qquad (4.153)$$

Equations (3.90) and (4.153) are combined to yield

$$36 \frac{C \Delta T_c}{\Delta H_m} \frac{\alpha \tau}{a^2} = 1 + 2 \frac{k_1}{k_m} \left(1 - \frac{a_1}{a_2} \right) + 2 \frac{a_1 k_1}{a_2 k_2} \tag{4.154}$$

By setting k_m and k_2 to infinity, the boundary conditions of both the unsteady and steady models will become identical. Then Eq. (4.154) reduces to the form of Eq. (4.148).

4.5.2 Microheat Transfer Due to the Firing of Action Potentials

Both nerve cells and muscle cells can be excited chemically, electrically, and mechanically to produce an action potential which is transmitted along their cell membranes. While neurons transmit nerve impulses, contraction is the specialized function of muscle cells.

4.5.2.1 Nerve cells. A motor neuron is a single nerve cell with a series of processes called dendrites which extend out from the cell body (soma) and arborize extensively (Fig. 4.21). It has a long fibrous axon which originates from a somewhat thickened area of the cell body called the axon hillock. A short distance from its origin, the axon acquires a sheath of myelin which is a protein-lipid complex made up of many layers of unit membrane. The myelin sheath envelopes the axon except at its ending and at periodic constrictions approximately 1 mm apart called the nodes of Ranvier. The axon is the single elongated cytoplasmic neuronal extension for conducting impulses away from the dendritic zone. The axon ends in a number of terminal buttons or axon telodendria. The cell body is often located at the dendritic zone end of the axon. However, it can be within the axon (e.g., auditory neurons) or attached to the side of the axon (e.g., bipolar neurons).

The nerve cell has a low threshold for excitation which may be electrical, chemical, or mechanical. The impulse which is the physicochemical disturbance created by these stimuli is normally transmitted along the axon to its termination at a constant amplitude and velocity. The whole sequence of potential changes within the cell membrane in response to an impulse, a stimulating current of threshold intensity, is called the action potential.

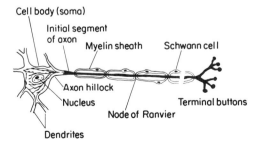

Figure 4.21 Motor neuron with myelinated axon.

As a result of the propagated disturbance, the nerve fiber is depolarized and subsequently, during the refractory periods, reestablishes its resting polarized condition. Depolarization of the membrane is analogous to the discharge of a condenser and can lead to heat production, while repolarization is like a recharging of the condenser and can lead to heat absorption. The rate of heat liberation involved in the passage of an impulse is very small and rapid and thus is difficult to measure accurately. The immediate rise of temperature for a single impulse at 20°C is about 7×10^{-8}°C (Hill, 1933). Two kinds of stimuli that can be applied to a nerve to study its heat production are single shock and tetanus (response to rapidly repeated stimulation).

The rate of heat production is roughly a measure of the rate of metabolism in the nerve, since heat is always released as a by-product of the chemical reactions of energy metabolism. There are three kinds of heat production in nerve activities: resting, initial, and delayed or recovery heat. Resting heat (heat given off at rest) is the external manifestation of the basal metabolic processes. Initial heat is the heat produced in excess of resting heat during excitation. It refers to a burst of heat production immediately associated with the passage of the stimulus. Recovery heat, the heat liberated by the metabolic processes to restore the nerve to its preexcitation state, is associated with restitutive reactions. Recovery heat may be divided into a very rapid A component, $A \exp(-at)$, which is half completed in 2–3 s, and a slow B component, $B \exp(-bt)$, half completed in 4–5 min. The rate of recovery heat production at time t after a stimulus is

$$y = A \exp(-at) + B \exp(-bt) \qquad (4.155)$$

In a stimulus of duration T s, successive impulses may be summed together with respect to the heat they produce. If, during constant stimulation, these impulses are of constant size, then Eq. (4.155) may be deduced for the rate of recovery heat production Y as follows:

$$Y = \int_0^t y \, dt = \frac{A(1 - e^{-at})}{a} + \frac{B(1 - e^{-bt})}{b} \qquad (4.156)$$

during the stimulus, $0 \leq t \leq T$, and

$$Y = A\frac{(e^{-a(t-T)} - e^{-at})}{a} + \frac{B(e^{-b(t-T)} - e^{-bt})}{b} \qquad (4.157)$$

after the stimulus, $t > T$. Equations (4.156) and (4.157), with appropriate values of A, B, a, b, fit well with the recording of heat production due to T s stimulation, for example Fig. 4.22 of $T = 16$ s. In the figure, there is a rapid rise at the beginning of stimulation and rapid decline at the end, both of which may be attributed to the initial heat. The slower rise in heat production rate is due to the cumulative effect of the recovery heat of preceding impulses, plus a steady component of initial heat which is considered to remain constant.

With a single shock, the initial heat is composite, being the result of an initial burst of heat production which lasts about 100 ms followed by a slower

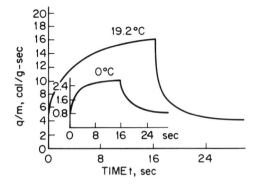

Figure 4.22 Extra heat production of a nerve during and after a 16-s tetanus.

absorption of heat of approximately the same total magnitude and lasting about 300 ms at 0°C (Abbot et al., 1958). The heat per impulse is estimated to be 14×10^{-6} cal/g produced and -12×10^{-6} cal/g absorbed.

Figure 4.23 shows the relationship of the rate of heat production in a nerve, q''', to the number of impulses transmitted by the nerve, n. It is observed that q''' increases exponentially with n. The process of recharging the nerve fiber membrane is caused by this increased heat derived from ATP as an energy source.

A theoretical model is developed to determine the timewise variation of axon temperature assuming the axon is an infinite cylinder with internal heat generation. The heat liberated is transferred from the axon surface into a medium at initial axon temperature T_0. The heat equation is, from Eq. (B.28),

$$\frac{1}{\alpha} \frac{\partial T}{\partial t} = \frac{1}{r} \frac{\partial}{\partial r} \left(r \frac{\partial T}{\partial r} \right) + \frac{q'''}{k} \quad (4.158)$$

subject to the initial and boundary conditions

$$T(r, 0) = T_o \qquad \frac{\partial T(0, t)}{\partial r} = 0$$

$$\frac{\partial T(a, t)}{\partial r} = -\frac{h}{k} [T(a, t) - T_o] \quad (4.159)$$

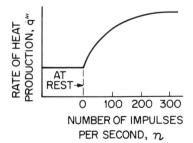

Figure 4.23 Rate of heat production in a nerve at rest and at progressively increasing rates of stimulation.

in which h denotes the heat transfer coefficient between the axon surface and the environment and k is the thermal conductivity of the axon.

$$q'''\rho = i^2 R^*$$

signifies the rate of heat liberated per impulse and is estimated to be 1.4×10^{-5} cal/g per pulse. i and R^* represent the current and electrical resistance in the axon, respectively. On temporal recordings, the current produced by an action potential i is of the spike form. The period of each impulse is about 3 ms with approximately 0.4 ms duration of the spike portion in the current produced by a series of action potentials. By averaging the rate of heat production $i^2 R^*$ with respect to time and treating it as a constant q''', the axon surface temperature is found to be

$$\frac{T - T_0}{q'''a/2h} = 1 - 4 \left(\frac{ha}{k}\right)^2 \sum_{n=1}^{\infty} \frac{\exp\left[-(\alpha t/a^2)(\beta_n a)^2\right]}{(\beta_n a)^2[(ha/k)^2 + (\beta_n a)^2]} \qquad (4.160)$$

where β_n are the positive roots of

$$\beta a J_1(\beta a) = \frac{ha}{\beta} J_0(\beta a)$$

The above equation indicates that the axon surface temperature varies as an exponential function of time. The solution appropriate to q''' liberated by the "spike" potential may be obtained through numerical integration of Eqs. (4.158) and (4.159).

4.5.2.2 Muscle cells. A muscle cell is activated by the action potential to produce a rapid and reversible shortening and relaxation. The muscle is generally classified into three types; skeletal (or voluntary), cardiac, and smooth. Skeletal muscle is made up of individual muscle fibers arranged in parallel between the tendinous ends. Each muscle fiber is a single cell, some 10–100 μm thick, multinucleated, from a few millimeters to several centimeters in length, and cylindrical in shape. It has well-developed cross-striations and is generally under voluntary control. Cardiac muscle also has cross-striations. It is functionally syncytial and contracts rhythmically even when denervated. Smooth muscle, present in the walls of blood vessels and most hollow viscera, lacks cross-striations and has semirhythmic contractile activity.

The muscle is a machine for converting chemical energy into mechanical energy. The immediate source of energy required in muscle contraction is ATP, the energy-rich organic phosphate derivative, but the ultimate source is the intermediary metabolism of carbohydrate. The hydrolysis of ATP is associated with the release of a large amount of energy. This energy appears in work done by the muscle, in energy-rich phosphate bonds formed for later use, and in heat. The overall efficiency of skeletal muscle is approximately 20% in lifting a weight during isotonic contraction and 0% during isometric contraction. Energy storage in phosphate bonds is only a small fraction. Therefore, heat production is considerable. With the use of appropriate thermocouples, the heat liberated can be measured accurately.

Like the nerve, the muscle has a resting heat, an initial heat, and a recovery heat. The resting heat is associated with the resting muscle and can be altered by stretching the muscle as well as changes in ionic strength in the surrounding fluids. In general, the period of evolution of initial heat can be divided into a shortening heat, a maintenance heat, and a relaxation heat. In a tetanus (i.e., under a constant load on the muscle), an initial outburst in heat production represents the phase of a shortening heat. The magnitude of the steep rise increases with the degree of shortening, as illustrated in Fig. 4.24a. The maintenance heat is represented by the constant slope portion of the curve. Since the weight lifted is the same, those curves run approximately parallel (as is also true for the isometric case). In the case of a single switch, an initial outburst in heat production, called the activation heat, occurs before shortening begins. It is then followed by a constant rate of heat production, with the curve almost parallel with shortening, as seen in Fig. 4.24b. This phase of thermal events is described as the shortening heat. In tetanus, the summated activation heats represent the maintenance heat. The curve for the relaxation heat shows a rapid upstroke if the muscle is allowed to lower its load. However, if the muscle is maintained in the isometric state, only a small hump appears. In the delayed

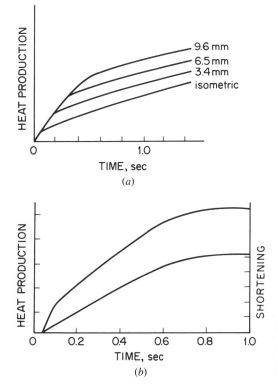

Figure 4.24 (*l*) Heat production during isotonic shortening from the start under constant load (tetanus). (*b*) Heat production (upper curve) and shortening (lower curve) in the isotonic twitch.

phase, the heat production rate falls abruptly followed by a prolonged recovery period. The recovery heat and the initial heat are about the same.

In either a tetanus or a single twitch, the shortening heat is a simple linear function of the shortening:

$$\text{Shortening heat} = ax \tag{4.161}$$

in which x denotes the shortening. The thermal constant a may be determined from purely mechanical measurements. A plot of the load against the velocity of contraction should yield a rectangular hyperbola with the asymptote of the load at $-a$.

The total energy released in a contraction, q''', whether a twitch or a tetanus, is given by

$$q''' = A + W + ax + h \tag{4.162}$$

where A is the activation (in a twitch) or maintenance (in a tetanus) heat; W, the work; and h, the relaxation heat. Experimental evidence indicates that $A + ax + h$ is virtually constant.

4.5.3 Cell Hypothermia

All living cells are exposed to some temperature variation in their environment. A temperature that is most favorable for one cell function, such as respiration, may not be most favorable for another, such as cell growth. Furthermore, a temperature that may appear favorable for a given function during brief exposure of a cell may prove harmful on longer exposure. The range of temperatures within which a living cell performs its life activities is called the biokinetic zone. It lies approximately between 10 and 45°C. In humans, the normal body temperature is 37°C.

The rate of thermochemical reactions in the cell, such as metabolism, increases with temperature up to a maximum. With further increase in temperature, the rate of activity declines as lethal thermal limits are approached. The temperature coefficient Q_{10} is the ratio of activity at a given temperature to the rate at a temperature 10°C below. The van't Hoff equation reads

$$Q_{10} = \left(\frac{k_2}{k_1}\right)^{10/(T_2 - T_1)} \tag{4.163}$$

in which k_2 is the rate of the reaction at temperature T_2 and k_1 is the rate at T_1. Arrhenius (1915) considered the rate of reaction to depend on the number of molecules N having energy equal to or in excess of the energy of activation E (the energy required for a given reaction). Utilizing the Maxwell-Boltzmann distribution law, he derived the equation

$$\frac{N}{N_0} = \exp\frac{-E}{RT} \tag{4.164}$$

where N_0 is the total number of molecules. Since k is proportional to N, Eq. (4.164) yields

$$\ln \frac{k_2}{k_1} = \frac{E}{R} \left(\frac{1}{T_1} - \frac{1}{T_2} \right) \tag{4.165}$$

Thus, E may be determined from the plot of the logarithm of the rate of the reaction against $1/T$, called the Arrhenius plot. Equation (4.164) can be used to determine the number of activated molecules N.

When unfavorable thermal conditions occur in the environment, cells may suspend life activities and become dormant. Their thermal resistance increases. In dormant states, cells become dehydrated. Their greater resistance to both hot and cold has not yet been fully explained. However, it may partially be the result of their dehydration, since the temperature required to denature proteins increases with a decrease in their water content.

Cells may become relatively inactive as the temperature falls to the lower limits they tolerate. Hypothermia results in decreased metabolism. When an organism is slowly exposed to freezing temperatures, water freezes outside the cells, thereby concentrating the solution surrounding the cells. As a consequence, water diffuses out of the cells into the bathing fluid resulting in an increase in the concentration of intracellular solutes, which lowers the freezing point inside the cell. The abnormally increased concentration of solutes is injurious to the cell. If the cells are slowly cooled to below $-40°C$, ice crystals outside the cells serve as nucleators through channels in the cell membrane and induce intracellular ice formation. However, if the intracellular water has been reduced to less than 10% of its original content, ice crystals will not form since this concentrated residue cannot crystallize. On the other hand, rapid freezing results in ice formation inside as well as outside cells. This may cause concentration of the dissolved materials to the point of injury. In addition, the intracellular ice formation may by itself disrupt cytoplasmic and nuclear structures beyond the point of recovery. This is especially the case if the cells are stored for a time, permitting growth of large ice crystals from a small size.

After either slow or rapid freezing, the nature of the thawing process is vitally important. Slow thawing allows the cell to be exposed to a high electrolyte concentration at a higher temperature than that applied during freezing so that proteins are more ready to denature. Growth of intracellular ice crystals during slow thawing may kill cells. Hence, even if the cell initially survives freezing, it might still be injured severely on thawing. Rapid thawing is therefore especially important.

Knowledge of cell dehydration and intracellular freezing, the two important phenomena associated with low temperature exposure which have a profound effect on cell survival, are well documented in the field of cryobiology. It is desirable (1) to have quantitative descriptions of the relationship between the amount of water remaining in a cell and its temperature and (2) to determine the circumstances under which the formation of intracellular ice crystals can occur.

Even in cells that are surrounded by ice, the intracellular water remains unfrozen, namely supercooled, at -5 to $-10°C$, because of the osmolal concentration of solutes in the protoplasm (Mazur, 1966). According to the Clausius-Clapeyron equation, the intracellular vapor pressure p_L, its temperature T and the extracellular vapor pressure p_S are interrelated as

$$\frac{d \ln (p_L/p_S)}{dT} = \frac{-L_f}{RT^2} \tag{4.166}$$

where L_f is the molar heat of fusion. The integrated form is

$$\ln \frac{p_L}{p_S} = \frac{L_f}{R} \frac{T_f - T}{T_f T} \tag{4.167}$$

in which T_f is the freezing point of the protoplasm. The molar free energy of fusion reads

$$\Delta G = RT \ln \frac{p_L}{p_S} \quad \text{or} \quad \Delta G = \frac{L_f(T_f - T)}{T_f} \tag{4.168}$$

Thus, the supercooled intracellular water has a higher vapor pressure and a higher free energy than the extracellular ice.

Since the mole fraction of water x_1 will change as water leaves the cell in response to the vapor pressure gradient, Eq. (4.166) should be rewritten as

$$\frac{d \ln (p_S/p_L)}{dT} = \frac{L_f}{RT^2} - \frac{d \ln x_1}{dT} \tag{4.169}$$

By expressing x_1 in terms of the volume of water in the cell V, the number of osmoles of solute in the cell n_2, and the molar volume of water V_1, Eq. (4.169) can be rewritten as

$$\frac{d \ln (p_S/p_L)}{dT} = \frac{L_f}{RT^2} - \frac{n_2 V_1}{V + n_2 V_1} \frac{dV}{dT} \tag{4.170}$$

The rate at which water leaves the cell is

$$\frac{dV}{dt} = \frac{KART}{V_1} \ln \frac{p_S}{p_L} \tag{4.171}$$

where K denotes the permeability constant and A is the area of the cell surface. K is usually an exponential function of T above $0°C$.

$$K = K_g \exp [b(T - T_g)] \tag{4.172}$$

where K_g represents the known permeability constant at temperature T_g and b is its temperature coefficient. It is assumed that Eq. (4.172) is applicable for T below $0°C$. With the assumption of constant cooling rate $dT/dt = B$, Eqs.

(4.170)–(4.172) can be combined to eliminate t and P_S/P_L to yield

$$T \exp \left[b(T_g - T) \right] \frac{d^2V}{dT^2}$$

$$- \left\{ (bT + 1) \exp \left[b(T_g - T) \right] - \frac{ARk_g nT^2}{B(V + nV)V} \right\} \frac{dV}{dT} = \frac{L_f AK_g}{Bv} \quad (4.173)$$

which expresses the volume of intracellular water as a function of temperature. Initially, the cell of volume V_i contains supercooled water at the freezing point of protoplasm and is externally surrounded by ice:

$$T = T_f \qquad V = V_i \qquad \text{and} \qquad T = T_f \qquad \frac{dV}{dT} = 0 \qquad (4.174)$$

Analytic solutions are obtained for four limiting cases in terms of the dimensionless parameters

$$\theta = \frac{T_f - T}{T_f} \qquad \psi = \frac{V_i - V}{V_i} \qquad \beta = bT_f \qquad \gamma = \frac{\Delta C}{R} \qquad (4.175)$$

$$\eta = \frac{nV}{V_i} \qquad \epsilon = - \frac{ARk_g T_f^2}{qvV_i} \exp \left[b(T_f - T_g) \right]$$

where ΔC denotes the difference between molar specific heats at constant pressure of water and ice. The magnitude of ψ indicates the extent of dehydration and $L_f = T\Delta C$. Under the condition of very small θ necessary for cell survival, one gets

1. For $\eta = 0$ (cell containing only water and no solute),

$$\psi = \frac{\epsilon\gamma\{1 + \beta \exp \left[-(\beta + 1)\theta \right] - (\beta + 1) \exp \left(-\beta\theta \right)\}}{\beta^2 + \beta} \qquad (4.176)$$

2. For $\psi \ll 1$ (small dehydration resulting from freezing process),

$$\psi = \frac{\epsilon\gamma[(\delta - \beta) + \beta \exp \left(-\delta\theta \right) - \delta \exp \left(-\beta\theta \right)]}{\beta\delta} (\delta - \beta) \qquad (4.177)$$

where $\delta = (\beta + 1) + \eta\epsilon/(1 + \eta)$.
3. For $\epsilon \ll 1$ (high cooling rate),

$$\psi = \frac{\epsilon\gamma\{1 + \beta \exp \left[-(\beta + 1)\theta \right] - (\beta + 1) \exp \left(-\beta\theta \right)\}}{\beta^2 + \beta} \qquad (4.178)$$

4. For $\epsilon \gg 1$ (infinitesimally slow cooling rate),

$$\psi = \frac{1 - (1 - \theta)^\gamma}{1 - (1 - \theta)^\gamma/(1 + \eta)} \qquad (4.179)$$

When η approaches zero, Eq. (4.177) reduces to Eq. (4.176). Equations (4.176) and (4.177) are identical indicating that for high cooling rates the cell behaves as though it contains only water.

Typical values for the various parameters including the constants are listed in Table 4.4. Figure 4.25 depicts the result for the case of $\psi \ll 1$ with $\epsilon = 1.236 \times 10^3$. Equation (4.177) agrees well with Mazur's (1963) numerical solutions within the stated limits of $\theta \ll 1$ and $\psi \ll 1$.

Many physiologists believe that the plasma membrane contains water-filled channels or pores (Ling and Tien, 1969) based on the discrepancies between the value of the permeability constants for water derived from osmotic flow across cell membranes and those derived from the rate of diffusion of isotropic water under zero osmotic gradient. Assuming these channels are of cylindrical shape, the equilibrium of three interfacial tensions between ice and water, between water and the capillary wall, and between ice and the capillary wall yields an expression for the freezing temperature of water in the capillary, θ_c. This expression is combined with Eq. (4.177) to yield

$$\theta_c = \frac{\ln [\delta(\beta - \alpha)/\gamma]}{\beta - \alpha} \tag{4.180}$$

where $\delta = 2v\sigma \cos \phi/aRT_f$; σ is the interfacial tension at the ice-water interface; and a is the pore radius. The temperature of lethal cell dehydration, θ_v, corresponds to the value of θ for which ψ is equal to ψ_c. $(1 - \psi_c)$ is the ratio

Table 4.4 Typical values for variables, parameters, and constants in cell freezing and dehydration

Parameter	Value
Intracellular water volume V	Dependent variable, μm^3
Cell temperature T	242–272 K
Molality of protoplasm m	0.5 mol/kg of water
Freezing point of cytoplasm T_f	272 K
Three initial volumes of internal water at $T = T_f$, V_i	0.42, 88, 6920 μm^3
Osmoles of solute in cell n_2 for the three values of V_i	2.1×10^{-16}, 4.4×10^{-14}, 3.45×10^{-12} moles
Temperature coefficient of permeability constant b	0.0325 degree^{-1}
Area of cell surface A for the three values of V_i	3.04, 107, 1964 μm^2
Permeability constant at T_g, K_g, for the three values of V_i	0.15, 0.3, 3 $\mu m^3/(\mu m^2$ min atm)
T_g	293 K
Rate of temperature change B	Dependent variable, K/min
Gas constant R	82.057×10^{12} μm^3 atm/(mol K)
Molar volume of pure water, V_1^0	18×10^{12} μm^3/mol
Molar heat of fusion of ice L_f	5.95×10^{16} μm^3 atm/mol

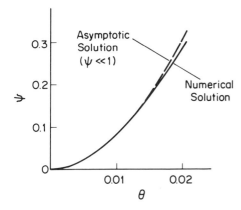

Figure 4.25 Supercooled intracellular water remaining at various temperatures in a spherical cell with $\epsilon = 1.236 \times 10^3$.

between the volume of water at the critical solute concentration and the initial volume of intracellular water V_i. For $\theta \ll 1$, ψ_c is essentially constant and equal to 0.8 for some cells. Equation (4.179) with $\psi = \psi_c$ and $\theta = \theta_v$ gives

$$\theta_v = 1 - \left[(1 - \psi_c) \frac{1 + \eta}{(1 - \psi_c) + \eta}\right]^{1/\gamma} \tag{4.181}$$

Since survival is to avoid both the formation of intracellular ice crystals and the level of lethal cell dehydration, cooling must cease at some θ less than the smaller of the two values, θ_v and θ_c. Theoretical predictions agree well with the experimental results (Ling and Tien, 1969).

4.5.4 Cell Hyperthermia

It is common knowledge that the activities of organisms are primarily catalyzed by enzymes which are made up of proteins. Many proteins are altered by heat (see Section 4.5.5). The high temperature required to kill a living cell, that is, the lethal temperature, depends on (1) the temperature range at which it has been previously adapted, (2) the length of exposure to that high temperature, and (3) the physiological state of the cell. Causes of heat death include (1) thermal inactivation of its enzymes, (2) derangements of the lipids of the cell, and (3) liberation of a coagulating enzyme by heat. The mechanisms of cell killing by hyperthermia are disclosed by exposing cell cultures to a range of elevated temperatures (43–48°C). Test data were correlated by the Arrhenius plot, as described in Section 4.5.3. Three major points were observed (Connor et al., 1977):

1. The Arrhenius plot shows a constant slope for all cell lines of five different cells beginning at a temperature of 43°C. The activation energies E evaluated from these lines lie in the range of 150 kcal for all cell lines at temperatures above 43°C. Activation energies in this range correspond to denaturation of enzymes and proteins.

2. The Arrhenius plot shows different slopes below 43°C compared to higher temperatures. These results may indicate either a different mechanism of cell inactivation at temperatures below 43°C or a similar mode of inactivation requiring multiple interactions.
3. Different cell lines have widely different sensitivities even though they have similar activation energies for thermal cell killing.

Hyperthermia affects the biochemical properties of normal cells in several ways (Conner et al., 1977): it inhibits cellular respiration and synthesis of DNA, RNA and protein; it inhibits the maturation of preribosomal RNA; it produces extended blockage of progression through the cell cycle; and it causes cell membrane changes.

Hyperthermia may affect respiration in tumor cells more severely than in normal cells. The effect is most pronounced in a very narrow range of temperatures, 41.5–42.0°C. Normal human diploid cells are less sensitive to treatment with 43°C than their transformed counterparts.

No study, theoretical or experimental, has been reported on heat transfer in cells during hyperthermia. The following analytic model will be useful in predicting temperature variations in cells. Assuming a cell to be spherical under external heating, the heat equation (4.143) may be used subject to the same initial and boundary conditions as Eq. (4.144), except for the last expression. For the case of constant heat flux q'' into the cell and neglecting the effect of heat production in the cell $\dot{q}$, Carslaw and Jaeger (1959) give

$$\frac{T - T_0}{aq''/k} = 3\frac{\alpha t}{a^2} + \frac{1}{2}\left(\frac{r}{a}\right)^2 - \frac{3}{10} - 2\frac{a}{r}\sum \frac{\sin(r\beta_n/a)}{\beta_n^2 \sin\beta_n} \exp\frac{-\alpha\beta_n^2 t}{a^2} \quad (4.182)$$

where β_n, $n = 1, 2, \ldots$, are the positive roots of $\tan\beta_n = \beta_n$. Numerical solutions are given in Fig. 4.26 for various values of $\alpha t/a^2$. This corresponds

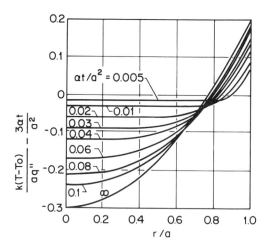

Figure 4.26 Temperature variation in a cell due to constant surface heat flux q''.

to the case where a cell is heated by radiation such as electromagnetic or ultrasonic waves.

In the case where a cell is heated in a medium at temperature T_∞ with the heat transfer coefficient between the cell surface and the medium h, the last expression in Eq. (4.144) should read

$$\frac{\partial T(a, t)}{\partial t} + H[T(a, t) - T_\infty] = 0$$

where H denotes h/k. The solution of Eq. (4.143) is given in Carslaw and Jaeger (1959) as

$$\frac{T - T_0}{T_\infty - T_0} = 2\frac{H}{r} \sum_{n=1}^{\infty} \frac{(a\beta_n)^2 + (aH - 1)^2}{\beta_n^2[a^2\beta_n^2 + aH(aH - 1)]} \sin a\beta_n \sin r\beta_n \exp\left(-\alpha\beta_n^2 t\right)$$

(4.183)

where β_n, $n = 1, 2$, are the roots of

$$a\beta \cot a\beta + aH - 1 = 0$$

4.5.5 Thermal Denaturation of Macromolecules (Proteins)

Changes in the specific properties that characterize the protein are called denaturation. Most studies on denaturation concern globular protein soluble in water, and almost all denaturing treatments are conducted on proteins in aqueous solutions. One of the most common methods of denaturing proteins is to heat them in solution.

The temperature coefficient Q_{10} for denaturation of protein is on a different order of magnitude from that for the thermochemical reactions. Most thermochemical reactions which obey the van't Hoff rule, Eq. (4.163), are characterized by a Q_{10} of 2.3. The coefficient is very high, being higher at the upper range of temperatures employed to denature the proteins. For example, Q_{10} is 13.8 in the temperature range of 60–70.4°C for protein coagulation of hemoglobin. In comparison, reactions in inanimated systems and reactions in organisms, with the exception of heat killing, have the coefficients on the order of 1–3 in temperatures up to 50°C. One should note that proteins are highly organized molecules and that denaturation involves the change from this highly organized structure to a much more random one. Therefore, even a small elevation in temperature in the range of denaturation has a marked effect out of proportion to the increase in the kinetic energy of the system.

Extremely high temperature coefficients are also observed for thermal death of cells, ranging from 10 to 1000, depending on the kinds of cells (Giese, 1968). These findings support the hypothesis that heat death of organisms is attributable to the denaturation of certain critical proteins (Johnson et al., 1954). So far, no model has been proposed to describe heat transfer associated with the process of protein denaturation.

4.6 REGULATION OF BODY TEMPERATURE AND CLOTHING

How are the ranges of normal thermal environment, hypothermia, and hyperthermia defined? Figure 4.27 illustrates the relationship between heat production and ambient temperature. There is a region of temperature, BC, called the neutral zone, over which the metabolic rate remains constant as the ambient temperature falls. The heat production in this zone is the basal metabolism, namely the minimum energy requirements at rest. *B* is the lower critical temperature below which heat production rises so that human beings can maintain their body temperature at a constant. The rise is proportional to the difference between the core temperature and the ambient temperature. Eventually point *A* is reached when the body temperature can no longer be maintained and hypothermia occurs. Line *AB* can be extrapolated to the core temperature E, say 38°C. Line *BE* is called the curve of thermogenesis. *C* is the upper critical temperature beyond which the metabolic rate rises mainly due to the increased respiration that often accompanies thermal stress. *D* is the temperature at which hyperthermia sets in. The temperature difference between *C* and *D* is very small, indicating that adjustment to a change in environmental temperature is mostly an adjustment to cold. Hyperthermia of 5°C is fatal.

4.6.1 Laws Pertinent to Thermoregulation

Thermoregulation may, in general, be divided into chemical and physical regulation. Chemical regulation deals with the mechanisms for the increase in heat production to cold, whereas thermal regulation deals with the mechanisms available for modifying the heat loss. In a human being, the latter is most highly developed, while chemical regulation contributes very little to the uniformity of body temperature. In addition, both physiological and nervous mechanisms are also brought into play in response to a change of ambient temperature within the range of accommodation.

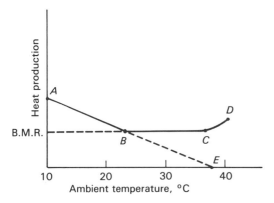

Figure 4.27 Heat production versus ambient temperature, *B*, lower critical temperature; *BC*, neutral zone; B.M.R., basal metabolic rate; *C*, upper critical temperature; *E*, body temperature.

The formulation of the problem of thermoregulation involves several important physiological and physical laws: homoisothermy, critical tissue temperature, van't Hoff law (see Section 4.5), conservation principles of energy (i.e., first law of thermodynamics) and mass, transport rate equations, and homeostasis.

4.6.1.1 Homoisothermy. Homoisothermy means uniformity of body temperature. Even though the environmental conditions undergo wide variations, body temperature, referring to the deep body temperature or the core temperature, remains fairly constant. The rectal temperature T_r represents the core temperature (which is occasionally taken in the mouth or axilla) because heat is better conserved in the rectum. In a normal human being, it is at 37°C under resting condition and at 39°C during strenuous exercise, varying ±1°C in direct proportion to oxygen consumption. Under resting conditions, the mean body temperature T_b may be evaluated from the equation for (subjects wearing underwear or indoor exercise clothing)

$$T_b = 0.65T_r + 0.35T_s \tag{4.184a}$$

For exposed nude subjects:

$$T_b = 0.8T_r + 0.2T_s \tag{4.184b}$$

For normally clad subjects:

$$T_b = 0.7T_r + 0.3T_s \tag{4.184c}$$

where T_s is the skin temperature.

The condition of homoisothermy must involve a delicate balance between heat production and heat loss. The balance may be achieved by independent variations in heat production and heat loss in order to meet the demands of the prevailing physical conditions.

4.6.1.2 Critical tissue temperature. If the body temperature rises as the result of inadequate loss of heat, sweating continues violently. In the case of fever, the rise of internal temperature during the first clinical phase occurs due to massive peripheral vasoconstriction. However, an increase in the mean body temperature causes heat production to go up, contributing to a further increase in the mean body temperature. Shivering in some fever conditions such as malaria will further increase the internal temperature.

It is known that damage is done to the tissue and the rates of metabolism and heat production start to fall when the temperature of the tissue exceeds a certain value between 40 and 45°C. The upper limit of temperature for survival is probably 42°C or below for the brain tissue, depending on the length of exposure time.

4.6.1.3 Conservation principles of heat and mass. The living body is also subject to the first law of thermodynamics which states that

$$q_{in} \quad - \quad q_{out} \quad + \quad \dot{q} \quad = \begin{array}{l} \text{Time-rate change} \\ \text{of internal energy} \\ \text{inside the body} \end{array} \quad (4.185)$$

Rate of heat transfer from the environment into the body	Rate of heat transfer out of the body into the environment	Rate of heat production within the body

Homoisothermy in humans requires that the time-rate change of internal energy should be either provisional or minimal. In a normal person, complex physiological actions immediately follow the change and restore the body to its normal core temperature, an excellent example of an integrative homeostatic mechanism. Therefore, the right-side term of the above equation can be treated as zero. This is not the case, however, in a sick person suffering from severe burn, fever, hyperthermia, heat illness, hypothermia, cold exposure, or under fever therapy or surgical hypothermia.

4.6.1.4 Heat and water production. Metabolism refers to the burning of foods, for example carbohydrates, fat, and protein. This combustion uses oxygen and produces carbon dioxide and water in proportion to the amount of food consumed. The heat production of the body may be measured by either direct or indirect calorimetry. The principle of direct calorimetry is to place the subject inside a large calorimeter for measuring total heat output. In indirect calorimetry, the heat production is calculated from the gaseous exchanges: the oxygen consumption and the carbon dioxide production. The ratio of the volume of carbon dioxide produced to that of oxygen used is called the respiratory quotient (R.Q.). It is the indicator of the dietary factor. For example, when glucose is oxidized, 1 mole (180 g) of glucose is oxidized by 6 moles (6×22.4 L) of oxygen. As a result, 6 moles of carbon dioxide are produced and 674 kcal of heat are liberated. The R.Q. is obviously 1.0. The body uses three kinds of food substances, fat, protein, and carbohydrate. Table 4.5 can be used to calculate the number of kilocalories produced per liter of oxygen for any desired ratio of carbohydrate and fat in the diet (Davson and Eggleton, 1962). When protein is burned 1 g protein will require 0.91 L O_2 and will produce 0.76 L CO_2. Therefore, its R.Q. is 0.84. The heat production amounts to 4.2 kcal/g protein, or 4.62 kcal/L O_2.

The amount of protein can be derived from the nitrogen excreted in the urine. Assuming that the protein contains 16% of N, the amount of N multiplied by 6.25 gives the amount of protein. Using the above figures, one can calculate the amount of O_2 and CO_2 associated with protein metabolism and also the heat production from protein metabolism. The subtraction of the total metabolic O_2 and CO_2 by the amounts of O_2 and CO_2 associated with protein metabolism

Table 4.5 Respiratory quotients and metabolic heat production of food substances

Consumption of 1 L O_2 is equivalent to			
Nonprotein R.Q.	Carbohydrate, g	Fat, g	kcal
1.00	1.232	0.000	5.047
0.95	1.010	0.091	4.985
0.90	0.793	0.180	4.924
0.85	0.580	0.267	4.862
0.80	0.375	0.350	4.801
0.75	0.173	0.433	4.739
0.707	0.000	0.502	4.686
Protein R.Q.	Protein, g	kcal	
0.84	0.909	4.62	

yields the nonprotein gaseous exchanges from which the nonprotein R.Q. is obtained. Then, by using Table 4.5, one can calculate the metabolic heat production for each liter of nonprotein oxygen utilization and subsequently the total heat production by adding the calories produced by protein metabolism.

This procedure is quite laborious and thus is not very often employed. In practice, the protein in the diet is ignored in the calculation of heat production. The gaseous exchange is treated as if only carbohydrate and fat were present. Since the protein has the R.Q. between those of carbohydrate and fat, the error in ignoring the protein is quite negligible. For example, a subject has a respiratory exchange at the rate of 302 cm^3 O_2 and 257 cm^3 CO_2 per minute. The corresponding R.Q. is 0.85 and the metabolic heat production is 4.862 kcal (from Table 4.5). For the oxygen usage of 18.12 L/h, the heat production rate is 57.6 kcal/h.

In general, the mechanical efficiency of the body is as low as that of an automotive engine, at most about 20%. In other words, at least 80% of the energy produced by metabolism has to be dissipated from the body as heat. When the subject is at rest, almost all the energy is converted into heat according to the first law of thermodynamics.

When the body is at rest, the heat produced at the basel metabolic level is derived about 38% from the respiratory muscle with the balance from various internal organs, particularly liver, spleen, and kidney. Heat generated in the body core must be transferred over a certain distance to the skin for eventual dissipation to the environment. In an active body with voluntary muscles doing external work such as strenuous exercise, the metabolism may increase to more than 10 times the basal rate with about 80% of the accompanying heat production in peripheral muscles close to the skin. Heat flow paths are thus much shortened for faster dissipation. However, in a human being doing extremely hard work, efficiency is limited not by muscular capability but by the rapidness of heat dissipation to maintain homoisotherm of the body.

4.6.1.5 Heat and mass transfer between skin and environment. For quantitative expressions of heat exchange and evaporative heat loss to the environment, the following transport-rate equations are employed:

Convection The rate of heat exchange q_c between the skin and the environment by convective mechanism (i.e., molecular diffusion and mixing in fluid) can be predicted by Newton's law of cooling, Eq. (1.26):

$$q_c = h_c A_s (T_s - T_a)$$

Here, A_s is the skin surface area and h_c is the convective heat transfer coefficient. When $T_s > T_a$ (or when q_c is positive), heat is lost from the skin to the environment and vice versa. The magnitude of h_c depends on the method of induction of air motion (natural or forced convection), flow conditions, geometry, and position of the surface, etc. Equations for evaluation of h_c are available in standard textbooks on heat transfer.

Convective coefficients for a nude subject in the common immobile postures calculated by the appropriate nondimensional equations of heat transfer using man-equivalent geometric shapes (flat plate for supine, vertical cylinder for standing, and sphere for seated) as thermal models, agree with those determined by physiological tests within 10% (9% for supine; 5% for standing; and 4% for seated; Hardy et al., 1970).

When air velocities are below about 30 m/min, a condition of either pure free convection or a combination of both free and forced convection occurs. Over the range of air velocities from 0 to 30 m/min or $(T_s - T_a)$ from 1 to 10°C, free convection dominates the convective transfer and the contribution of forced convection is found to be less than 10% (Hardy et al., 1970).

Radiation Radiation is the transfer of heat from the body surface by means of electromagnetic waves. Its rate, radiated from a black body (a perfect radiator) is described by the Stefan-Boltzmann law

$$q_r = \sigma A T^4$$

in which σ, the Stefan-Boltzmann constant, is equal to 5.67×10^{-8} W/m²K⁴; A is the radiating surface area; and T is the surface temperature on the absolute scale. For radiation from nonblack surfaces, its rate is given by

$$q_r = \epsilon \sigma A T^4$$

Here, ϵ is the emissivity of the surface defined as the ratio of its emissive power $(E = q_r/A)$ to that of a black body E_b. The reflectivity ρ (i.e., the fraction of the incident radiation reflected from the surface) is equal to $1 - \epsilon$ if the surface is opaque (does not transmit radiation).

The human skin and most clothing are black or almost so for some radiations (for example, in the infrared zone) and not for others (for example, in the visible range), as shown in Fig. 4.28. In the infrared range, the emissivity of human skin, regardless of its color and pigmentation, is about 0.99, which is sufficiently close to unity for all practical purposes. However, the reflective power of the human skin depends on skin color and varies with the wavelength of the incident radiation. Both black and caucasian (white) skin are almost black in the invisible infrared range but are not in the visible range. Figures 4.29*a* and 4.29*b* are a summary of the spectrographic results of several investigators for white and black skin, respectively (Boehm and Tuft, 1971). The maximum and minimum curves indicate an envelope of data obtained by those investigators. Most data fall within a band of ±5% of the average curve. A large portion of the minimum curve is composed from data obtained from excised skin, while the majority of the maximum curve is made up of readings taken from body parts seldom exposed to sunlight. The total (i.e., contributions

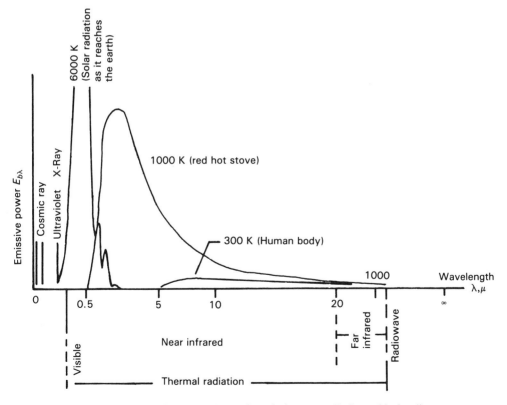

Figure 4.28 Spectral distribution monochromatic emissive power E_b for an ideal radiator.

of all wavelengths) reflectivity is given by

$$\rho = \frac{\int_0^\infty \rho_\lambda G_\lambda d\lambda}{\int_0^\infty G_\lambda d\lambda} \tag{4.186}$$

where ρ_λ is the average, angular-hemispheric reflectivity and G_λ is the spectral energy incident upon the skin. For the case of black-body or gray-body radiation incident upon the skin, the substitution of a ρ_λ in Figs 4.29a and 4.29b into Eq. (4.186) yields the average, total, angular-hemispherical reflectivity of white and black skin, as illustrated in Fig. 4.29c. It is seen that white skin reflects about 40% of the solar radiation, whereas the dark skin of blacks reflects about 18% of these rays.

In evaluating the net radiative heat exchange between the skin and the environment, the emissivity of the environment ϵ_e is an important factor. For radiation between two parallel flat surfaces, the gray-body shape factor takes the form of $F_{s-e} = 1/(1/\epsilon_s + 1/\epsilon_e - 1)$ if the end effects are neglected, where subscripts s and e denote the skin and environment, respectively. The rate of radiative heat transfer between the skin and the environment can be expressed as [Eq. (4.5)]

$$q_{rs-e} = F_{s-e} A_s \sigma (T_s^4 - T_e^4) \tag{4.187}$$

For indoor environments in which the walls are at air temperature and are painted, ϵ_e is approximately unity for practical purposes. However, for most environments, a ϵ_e of 0.95 is a safe value. Since both emissivities, ϵ_s and ϵ_e, are approximately or very close to unity, $F_{s-e} = \epsilon_s \epsilon_e/(\epsilon_e + \epsilon_s - \epsilon_s \epsilon_e)$ may be approximated to be $\epsilon_s \epsilon_e$. In other words, q_{rs-e} may be simplified

$$q_{rs-e} = \epsilon_s \epsilon_e A_s \sigma (T_s^4 - T_e^4)$$

The above expression has been extensively used in the measurement of the total emissivity of the skin with radiometric type devices, where ϵ_e and T_e denote the emissivity and temperature of the radiometer receiver. For application in calculating the radiation heat transfer between a nude man and the environment, this equation may be reduced to

$$q_{rman-env} = f A_b \epsilon_e \sigma (T_s^4 - T_e^4)$$

where $\epsilon_s = 1$ is imposed; f is the ratio of effective radiating surface area to the DuBois surface A_b. Here, $f = 0.78$ for lying in anatomic position, $f = 0.85$ in a spread-eagle position, $f = 0.71$–0.75 for sitting in a chair in a semireclining position. T_s, the average skin temperature, is 32.02°C based on area fractions as shown in Table 4.6.

In the presence of the sun, the actual radiant heat loss from man is

$$q_r = q_{rman-env} - (1 - \rho_s) E_{bsun} f_s A_b$$

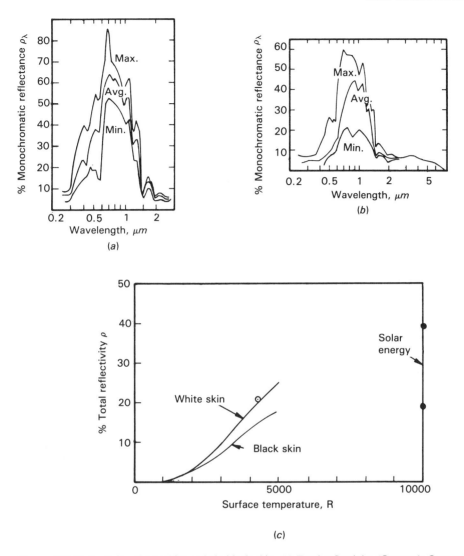

Figure 4.29 Reflectivity of (*a*) white and (*b*) black skin. (*c*) Total reflectivity (○ tested; ● tested and calculated). (Boehm and Tuft, 1971)

Here, ρ_s is the reflectivity of skin for solar radiation, $E_{b\text{sun}}$ is the emissive power of solar radiation and f_s is the fraction of the body area exposed to the sun. This equation needs further modification in the presence of appreciable reflection of the solar radiation from the surroundings. It should read

$$q_r = q_{r\text{man}-\text{env}} - (1 - \rho_s)(f_s + \rho_e f_e)E_{b\text{sun}}A_b$$

where f_e is the body area exposed to this reflected radiation.

Table 4.6 Skin temperature and area fraction for body segments

Body segment	Skin temperature, °C	Area fraction	Product
Head	35.0	0.07	2.45
Arms	33.3	0.14	4.66
Hands	34.1	0.05	1.71
Feet	30.6	0.07	2.14
Legs	31.8	0.13	4.13
Thighs	31.9	0.19	6.06
Trunk	33.9	0.35	11.87
Whole body, average	32.02	1.00	32.02°C

Combined action of convection and radiation Equation (4.187) may be simplified (linearized) to read

$$q_r = h_r A_s (T_s - T_e)$$

which is Eq. (1.26). Here,

$$h_r = F_{s-e} \sigma (T_s^3 + T_s^2 T_e + T_s T_e^2 + T_e^3)$$

h_r is called the radiative heat transfer coefficient.

Convection and radiation may be combined to yield

$$q_c + q_r = h A_s (T_s - T_e) \qquad (4.188)$$

where $h = h_c + h_r$ is the combined heat transfer coefficient. The magnitude of h can be obtained by evaluating h_c and h_r, individually utilizing the information available for convection and radiation, respectively. However, several empirical equations have been proposed to predict the magnitude of h in air using the expression (Hardy et al., 1970)

$$h = a + bv^n \qquad \text{[kcal/n m}^2 \text{ °C]} \qquad (4.189)$$

where v is the air velocity in m/s and a, b, n are experimentally determined constants as given in Table 4.7. $v = 0$ corresponds to the case of natural convection in still air.

Conduction and mass diffusion to environment Conduction and mass diffusion refer to the molecular diffusion of heat and mass, respectively. The diffusion rates q_k and N can be expressed in differential form, Eqs. (1.25) and (1.21), as

Fourier conduction law: $q_k = -kA \, dT/dx$

Fick's first law of mass diffusion: $N = -DA \, dC/dx$; or in finite difference form, Eqs. (1.26) and (1.22), $q_k = kA\Delta T/\Delta x$ and $N = DA\Delta C/\Delta x$, respectively.

Here, k is the thermal conductivity, D is the mass diffusivity, dT/dx and dC/dx are, respectively, the temperature and concentration gradients in the flow direction x, and ΔT and ΔC are, respectively, the temperature and concentration differences between two points at a distance of Δx.

As characterized by the low values of k and D, heat and mass losses by molecular diffusion from the skin to the air are very small compared with those lost by convection. However, significant heat loss occurs when the body comes into contact with a material of high thermal conductivity, for example, sitting on a marble slab or immersion in water.

Insensible evaporation Water produced as a by-product of a metabolic reaction in the tissues is dissipated from the skin and through the respiratory tract to the environment as a vapor. Upon evaporation, which is a phase change from liquid to vapor, a large amount of heat is removed from the body. This is an important mechanism in body temperature regulation. The heat of 576 kcal/kg is absorbed or removed from the surface of evaporation.

At an ambient temperature below 25°C, a resting, lightly clad individual does not sweat (sensible perspiration) and the evaporative loss is entirely due to the loss of water vapor from the skin and lungs (insensible perspiration). Little sweating occurs at temperatures below 28°C. Sweating definitely begins at 29°C and its role in thermal regulation of the body becomes increasingly

Table 4.7 Combined convective and radiative heat transfer coefficient in air

Experimental conditions	Results*		
	a	b	n
Subjects lying on a table	4.85	6.3	0.5
	2.7	6.8	0.5
Semireclining, lightly clothed subjects	3.6	10.9	0.5
Nude standing subjects	5.7	7.5	0.5
	4.5	9.4	0.5
	6.9†	5.6	0.67
Nude seated objects, still air	6.00	0	0
	6.30		
Nude reclining subjects	6.9	7.5	0.67
Nude reclining subjects, still air	6.1	0	0
Nude reclining subjects	6.9‡	7.5	0.67
Nude seated subjects	6.4§	5.6	0.67

*Results under three conditions given: (†) air flow parallel to the main body axis; (‡) air flow perpendicular to the main body axis; and (§) air in turbulent flow.

Source: Hardy et al. (1970).

important with rising ambient temperatures. When the air temperature reaches 35°C or higher, practically all the heat loss is by evaporation. The mean skin and body temperatures at which sweat secretion occurs are about 34°C and 37.2°C, respectively.

The rate of water evaporated from the skin to the ambient air can be expressed as

$$N = h_D A_w (\rho_{ws} - \rho_{wa}) \quad \text{(kg/h)}$$

where A_w is the area of the skin surface for evaporation in square meters and h_D is the mass transfer coefficient in m/hr. ρ_{ws} and ρ_{wa} are the mass densities or concentrations of water vapor at skin surface and in ambient air, respectively in kilograms per cubic meter. The mass transfer is accompanied simultaneously by phase change of water from liquid to vapor, which requires the latent heat of vaporization L in kilocalories per kilogram to be supplied from the skin. The evaporative heat loss from the skin in kilocalories per hour is

$$q_e = NL = h_D A_w (\rho_{ws} - \rho_{wa}) L \quad (4.190)$$

Within the usual range of physiological conditions, the skin-air temperature difference $(T_s - T_a)$ is small compared to the film temperature, $T_f = (T_s + T_a)/2$, and the density of the air-vapor mixture ρ_a is essentially constant. Treating the vapor as an ideal gas, the equation of state can be utilized to rewrite Eq. (4.190) in terms of partial pressure difference as

$$q_e = h_e A_w (p_{ws} - p_{wa}) \quad (4.191)$$

where h_e is the evaporative heat transfer coefficient defined as

$$h_e = \frac{h_D L}{R_w T_f} \quad \text{(kcal/h m}^2 \text{ mmHg)} \quad (4.192)$$

p_{ws} and p_{wa} are the partial pressures of water vapor at skin surface and in ambient air, respectively in mmHg. R_w is the gas constant for water vapor, 47 kgf m/kgm K or 3.47 mmHg m³/kgm K, where kgf is kilograms of force and is equal to kgm, kilograms of mass, times acceleration. The evaporation rate m and the skin area A_w are measured, while the partial mass densities (ρ_{ws}, ρ_{wa}) and partial pressures (p_{ws}, p_{wa}) are determined by the psychrometric conditions at the skin surface and in the ambient air at the respective temperatures T_s and T_a.

For any given skin area, temperature, and environmental conditions, the evaporative heat loss can be calculated if the value of the mass transfer coefficient h_D can be determined. In gases (e.g., moist air), where the partial pressure differences are small, which is the case in most physiological situations, the heat and mass transfer analogy gives a close approximation of h_D:

$$\frac{h_D}{h_e} = \frac{(\text{Pr/Sc})^{2/3}}{\rho_{am} C_{pm}} \quad (4.193)$$

This equation is valid for both laminar and turbulent flow. Sc is the Schmidt number for the diffusion of water vapor in air and takes a value of about 0.60 over the air temperature range of 0–50°C under 1 atm. Pr is the Prandtl number of dry air, which is 0.72. ρ_{am} and C_{pm} are the density and specific heat of moist air, respectively. C_{pm} is defined as

$$C_{pm} = \frac{C_{pa} + WC_{pw}}{1 + W} \quad \text{(kcal/kg °C)} \tag{4.194}$$

where

$$W = \frac{18p_w}{29p_a} \tag{4.195}$$

is the absolute humidity. p_w and p_a are the partial pressures of water vapor and dry air in the ambient air, respectively. C_{pa}, the specific heat of dry air, is 0.240 cal/g °C, while C_{pw}, the specific heat of water vapor, takes a value of 0.441 cal/g °C. Over the usual physiological ranges, ($T_a = 0$–50°C and relative humidity $\phi = 0$–100%), the product $\rho_{am}C_{pm}$ may vary from 0.25 to 0.32 kcal/m³ °C. For any skin surface, when the size, shape, air-stream orientation and air velocity are known, the convective heat transfer coefficient h_e in Eq. (4.193) for both free and forced convection can be determined by appropriate formulas available in standard textbooks on heat transfer.

4.6.1.6 Physiological adaptations to hot and cold environments. In the regulation of the body temperature, the chemical and physical processes described in Section 4.6.1.4 are modified by three physiological phenomena: changes in cutaneous blood circulation, sweating, and shivering. These adaptive reactions are activated by the reflex mechanism. A scheme illustrating the main nervous pathways in the reflex mechanism of temperature regulation is presented in Fig. 4.30. Thermal receptors are situated in the skin and in the brain. The hypothalamus has an area which is thermosensitive. Efferent pathways are in the spinal cord and sympathetic chain. Efferent arcs and mechanisms involve all the functions of the tissues and all the organs.

Vasomotor responses A resting, lightly clad human being can easily establish thermal equilibrium in stall air at about 27°C. When the air temperature T_a falls below this, the skin vessels constrict. Vasoconstriction slows the flow and diverts the venous return from the superficial toward a deeper vein. On the other hand, when T_a rises above 27 °C, cutaneous vasodilation occurs. Blood is then diverted from the core of the body to the surface. Vasomotor regulation occurs in the range of environmental temperatures between 25 and 29 °C.

Consider a layer of tissue in which blood vessels are distributed. If one side of the tissue (denoted by subscript 1) is at a higher temperature than the other (subscript 2), heat transfer will take place by both conduction through the tissue and convection through the blood vessels. Equations (1.25), (1.21), and (1.26), (1.22) determine the rates by the conduction mechanism locally and

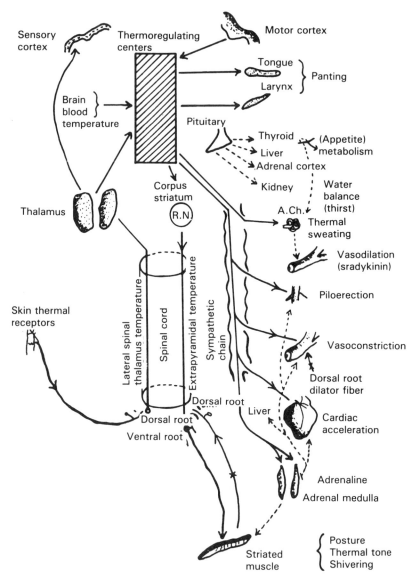

Figure 4.30 Main nervous pathways in the reflex mechanism of temperature regulation. R.N. = red nucleus; --- = humoral agent. The final effector systems are listed on the right-hand side (Burton, 1953).

across the tissue layer, respectively. The convective rate transferred by the blood in each vessel with a poor thermal-conducting wall is

$$q_c = \rho_b v_b A_c C_b (T_{bl} - T_{b2}) \qquad (4.196)$$

where ρ_b, v_b, C_b, and T_b are the density, velocity, specific heat, and temperature of the blood, respectively, and A_c is the cross-sectional area of the vessel. If there is a bunch of blood vessels lying in parallel in the tissue layer, the total convective rate contributed by the blood flow is the summation of the q_c's of all vessels. Equation (4.196) indicates the proportional relationship between the q_c's of all A_c. Therefore, when the skin vessels dilate, heat is diverted from the tissues to the surface, contributing to a rise in the skin temperature. As a result, the rate of heat loss to the environment increases due to an increase in the temperature difference between the skin surface and the surroundings. When the skin vessels constrict, heat is conserved. The extent of alteration in any part of the peripheral circulation depends on the local temperature, in particular the skin temperature, and the deep body temperature.

Sweating (sensible evaporation) Sweat secretion begins at an air temperature of about 31°C for resting nude subjects and 29°C for lightly clad persons whose average skin temperature is about 34°C. When the skin surface is completely wetted by sweating, its temperature is at the saturated value, T_{sat}, corresponding to the water vapor pressure, p_{ws}. As water evaporates from the wetted skin to ambient air, the latent heat ($L = 0.576$ kcal/g) is removed from the skin surface. If sweat runs off without evaporation, it has no cooling effct.

The rate of evaporation and the accompanying heat loss can be predicted by the same equations for insensible perspiration provided that the skin surface temperature T_s is replaced by the saturation temperature T_{sat}.

Over the air temperature range of 40–60°C and air velocity range of 3–300 m/min, the maximum skin evaporative heat transfer coefficient for a completely sweating (over the entire body) nude human in supine, standing, and seated postures can be predicted (Hardy et al., 1970) within an estimated accuracy of ±10% by

$$h_e = 2.26h_c \qquad (4.197)$$

Within these ranges of air temperature and velocity, h_e varies from 3.5 to 35 kcal/h m^2 mmHg. When localized sweating occurs on whole body segments such as the thighs, arms, legs, or head, the evaporative heat transfer coefficients are 10–90% higher than those obtained when the entire body is wet, depending on the mean segment diameter. However, they are approximately the same magnitude as those for the fully wetted trunk.

Muscular activity When voluntary muscles are used for the exercise or to do external work, the metabolic rate increases greatly, more than 10 times the basal level in the case of severe exercise. If one is unable to dissipate all the heat produced by metabolism, one's muscular activity or efficiency may be limited by difficulties of heat dissipation. This will be less a problem in cold environments where heat can be removed with relative ease. During muscular activity, almost 80% of the total metabolic heat is produced in the peripheral muscles closed to the skin. Hence, heat dissipation is easily promoted.

When the ambient temperature falls below 25°C, the resting nude body begins to lose more heat than it produces. As a response to the cold environment, heat production is increased in the body. If this increase still cannot offset the loss, more heat must be produced in order to maintain a normal body temperature. This is accomplished by shivering, a phenomenon of converting work to heat. This is accompanied by increases in oxygen consumption, respiration, and blood pressure—the typical reaction to cold. The environmental temperature corresponding to the onset of shivering is called the critical temperature. It differs with species. A nude human has a critical temperature of about 29°C.

4.6.1.7 Conduction and mass diffusion in tissues. In the body, heat produced by metabolic reaction has to be dissipated to the environment. The heat is transferred from the deep tissues to the skin by conduction as well as convection through blood vessels. The rate of conduction through the tissues can be predicted by Eqs. (1.25) and (1.26). There is a temperature gradient from within to outward. Its steepness depends on the thermal conductivity of the tissues. For example, epithelium and fat have low thermal conductivity. Therefore, the gradient through the skin and subcutaneous tissue could be quite steep. However, the circulation of the blood tends to reduce differences of temperature and accordingly the gradient may be reduced.

In 1934, Burton introduced the concept of the thermal conductance of tissues, which has been adopted in medical science. Conductance is nothing but the conductive heat transfer coeefficient as defined by Eq. (1.26) in which q_k is replaced by the rate of metabolic heat loss through the skin. ΔT is the temperature difference between the two surfaces of the tissue layer, although Burton employed the difference between the rectal temperature and the mean skin temperature. Since the contributions of both conduction through the tissue and convection by blood perfusion are included in the definition, conductance becomes dependent not merely on the tissue geometry and temperature but most significantly on the rate of cutaneous blood flow. Consequently, convection from vasodilatation and vasoconstriction induced by changes in the air

Table 4.8 Thermal conductivity of some body elements

Body element	Temperature range, °C	Thermal conductivity, cal/cm²/min°C	
		Male	Female
Forearm	15–38	0.065	0.055
Palm	15–38	0.114	0.136
Fat	–	0.032	0.032
Muscle (excised)	–	0.086	0.086
Gelatin (excised)	–	0.086	0.086

Source: Burton (1934).

temperature can significantly affect the magnitude of conductance. Even more inaccurate is the determination of the thermal conductivity of the tissues based on the concept of conductance. Thermal conductivity is a physical property and as such should depend on only temperature and pressure. Lack of a sound physical concept of thermal conductivity has led to a substantial deviation in test data and inconsistency in the interpretation of observed phenomena often found in medical publications. Table 4.8 presents the thermal conductivity of various tissues in body segments.

The rate of mass diffusion through the tissues can be determined by Eqs. (1.21) and (1.22).

4.6.1.8 Heat storage. The human body itself is a heat reservoir. Heat may be stored by body water with its high heat capacity. This heat reservoir is provided against thermal demands or supplies so that regulatory changes do not induce abrupt disturbance of heat balance in the body. The amount of heat stored S, the difference between the heat produced and that being dissipated, can be determined by the heat balance. As heat is stored, the mean body temperature rises above a normal basal value. The change in mean body temperature ΔT_b can be evaluated using the expression

$$\Delta T_b = \frac{S}{m_b C_{pB}} \tag{4.198}$$

where m_b denotes the mass of the entire body. The specific heats of water and fat are 1.0 and 0.5 cal/g °C, respectively. With S measured by direct or indirect calorimetry and the corresponding ΔT_b known, the above equation can be used to determine C_{pB} of the body, which is 0.83 cal/g. The product of mass and specific heat, $m_b C_{pB}$, is called the heat capacity of the body. The magnitude of heat capacity indicates the capacity of heat storage in the body.

A 70-kg man has a heat capacity of 58 kcal/°C. If his mean body temperature rises by 2°C in exercise, he has 116 kcal of heat stored which has to be later dissipated to restore thermal balance. The mean temperature of the body can vary considerably with additions to or removals from the heat reservoir. The deep body temperature or the core temperature represents the mean value with respect to the mean body temperature flactuations. The function of the heat reservoir is the damp down the effects of fluctuation in heat production and removal. The deep body temperature is therefore not a fixed value but varies in a limited range, depending upon the person's physiological status.

4.6.1.9 Homeostasis. Homeostasis is the maintenance of static, constant, or equilibrium conditions in the internal environment. Essentially all the organs and tissues of the body perform functions that help to maintain these equilibrium conditions. Thermal regulation is an integrative homeostasis in which practically every aspect of physiological control is involved. Examples include control of blood circulation by sympathetic and parasympathetic systems, hormonal control, the function of external secretions such as sweat, behavioral reactions like posture, the action of a brain center, and the activity of voluntary muscle. All this in combination with the physical laws of heat transfer and Arrhenius' fundamental physicochemical laws have contributed to the achievement of homoisothermy in man.

Fever is evidenced by an elevated central temperature. In moderate fevers, the homeostatic mechanism is not impaired except that the temperature is at a new level higher than normal. Fever is characterized by three clinical phases. First is the rising temperature phase during which massive peripheral vasoconstriction occurs. Heat is conserved in the body yet the patient feels cold. The sensation of cold actually follows the rise of internal temperature. In the second or the steady phase, the patient's temperature remains steady for days (the duration depends on the type of infection) at a higher level. Heat is produced at a rate of about 13% higher than the basal value for every degree Celsius rise in body temperature. The continuation of this higher metabolic rate is sustained either by dietary compensation (such as consumption of extra protein in typhoid fever) or at the expense of body tissues. The higher heat production is balanced by a higher heat dissipation from the body, however, at no change in sweat loss. Thermal regulation still proceeds quite accurately at the higher temperature level in response to exercise or variation in environmental conditions. Hence, the use of cold compresses during fever depends on the environmental temperature and humidity. The third and final stage is the falling temperature phase during which profuse sweating and peripheral vasodilation occur. All routes of heat dissipation are put in use, yet the patient complains of feeling hot. A decrease in fluid intake rather than an increase in water loss results in dehydration with mild fever.

4.6.2 Thermoregulation

In order to maintain the homoisothermy within the normal range of temperature, changes in circulation, respiration, and metabolism are involved to counteract the effects of thermal interactions between the human body and its environment. The interactions include both mechanical and thermal types. Hence, one must consider (1) heat production by metabolism (q'''); (2) heat exchange with the environment through physical processes such as conduction (q_k), convection (q_c), radiation (q_r), and evaporation (q_e); (3) physiological adaptations through the variations of peripheral vasomotor tone, sweating, and muscular activity such as work and shivering (dW/dt); and (4) heat storage (dE/dt). Item (1) includes normal as well as fever situations. The nature of the interactions can be established by treating the human body as a control volume interacting with the environment across a control surface. Application of the principle of conservation of energy [Eq. (4.185)] yields

$$\pm q_k \pm q_c \pm q_r - q_e + q''' - \frac{dW}{dt} = \pm\frac{dE}{dt} \qquad (4.199)$$

in which the $+$ and $-$ signs indicate the heat gain and loss of the body. Equation (4.199) represents the overall heat balance formulation and is applicable to systemic thermoregulation.

4.6.3 Clothing

The principal function of the coats of homeothermic animals and the clothing of human beings is the regulation for heat transfer between the body and its environment. This is crucial for survival and comfort. The studies of clothing were prompted by the problems of clothing military personnel during World War II for a wide variety of severe climates, from the tropics to the arctic, from submarines to airplanes (e.g., Newburgh, 1968). Afterward, the air conditioning of buildings for human comfort led to the research of its relationship to clothing and work. Efforts have also been directed toward studies of clothing requirements under severe environments such as diving, arctic exploration, and space flight. However, our everyday clothing is determined by fashion as well as function.

At steady state, metabolic heat production is balanced with heat losses at the body surface by convection, radiation, and evaporation. With the presence of a coat, the heat lost to the environment is attenuated. Evaporative heat transfer through the coat is considered separately by the difference of water vapor concentrations, C_s and C_{ct}, at the skin surface and in the air outside the coat, respectively. The sensible (combined convective and radiative) heat flux drives through the coat thermal resistance δ/k. Hence, the first law of thermodynamics leads to

$$\frac{q'''}{A} = \frac{T_s - T_{ct}}{\delta/k} + \frac{(C_s - C_{ct})}{\delta/D} h_{fg} \qquad (4.200)$$

where q'''/A signifies the metabolic heat flux; T_s, the skin temperature; T_{ct}, the temperature of the coat surface; δ, the coat thickness; k, the thermal conductivity of the coat; D, the mass diffusivity of water vapor through air; and h_{fg}, the latent heat of vaporization of water. The quantity δ/k is called the insulation I_s:

$$I_s = \frac{\delta}{K} \qquad (4.201)$$

Historically, an empirical unit of insulation called "clo" has been used. One clo is equal to 0.155 m² K/W. Measured values of typical garments and fabric insulation are presented in Table 4.9. It is interesting to note that insulation manufacturers called δ/k the R value and use it for measured values of house insulation.

Once heat dissipation to the environment is changed by insulation and environmental stress, metabolic heat production will respond to cope with the new situation. Figure 4.31 illustrates the relationships between metabolic heat

Table 4.9 Thermal insulation of garments and fabrics

Garment or fabric	Insulation, $\times 10^3$, m^2 K/W
Socks	1.5–5
Light underwear	8
Winter underwear (cotton briefs and short-sleeve vest)	30
Shirt or blouse (short sleeve)	30–40
Shirt or blouse (long sleeve)	39–45
Light trousers	40
Heavy trousers	50
Sweater (long sleeve)	26–57
Jacket	26–57
Heavy jacket or quilted anorak	76
Boiler suit or long winter underwear	77–116
Tights	1.5
Shirt	15–34
Light dress	26
Winter dress	100
Track suit	77
Arctic combat assembly	666
Cotton poplin shirting (0.5-mm thick)	8
Wool serge (1.0)	17
Light woollen (2.2)	46
Heavy woollen coating (4.3)	93
Cellular woollen blanket (6.6)	147
Anorak lining (raised wool) (8.3)	201
Mohair pile (12.7)	310

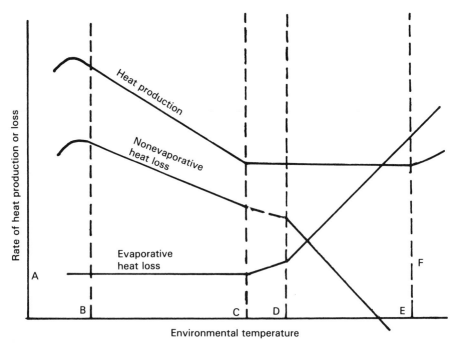

Figure 4.31 Relationship between heat production and evaporative and nonevaporative heat loss (Fazzio and Jacobs, 1972).

production, and evaporative, and nonevaporative heat loss. The abscissa is the environmental temperature T_a. A is the zone of hypothermia, while F is the zone of hyperthermia. B and E represent the temperatures of incipient hypothermia and hyperthermia, respectively. BE is the thermoregulatory range. The environmental temperature either below B or above E is lethal and must be avoided. C is called the critical temperature, T_{cr}, below which metabolic heat production increases significantly in response to cold stress. D is the temperature beyond which evaporative (or insensible) heat loss increases sharply in response to heat stress. CD is the thermoneutral zone in which metabolism is a minimum and the least amount of effort is required for thermoregulation. Most homeoisotherms have a limited width of this zone which may vary depending upon their size and insulation. CE corresponds to the zone of minimal metabolism.

From the viewpoint of survival, the lower B is, the longer a homeotherm may survive. The wider the thermoneutral zone CD is, the greater the range of comfort. These two goals may be simultaneously achieved through increasing insulation; for example, at steady state, metabolic heat is transferred through the insulation by conduction (evaporative heat loss is due to mass transfer cooling and is ignored for convenience) and then from the outside surface of the insulation (at a temperature T_{ct}) into the environment (at T_a) by convection and radiation. When the environmental temperature is at the critical temperature T_{cr}, the sensible heat flux through the coat into the environment is approximated to be the metabolic heat flux at the minimum rate $(q'''/A)_{min}$, neglecting evaporation,

$$\left(\frac{q'''}{A}\right)_{min} \cong \frac{T_s - T_{cr}}{\delta/k + h_r h_c/(h_r + h_c)} \tag{4.202}$$

Since in most coated homeotherms, δ/k is substantially larger than $h_r h_c/(h_r + h_c)$, the above equation yields

$$T_{cr} \cong T_s - \left(\frac{q'''}{A}\right)_{min} \frac{\delta}{k} \tag{4.203}$$

In human beings, the minimal metabolism corresponds to the basal metabolic rate. Therefore, T_{cr} is determined largely by coat insulation. Older people, whose basal metabolic rates are lower than those of younger people, need garments of higher insulation values to be comfortable at comparable levels of T_{cr}.

4.7 THERMAL MODELING

Treating biomaterial as an isotropic, homogeneous continuum permeated by a vascular network and taking into account the effects of tissue perfusion, metabolism, and heat production due to absorption of radiation fluxes such as electromagnetic and ultrasonic diathermy, the first law of thermodynamics yields the bioheat transfer equation (4.54). The expression can be applied to the thermal modeling of regional thermoregulation as well as the measurement of the thermal properties of biomaterials. Its solution, when combined with the experimental data, yields the thermal properties (Cravalho, 1972).

4.7.1 Systemic Thermoregulation

Systemic thermoregulation is a well-documented phenomenon. A review of mathematical models of the human thermal system is presented in Fan et al. (1971). Other important references include Hardy (1963), Newburgh (1968), Hardy et al. (1970), and Shitzer and Eberhart (1985). All standard textbooks on human physiology, to a certain extent, cover the topic of systemic thermoregulation (see for example Davson, 1970).

A simple thermal model of systemic thermoregulation uses only Eq. (4.199). In the evolution course of thermal modeling, the concepts of "core and shell," "concentric cylinders," and "infinite slab" were introduced. Wissler (1961, 1963) made an even closer simulation to the human body in his six-cylinder-elements model consisting of the trunk, the head, two arms and two legs. Each element is assumed to be a homogeneous cylinder of bone and tissue which is coverd with a layer of fat and skin. Metabolic reactions are assumed to generate heat uniformly in each section of the cylinder, but the rates are not necessarily equal. The capillary beds of each section are uniformly supplied with arterial blood at a temperature T_b. Then equations similar to Eq. (4.54) are obtained, subject to the appropriate initial and boundary conditions. Numerical results can be obtained for either steady or transient state temperature distribution and heat losses.

4.7.2 Regional Thermoregulation

Skin functions as temporary or permanent storage of certain materials such as fat, extracellular fluid, glucose, salts, and sugar. Skin also serves as a barrier for hindering or delaying penetration of water, electrolytes, or bacteria. Additionally, skin plays an important role in the regulation of regional temperature as well as the whole body temperature, since heat loss from the body must occur almost entirely from its surface.

Three routes are available by which direct heat exchange between the skin and the environment may take place: conduction, convection, and radiation. Heat is transferred from the skin to the environment when the skin temperature exceeds the ambient temperature T_a, and vice versa. When the two temperatures are equal, there is no driving potential and thus heat exchange by any of the three modes ceases. Therefore, when the environmental temperature reaches or exceeds the skin temperature, evaporation of water from the skin and air passage becomes the only way of dissipating body heat. The process takes place continuously at all environmental temperatures. Water has a high latent heat of evaporation, 0.58 kcal/g at 33°C. Evaporation from the wet surfaces of the lungs and respiratory passages, which is called insensible perspiration, may be increased by increasing lung ventilation. However, adjustments of evaporation in this way are limited as compared with the range of adjustment by the sweat glands in the skin.

There are two ways by which water migrates through the skin: insensible perspiration and sweating. Insensible perspiration involves the evaporation of water which diffuses through the epidermis from the deeper layers of the skin. The loss of water from the lungs and skin are approximately in proportions of 1:3 to 2:3. About 400–500 g of perspiration may be lost in a day but this varies with the ambient conditions, metabolic rates, blood perfusion rates, etc. Evaporative heat loss from the skin amounts to about 14% of the total resting heat loss. When the air temperature T_a exceeds about 27°C, blood is diverted from the core of the body to the surface and heat loss to the environment is increased. On the other hand, if T_a falls below this value, the skin vessels constrict and heat is conserved. This vasomotor regulation takes place for T_a in the range of 25–29°C.

For a person at rest, sweating begins at a T_a of about 30°C. When the skin is completely wetted by sweat, insensible perspiration from the skin ceases.

The following subsections discuss the transport phenomena from nude skin and in sauna and steambaths.

4.7.2.1 Nude skin (Chao et al., 1973a). The skin is a composite of morphologically distinct tissues. When studying heat transfer through skin, all living layers of the skin are treated as a homogeneous, isotropic tissue, except for the stratum corneum which is a layer of dead tissue and is included in the skin surface. The coordinate system is fixed on the skin surface with y measuring depth, as shown in Fig. 4.32. The skin (denoted by subscript 1) is above a layer of subcutaneous tissue (subscript 2). T_i is the temperature, while $\dot{V}_i$, q_i''' and m_i''' represent the rates of capillary blood circulation, metabolic heat generation, and water production resulting from metabolic reaction and dehydration, re-

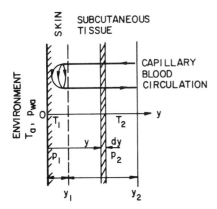

Figure 4.32 Analytic model for nude skin.

spectively, where $i = 1, 2$. Subscripts a and, b refer to the environment and body, respectively. We will look at three examples of heat transfer.

Heat transfer from dry skin in a cool environment, in which insensible perspiration is negligible Equation (4.54) is reduced to

$$(wc_p)_{bi}(T_i^* - T_i) + k_i \frac{d^2T_i}{dy^2} + q_i''' = 0 \tag{4.204}$$

in which $T_1^* = T_2(y_1)$ and $T_2^* = T_b$, the temperature at the deepest subcutaneous tissue. The appropriate boundary conditions are

$$k_1 \frac{dT_1(0)}{dy} = h[T_1(0) - T_a] \tag{4.205a}$$

$$T_1(y_1) = T_2(y_2) \qquad k_1 \frac{dT_1(y_1)}{dy} = k_2 \frac{dT_2(y_1)}{dy} \qquad T_2(y_2) = T_b \tag{4.205b}$$

where h is the heat transfer coefficient for the combined effects of convection and radiation.

Heat transfer accompanied by insensible perspiration The contribution of insensible perspiration becomes important when the environment temperature is between 20 and 29°C for resting clothed humans or between 20 and 31°C for resting nude humans. Equation (4.204) is then coupled with the mass equation governing water migration:

$$D_i \frac{d^2\rho_i}{dy^2} + m_i''' = 0 \tag{4.206}$$

in which D_i is the mass diffusivity of water in the tissues and ρ_i is the density of migrating water in the tissues. At the skin surface, $p_1(0) = \rho_1(0)RT_1(0)/M$, where R and M are gas constants of vapor and molecular weight. The appropriate boundary conditions include Eq. (4.116) with Eq. (4.115a) replaced by

$$k_1 \frac{dT_1(0)}{dy} = h[T_1(0) - T_a] + h_{fg}\dot{m}_1 \tag{4.207}$$

and

$$D_1 \frac{d\rho_1(0)}{dy} = \dot{m}_1 = \frac{h_D M[\rho_1(0) - p_a]}{R\bar{T}_a}$$

$$p_1(0) = p_{1s}[T_1(0)] \tag{4.208}$$

$$D_1 \frac{d\rho_1(y_1)}{dy} = D_2 \frac{d\rho_2(y_1)}{dy} \qquad \rho_1(y_1) = \rho_2(y_2)$$

where $\dot{m}$ is the rate of insensible perspiration, h_{fg} is the latent heat of water (also sweat), h_D is the mass transfer coefficient and p is the saturated vapor pressure corresponding to the skin temperature $T_1(0)$. $\overline{T}_a$ is the absolute temperature of air.

Heat transfer accompanied by sweat secretion (nondrip skin water loss) Sweating is initiated in the average nude human at rest in environmental heat stresses in which the skin attains a critical temperature of 34.5°C. With the skin surface covered by a layer of sweat, heat transfer is governed by Eq. (4.204) subject to the boundary conditions (4.205), except that Eq. (4.205a) is replaced by

$$k_1 \frac{dT_1(0)}{dy} = h(T_m - T_a) + h_{fg}\dot{m}_s + h_{fg}\dot{m}_1 \tag{4.209}$$

Here, $T_m = [T_1(0) + T_a]/2$ is the sweat-air interfacial temperature, m_s is the rate of sweating and p_{1s} is the saturated pressure of water vapor corresponding to T_m.

The above equations and boundary conditions are solved using a digital computer. Numerical results corresponding to the physiological and physical factors listed in Table 4.10 are graphically shown in Figs. 4.33 and 4.34. Figures 4.33a and 4.33b indicate temperature distributions in the skin and subcutaneous tissue in the arm of a resting nude human under ambient temperature T_a of 20 and 30°C, respectively, while Fig. 4.34 shows the corresponding heat and water losses to the environment.

The theory agrees well with the in vivo test data, particularly the skin surface temperature. It is found that as T_a increases in this range, both the skin

Table 4.10 Numerical values of physiological and physical factors

Factor	Skin	Subcutaneous tissue
C_p (cal/cm³ °C)	0.97	0.97
$\dot{w}/\rho$ (cm³/cm³ min)	0.0275	0.0275
q''' (cal/cm³ min)	0.005	0.01
k (cal/cm min °C)	0.03	0.06
D (cm²/min)	0.00402	0.00204
m''' (gm/cm³ min)	0.001	0.002
$h = 0.009$ cal/cm² min °C		
$h_D = 35.7$ cm/min		
$h_{fg} = 579$ cal/g at 25°C		
$R_w = 82.06$ cm³ atm/K		
p_{wa} for 40% relative humidity at T_a		
$U = 0.00497$ cal/cm² min °C		(corresponding to entrapped air layer = 2 mm and Apollo constant wear garment)
$U_D = 19/7$ cm/min		

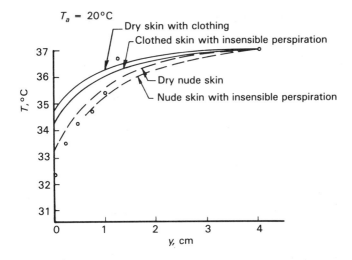

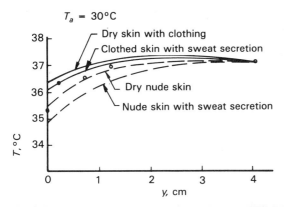

Figure 4.33 Temperature distributions in skin-subcutaneous tissues of forearm. (○) In vivo test data for nude skin.

temperature and the rate of water loss increase, while the rate of heat loss decreases.

4.7.2.2 Sauna and steambaths. The transient response of skin and tissue temperatures in sauna and steambaths is analyzed together with water loss from the skin. Changes in rectal temperature, blood perfusion rate, thermal conductivity, and rates of heat transfer with the ambient air as well as sweat secretion are taken into account.

Consider a nude subject staying in an environment with the air temperature T_a and the specific humidity ϕ. His physiological condition is in a homeostatic

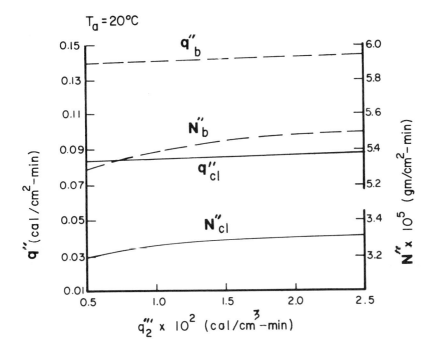

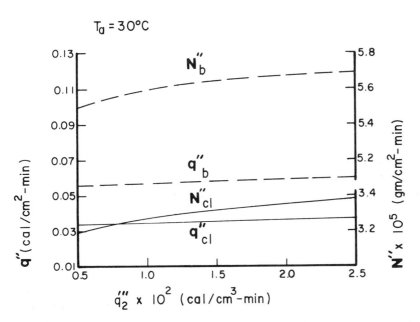

Figure 4.34 Rates of heat and water losses from nude skin surface (subscript *b*) and clothed skin surface (subscript *cl*) against metabolic heat generation.

state with the temperature distribution $T_i(y)$ and the density of the migrating water in the tissue ρ_i. The entering of the subject into a bath constitutes a sudden change in the ambient conditions including T_a and ϕ. The special features of a sauna include a high T_a and very low ϕ, while those of a steambath are lower T_a but very high ϕ. As a result of the sudden changes in T_a and ϕ, unsteady heat and water migration take place in the skin and subcutaneous tissues. Equation (4.54) takes the form of

$$(\dot{w}c_p)_{bi}(T_i^* - T_i) + k_i \frac{\partial^2 T_i}{\partial y^2} + q_i''' = (\rho c_p)_{ti} \frac{\partial T_i}{\partial t} \tag{4.210}$$

and the mass equation reads

$$D_i \frac{\partial^2 \rho_i}{\partial y^2} + m_i''' = \frac{\partial \rho_i}{\partial t} \tag{4.211}$$

Here, t is the time and the subscripts b and t refer to the blood and tissue, respectively. The initial conditions correspond to the distribution of T_i and ρ_i in nude skin. The appropriate boundary conditions are

$$k_1 \frac{\partial T_1(0, t)}{\partial y} = h[T_1(0, t) - T_a] = h_{fg}(\dot{m}_1 - \dot{m}_s)$$

$$T_1(y_1, t) = T_2(y_1, t) \qquad k_1 \frac{\partial T_1(y_1, t)}{\partial y} = k_2 \frac{\partial T_2(y_1, t)}{\partial y}$$

$$T_2(y_2, t) = T_b \tag{4.212}$$

$$D_1 \frac{\partial \rho_1(0, t)}{\partial y} = \dot{m}_1 = \frac{h_D M[p_1(0, t) - p_a]}{R\bar{T}_a}$$

$$p(0, t) = p_{1s}[T_1(0, t)] \qquad D_1 \frac{\partial \rho_1(y_1, t)}{\partial y} = D_2 \frac{\partial \rho_2(y_1, t)}{\partial y}$$

$$\rho_1(y_1, t) = \rho_2(y_1, t) \qquad \dot{m}_s = \frac{0.117(T_y - 36.6)[T_1(0, t) - 33.5]}{L}$$

where T_y is the tympanic temperature.

Numerical computations are performed for the case when a nude subject in a room at 25°C enters a sauna (at 90°C, $\phi = 0\%$) and steambath (at 45°C, $\phi = 100\%$). Results are shown in Fig. 4.35 and 4.36.

High ambient humidity, even at a lower temperature, constitutes a barrier for temperature regulation as seen in Fig. 4.35. The skin and subcutaneous temperatures are higher in the steambath case than those in the sauna case. The difference diminishes with depth, however, It is observed in Fig. 4.36 that the skin surface temperature, T_s, is lower for the sauna than for the steambath. The heat transfer rate, $q'' = -k_1 \partial T_1(0, t)/\partial y$, at the skin surface varies with time in an interesting manner: for both cases, heat is transferred from the ambient into the skin as indicated by a negative value during the first few

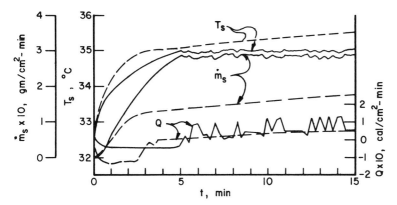

Figure 4.35 Transient skin temperature, heat, and water losses from $T_a = 25°C$ into sauna of 90°C (solid lines) and steambath of 45°C (broken lines).

minutes. Only after T_s becomes steady (in the sauna, T_s continues to fluctuate slightly about the sweat secretion temperature) or quasi-steady (steambath), is the heat flow reversed; that is, it flows from the skin surface to the ambient, as evidenced by a positive sign in q''. In the case of a sauna, the inward heat flow is lower in rate, but lasts longer. After the heat flow has turned outward, its rate fluctuates quite severely and irregularly due to very slight fluctuations in both T_s and m_s. In contrast, the outward heat flow rate in the steambath does not fluctuate since no oscillation in T_s or $\dot{m}_s$ is observed.

4.7.3 Nonclinical Applications

Applications of bioheat transfer can be grouped into two areas: (1) internal or clinical applications in which heat transfer takes place within the human body

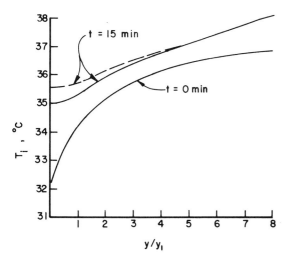

Figure 4.36 Transient temperature profiles in the forearm of a subject from $T_a = 25°C$ into sauna (solid lines) of 90°C and steambath (broken lines) of 45°C.

and (2) external or nonclinical applications in which the hardwares and techniques are developed to control, induce, or observe, by external means, certain thermal states in the human body.

Since humans first appeared on the face of the earth, they have been concerned with protecting themselves from an adverse thermal environment, as evidenced by their search for shelter and use of fire to modify their thermal environment. Civilization progressed with the development of the techniques and hardware to provide a thermal environment most compatible with the thermal requirements of the human body. The climax was reached in early 1970s when American astronauts, clad in space suits, walked on the moon's surface. An excellent monograph edited by Newburgh (1968) covered the human response to the climatic environment and clothing.

In sharp contrast to the great number of publications, covering both theory and in vivo tests available for heat and water losses from nude skins, there are virtually no studies on the transport phenomena from clothed regional skin.

The analytic model for nude skin, case b of Section 4.7.2.1, is extended to the case of clothed subjects (Chao et al., 1973b). A layer of entrapped still air is assumed between the skin surface and a clothing of certain thickness. The modified model describes regional heat and water migration from deep tissues through subcutaneous tissues, skin, unventilated entrapped air layer, and cloth to the environment. Three cases of heat transfer are studied.

Heat transfer from dry skin in a cool environment, in which insensible perspiration is negligible The same energy equations and boundary conditions, Eqs. (4.204) and (4.205), for nude skin can be applied to clothed subjects provided that the heat transfer coefficient h is replaced by the overall heat transfer coefficient between the skin surface and the environment, including contributions due to thermal resistances of the entrapped air layer and the cloth and the combined effects of convection and radiation between the cloth surface and the environment U.

Heat transfer accompanied by insensible perspiration The governing equations and the appropriate boundary conditions are identical with those for nude skin with the replacement of h_D by U_D. U_D is the overall mass transfer coefficient between the skin surface and the environment including contributions due to mass transfer resistances of the entrapped air layer, the cloth and the cloth-environment boundary layer, P_{ws} is the saturated vapor pressure corresponding to the skin temperature $T_i(0)$.

Heat transfer accompanied by sweat secretion (nondrip skin water loss) The governing equations and appropriate boundary conditions are identical with those of the nude skin case provided that h and h_D are replaced by U and U_D. The above equations and boundary conditions are solved using a digital computer for T_a ranging from 20 to 35°C, various air layer thicknesses, and different clothings. The physiological and physical factors corresponding to $T_a = 25°C$

are listed in Table 4.10. Numerical results are graphically illustrated in Figs. 4.33, 4.34, and 4.37.

Figure 4.37 demonstrates the variation of the clothed skin temperature with the ambient temperature at a basal metabolic level. The validity of the analytic model is borne out by good agreement between the theoretical skin temperatures and the corresponding in vivo test data for a nude body. The effect of clothing on the skin temperature, as indicated by the vertical distance between the solid and dotted lines, diminishes as the ambient temperature increases. Figures 4.33a and 4.33b show the temperature distributions in the skin ($0 < y < 0.5$ cm) and subcutaneous tissue ($0.5 < y < 4$ cm) in the arm of a resting man under the ambient temperatures of 20 and 30°C, respectively, while Fig. 4.34 indicates the corresponding heat and water losses to the environment form the nude skin and the cloth surface. Insensible perspiration and sweat secretion take place on nude skin surface in the ambient temperature range of 20–30°C. The theory agrees well with the in vivo test data for nude skin, particularly the skin surface temperature. It is seen in Fig. 4.34 that clothing contributes to substantial decreases in both heat and water losses to the environment: heat loss decreases by approximately 40 and 50%, while water loss is cut by approximately 53 and 70% at a T_a of 20 and 30°C, respectively. Clothing does not change the way the rate of metabolic heat generation in the subcutaneous tissue would affect heat and water losses from nude skin. It is also found, for both cases of nude and clothed skin, that as T_a increases in this range, both skin temperature and the rate of water loss increase, while the rate of heat loss decreases.

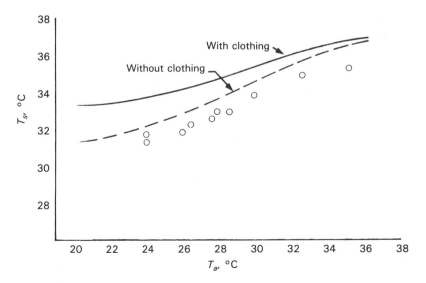

Figure 4.37 Relationship between skin temperatures and ambient temperature. ($\bigcirc$), In vivo test data for nude skin.

For all three cases, the skin temperatures are higher while both the heat and water losses to the environment are lower than the corresponding cases for nude skin. As T_a increases, both the skin temperature and the rate of water loss increases, while the rate of heat loss decreases. The model also predicts the temperature profiles in the skin and subcutaneous tissue layers.

4.8 THERMAL MEASUREMENTS

In studying thermal interaction with the environment, the temperature of the body surface is one of the most important variables controlling heat exchange with the surroundings.

4.8.1 Skin Temperature

4.8.1.1 Probe techniques. Skin temperatures are measured using thermocouples or thermistors. A number of these surface-contact probes can determine the temperature distribution over an area. The complex nature of skin temperature, and its variation with different environments and with exercise, made it difficult for probe techniques to provide detail over the body surface.

For applications where continuous temperature ranges are to be monitored, the choice is limited by practical considerations to three familiar transducers: thermistors, resistance thermometers, and thermocouples. Thermocouples consist of two dissimilar metal wires forming two junctions, namely reference (or cold) and measuring (or hot) junctions. The temperature difference between the two junctions is measured.

Thermistors and resistance thermometers are resistance-temperature devices which provide a direct indication of absolute temperature. Thermistors are essentially semiconductors which behave as thermal resistors, namely resistors with a high (usually negative) temperature coefficient of resistance. In some instances, for example, the resistance of a thermistor at room temperature may decrease by 6% for each degree Celsius increase of temperature. Thermistors operate as either self-heated or externally heated units. When externally heated, they convert changes in ambient or contact temperatures directly into corresponding changes in voltage or current. Self-heated units, on the other hand, utilize the heating effect of the current flowing through them to raise and control their temperature and thus their resistance. Because of their very large change in resistance versus temperature (no other transducers provide such a high degree of resolution or gain), both units are well-suited for precision measurement, control, or compensation of temperature.

Thermistors are composed of a sintered mixture of metallic oxides such as manganese, nickel, cobalt, copper, iron, and uranium. They can be made small in size and provide a high degree of stability over a long life span. Standard configurations include beads, probes, disks, washers, and rods.

4.8.1.2 Thermography. The essential elements of a thermograph are: (1) a lens system for focusing the incoming radiation, (2) an infrared detector which converts the focused radiation into an electrical signal by some physical process such as a thermoelectric effect or a photoconductive effect, depending on the type of detector, (3) an electronic package which processes the electrical signal for imaging or displaying purposes, and (4) an imaging or display unit which exhibits the processed data as the output of the device. The image recorded on a film is called a thermogram. For infrared thermographs in use, temperature resolution typically is on the order of 1°C. Receiver temperature sensitivity in microwave thermographs is on the order of 0.1°C. For the potential of thermography to be fully realized, one needs to develop more rapid and more sensitive instruments with higher resolution and more sophisticated thermal modeling techniques that can establish the relationship between temperature abnormalities recorded on the thermogram and the magnitude and location of the physiological defect.

A noninvasive diagnostic procedure known as thermography consists of making a photographic analog of the thermal state in biomaterials. This temperature map may be examined for patterns indicative of abnormal local heating or cooling. Potential diagnostic applications include detection of cancerous tumors, peripheral vascular disease, incipient strokes, local inflammations, breast cancer, and cerebrovascular blockage.

All objects emit thermal radiation due to the acceleration of their electrical charges by internal thermal motions. Radiation by human tissue in thermal equilibrium may be described by Planck's law for the emitted specific intensity:

$$I(v) = \epsilon(v)2hv^2[c^2(e^{hv/kT} - 1)] \tag{4.213}$$

in which v denotes the frequency; T is the temperature of the radiator; h, c, and k are, respectively, Planck's constant (6.62517×10^{-27} erg s), the speed of light, and Boltzmann's constant (1.3804×10^{-6} erg/K); and ϵ is the emissivity which depends on the dielectric properties of the emitting and receiving medium. For $hv \ll kT$, Eq. (4.213) transforms to the Rayleigh-Jeans expression as

$$I(v) = \frac{\epsilon 2kTv^2}{c^2}$$

which illustrates a linear relationship between the intensity and the temperature of the emitter. The frequency of maximum intensity can be predicted by Wien's displacement law:

$$\max v = 1.04 \times 10^{11} \, T, \text{ Hz}$$

This equation gives 3×10^{13} Hz (approximately 10 μm wavelength) of maximum intensity by the human body at 300 K. This has led to the operation of infrared thermographs at wavelengths near 10 μm. At a typical microwave frequency of 3 GHz (10 cm wavelength), the intensity is reduced by a factor of approximately 10^8 from its maximum, but radiation with this intensity can be easily detected by radiometers. Microwave thermography has been developed to apply this technique of microwave radiometry to the mapping of microwave emission from human tissues (Barrett and Meyers, 1975). Microwave thermography is different from infrared thermography in its coarser resolution and in its ability to sense the temperature of deeper tissue.

The introduction of a measuring probe into live tissue will inevitably cause pooling of blood around it, inflammation, etc., resulting in a distortion of the measured quantity. Therfore, a noninvasive technique of temperature measurements such as thermography is desirable. The functional principle is briefly described below.

All substances at a temperature higher than absolute zero emit electromagnetic waves (i.e., radiation) toward their surroundings and absorb electromagnetic radiation from their surroundings. The intensity I of electromagnetic radiation emitted by the human body at a temperature T is given by the Rayleigh-Jeans expression

$$I = \frac{2ekT}{\lambda^2}$$

where k and λ denote the Boltzmann constant and the wavelength, respectively. e is the emissivity which depends on the complex dielectric constant of the skin and subcutaneous layers. A subcutaneous "hot spot" of ΔT relative to the surrounding tissue temperature results in an increase of ΔI of the reradiation. ΔI is detected externally to construct a temperature-distribution map called the "thermograph." Thermography at centimeter, millimeter, and infrared wavelengths can be utilized for hyperthermic treatment, monitoring of

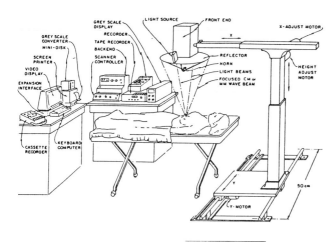

Figure 4.38 Breast scanning using image thermographs at centimeter and millimeter wavelengths.

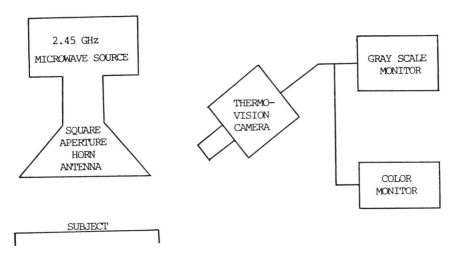

Figure 4.39 Experimental apparatus for thermographic visualization of electromagnetic diathermy.

arthritis, detection of tumors in the brain and the thyroid, and detection of breast cancer.

The centimeter, millimeter, and infrared wave emission from the human body is focused through a lens or large reflector into a horn (Fig. 4.38) (Edrich et al., 1980). Subsequently, the received power is amplified and detected by a highly sensitive radiometer. The scanner is slowly moved in raster fashion over the supine patient. The Data are stored and thereafter a gray-scale image of the temperature distribution is produced. Further data processing, smoothing, and display may be performed by means of a microcomputer with memory, a minidisk and a gray-scale converter (Edrich et al., 1980; Myers et al., 1980). The first medical thermograph was built by Dr. B. Barnes in the 1950s. A review of the relatively brief history of thermographic instruments is available in Baeu (1980).

An experimental apparatus for the thermographic visualization of electromagnetic diathermy is shown in Fig. 4.39. It consists of an electromagnetic source, a subject, and a thermovision camera system. Radiation is emitted from the 2.45-GHz, 400-W source to the subject using an appropriate waveguide and square-aperture horn applicator. The entire arrangement is enclosed in a microwave-absorbing anechoic chamber.

4.8.1.3 Tomographic thermography. The conventional thermography of a steady-state, two-dimensional image suffers from two major shortcomings. First, heat is transferred diffusively from the interior to the skin surface. Thus, the observed thermal image is substantially smeared. Second, considerable information is lost in the conventional thermographic image due to the three dimensionality of the interior structure. Tomographic thermography utilizes a time-dependent set of thermographic skin temperature measurements in an "inverse" bioheat transfer computation to reconstruct the interior perfusion and temperature patterns (Chen et al., 1977; Pantazatos and Chen, 1978; Chen and Panazatos, 1980). This technique needs some refinement in areas such as the computational scheme and noise in infrared detectors.

Infrared thermography has recently been used in a number of investigations, such as the skin temperature distribution at different environmental temperatures and during exercise, the effect of solar radiation on skin temperature, the temperature distributions of infants nursed in incubators, and evaluation of the mean skin temperature. Thermography produces a very detailed pattern of skin temperature at steady state as well as transient state.

4.8.2 Measurements of Boundary Layers over Body Surfaces

When a subject is at rest, either standing or lying, the temperature difference between the skin and the surroundings induces the body to exchange heat by natural convection. If the mean skin temperature is lower than the air temperature, the air adjacent to the skin surface will become heated by thermal diffusion and will rise due to buoyancy. This results in an envelope of air moving upward which surrounds the body. When the body is moving through the air or is exposed to the wind, the body exchanges heat by forced convection, which is affected by the mean air velocity, the nature of the flow, and the flow direction.

4.8.2.1 Probe techniques. The temperature and velocity profiles in the flow around the human body can be measured using thermocouples and hot wire anemometers. However, the complex shape of the body makes analysis difficult.

4.8.2.2 Optical methods. Optical methods including shadowgraphic, Schlieren, interferometric, and holographic techniques can be used in visualizing the convective boundary layer flow around the human body. However, due to its simplicity and good image quality, the Schlieren method is favored. The principle of operation of a typical Schlieren optical system is as follows: The system essentially consists of a parallel beam of light focused on a particular point. If any of the light rays in the parallel beam are bent by passing through the air with a different refractive index (due to a temperature difference), the focus of the beam is displaced from its original position. By arranging the undeviated beam so that it is focused through a colored filter and by letting the displaced beam be focused through a different color, it is possible to visualize heated air streams as one color against a background of another. Figure 4.40 illustrates a schematic of the Schlieren optical system employed to visualize the boundary layer flow over the whole body.

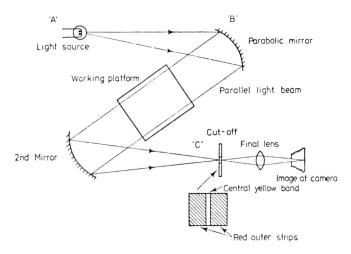

'A'
Light source

Working platform

'B'
Parabolic mirror

Parallel light beam

2nd Mirror

Cut-off
'C'
Final lens

Image at camera
Central yellow band

Red outer strips

Figure 4.40 Diagram of Schlieren optical system.

4.8.3 Thermal Property Measurements

Information about the thermal properties of biomaterials, including thermal conductivity, thermal diffusivity, specific heat, and metabolic heat generation, is becoming increasingly important for diagnostic and therapeutic medicine as well as for the understanding of general physiology. Thermal conductivity data, which are well characterized with respect to tissue type, temperature, and composition, are limited, while thermal diffusivity data are even more scarce. With thermal conductivity k, thermal diffusivity α, and density available, specific heat can be determined from the definition of α.

Thermistors are uniquely qualified as transducers in determining k because of (1) the high sensitivity to small variation in their own temperature and (2) their ability to operate in a self-heated mode. When a current of sufficient magnitude is passed through a thermistor, its temperature increases until the power in the circuit is balanced by the heat dissipated from the thermistor. When used in thermal conductivity measurements, two thermistors are normally connected in the adjacent legs of a Wheatstone bridge with the voltage applied at their junction. One thermistor is mounted in a static area to provide for temperature compensation while the other, called the sensing thermistor, is placed in the medium to be measured. Any difference in the thermal conductivity of this medium from that of the static area will change the rate at which heat is dissipated from the sensing thermistor, thus changing its temperature. This results in a bridge imbalance which can be calibrated in appropriate units. Valvano et al. (1981) combined the use of a self-heated spherical thermistor probe and the closed-form solution of the time-dependent probe-tissue coupled thermal model for the in vivo determination of thermal conductivity, thermal diffusivity, and perfusion of biomaterials. Earlier, Bowman et al. (1980) developed a technique capable of accurately characterizing the thermal properties of both normal and diseased tissue.

Spherical beads and cylindrical wires with pulse or step heat input (transient technique) are used in measuring the thermal properties of tumors as well as normal tissues (Chato, 1980; Chen and Rupinskas, 1977; Balasubramanian and Bowman, 1977). A comprehensive review of this subject was given by Bowman et al. (1980).

Measurement of the local metabolic heat generation rate has not been as successful. Techniques have been limited to direct calorimetry on small sample preparations or have been inferred from local measurement of oxygen consumption (Gibbs, 1978). Neglecting the conduction terms. Eq. (4.54) can be readily solved to yield an explicit measure of the local metabolic heat generation and perfusion rates. Eberhart et al. (1980) reviewed the heat clearance/bioheat transfer equation methods to determine the local heat generation and perfusion rates.

4.9 HYPOTHERMIA, HYPERTHERMIA, AND THERMAL LESION

4.9.1 Hypothermia

Hypothermia from a medical viewpoint refers to the reduction of body temperature, locally or generally, below normal in nonhibernating homeothermic animals. The application of hypothermia may be systemic or local and may vary in degree and duration. Its most important use is in prolonging the safe period of circulatory arrest so that intracardiac and neurological surgical procedures may be performed. It also helps alter the course of disease states varying from febrile illness to gastrointestinal hemorrhage to anoxic brain damage. The fundamentals in hypothermia are well documented, for example, in Newburgh (1968), Hardy (1963), and Lunardini (1981). Thermal modeling is well known and thus is not repeated here. Instead, subzero hypothermia, namely cryobiology and cryosurgery, is discussed.

The central aim of cryobiology is the successful preservation and storage of biomaterials, organs, and tissues, by freezing. In contrast, the objective in cryosurgery is to produce localized necrosis in tissues or organs. Research in cryobiology has concentrated on studies of the effects of temperatures on the in vitro behavior of cell suspensions (see Section 4.5). Despite considerable effort, success is still limited to relatively few biomaterials such as blood, sperm, skin, cornea, and certain tissue cultures. In cryosurgery, the biomaterial must be cooled at a rate to kill cells. However, progress in cryosurgery is hampered by complexity in the geometry of the material and different survival signatures of several cell types in the medium.

When a cryosurgical probe is inserted into a tissue and suddenly brought to a cryogenic temperature, a transient thermal field develops in the medium. An injury is produced when the tissue in the vicinity of the probe undergoes a phase change (i.e., freezing). The freezing front propagates through the tissue until a steady-state condition prevails. The phase interface at the steady state determines the maximum lesion size for a probe at a particular surface temperature. There are many factors affecting the rate of growth and the maximum lesion size, such as probe geometry and size, probe surface temperature, thermal properties of both the frozen (subscript 1) and unfrozen (subscript 2) tissues, metabolic heat production rate and blood perfusion rate in the unfrozen tissue, initial tissue temperature, and freezing temperature T_f.

Both hemispheric and cylindrical (assumed infinitely long to be one-dimensional geometry) cryoprobes may be employed to produce cryogenic lesions. It has been disclosed that the temprature field in the region directly below a hemispherical probe is approximately the same as found in a similar region around a sphere of the same size and temperature (Cooper, 1970). For both hemispheric and cylindrical probes, steady-state equations similar to Eq. (4.113) apply. They are subject to the following boundary conditions:

1. The temperature of the frozen tissue T_1, at the probe surface r_0, equals a specified probe surface temperature T_s.
2. Continuity of both the temperature and the heat flux exists at the phase interface r_{12}.
3. The unfrozen tissue temperature T_2 at a large distance from the probe surface remains unchanged at T_0. The steady-state temperature distributions and the location of the frozen front are obtained for both probes as (Cooper, 1970).

	Spherical cryoprobe	Cylindrical cryoprobe
Frozen tissue Θ_1	$1 + (\Theta_f - 1)r_{12}^* \dfrac{(1 - r^*)}{r^*}$	$(\Theta_f - 1)\dfrac{\ln r^*}{\ln r_{12}^*} + 1$
Unfrozen tissue Θ_2	$\dfrac{\Theta_f r_{12}^*}{r^*} \exp\left[-b(r^* - r_{12}^*)\right]$	$\dfrac{\Theta_f K_0(br^*)}{K_0(br_{12}^*)}$
Location of phase front r_{12}^*	$\dfrac{b-1}{2b} + \left[\dfrac{(1-b)^2}{4b^2} + \dfrac{1+a}{b}\right]^{1/2}$	$a = br_{12}^* \ln r_{12}^* \dfrac{K_1(br_{12}^*)}{K_a(br_{12}^*)}$

where

$$\Theta_1 = \frac{T_f - T_0}{T_s - T_0} \qquad b^2 = \frac{m_B C_{pB} r_0^2}{k_2} \qquad \Theta_2 = \frac{T_2 - T_0}{T_s - T_0} \qquad a = \frac{k_1(T_f - T_s)}{k_2(T_f - T_0)}$$

$$\Theta_f = \frac{T_f - T_0}{T_s - T_0} \qquad *r = \frac{r}{r_0} \qquad r_{12}^* = \frac{r_{12}}{r_0} \qquad T_0 = T_B + \frac{q_m}{m_B C_{pB}}$$

T_B denotes the arterial systemic blood temperature; K_0 and K_1 are modified Bessel functions. The quantity b^2 may be written in the form $(m_B C_{pB} r_0)/(k_2/r_0)$. The term k_2/r_0 represents conductance through the tissue, while the term $m_B C_{pB} r_0$

signifies convective effects due to blood flow. Thus, when $b \ll 1$, the conductive mode of heat transfer dominates blood flow effects, while $b \gg 1$ represents the opposite situation. Figures 4.41 and 4.43 predict lesion sizes for the spherical and cylindrical cryoprobes, respectively.

In addition to the ultimate lesion size, knowledge of the rate of lesion growth may be used by the cryosurgeon to estimate the amount of time required to reach a steady-state condition. Equations similar to (4.54) with the transient terms are solved numerically with the latent heat of fusion (L) condition incorporated in the boundary conditions. A typical example showing the rate of lesion growth is shown in Fig. 4.43 for the spherical probe.

Clinical applications of cryosurgery have been extended to such areas as cryoimmunology, neurosurgery, ophthalmology, otology, gynecology, and dermatology. There are numerous publications on cryobiology and cryosurgery including Meryman (1966), Smith (1970), Wolstenholme and O'Connor (1970), Rand et al. (1968), Zacarian (1968), Asahina (1966), Cooper (1970), von Leden and Cahan (1971), Holden (1975), and Ablin (1980).

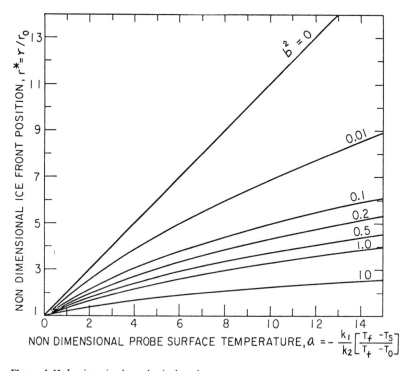

Figure 4.41 Lesion size by spherical probe.

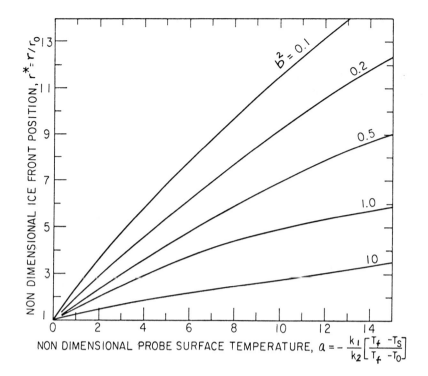

Figure 4.42 Cylindrical probe lesion size.

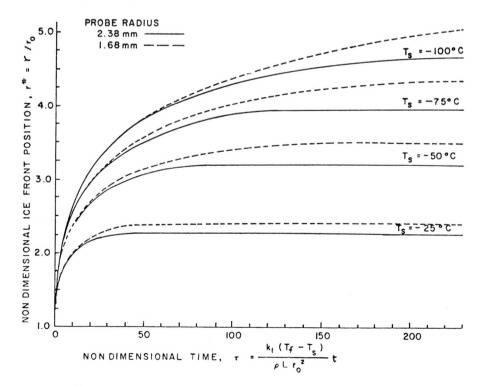

Figure 4.43 Position of ice front with time for spherical probes.

4.9.2 Hyperthermia

Two types of hyperthermia can be administered: systemic and localized. The history of systemic hyperthermia in humans was summarized by Miller et al. (1977). New interest has been stimulated in the use of localized hyperthermia as a potential anticancer agent. Laboratory reports (Miller et al., 1977 and Sugahara & Abe, 1984) indicate that (1) tumor cells may be more sensitive to heat than normal tissue; (2) hyperthermia enhances response to irritation and can increase the therapeutic ratio; (3) cells are most sensitive to hyperthermia during the S phase, a time when they are resistant to ionizing radiation; (4) the oxygen effect is absent for hyperthermic cell killing, and radiation effects are less oxygen-dependent when potentiated by heat treatment; and (5) biological damage occurs more rapidly at temperatures above 43°C.

Localized hyperthermia may be produced through a number of heating techniques including microwave, ultrasound, convection (including regional perfusion, fluid immersion, and irrigation), and radio frequency (100–1000 kHz) current fields. The use of localized hyperthermia for treatment of malignant disease in humans depends on the feasibility of producing homogeneous heating in selected volumes within the body. Ideal heating would result in a uniformly elevated temperature within the volume of interest with essentially no heating outside it. Only microwave and ultrasonic techniques are most suitable for heating deep underlying tissues. The former is more suitable for heating tissues with high water content, particularly muscles, while the latter can easily heat up bones and nerves but not muscles. Being a kind of electromagnetic radiation phenomenon, the microwave technique is more effective in exothermic function and less damaging to nerve transmission function than the ultrasonic wave heating which is produced by longitudinal mechanical vibration.

Before hyperthermia can be used widely in humans, it is necessary to know normal tissue tolerances of hyperthermia (biological aspects) and thermal distribution and dose in normal tissues during the production of localized hyperthermia (physical aspects). Figure 4.44 summarizes the physical considerations of hyperthermia in humans including the methods of heating, hyperthermia dosimetry, and temperature measurement and control.

Some studies have been conducted on ultrasonic diathermy, see for example Lehmann et al. (1959), Schwan (1957), and Schwan et al. (1953).

The evolution of therapeutic heating with electromagnetic waves is well documented (Guy et al., 1974). Johnson and Guy (1972) present an extensive survey of the literature pertinent to the biological effects of electromagnetic waves. Miller et al. (1977), Connor et al. (1977), and Lou et al. (1988) summarize the current progress toward the application of hyperthermia to humans for cancer therapy. Earlier works directed contributed greatly to recent developments in the use of microwave diathermy for hyperthermia: measurements of the dielectric properties of living tissues and absorption characteristics of microwaves in the body over a frequency range from dc to 30 GHz by Schwan; in vivo tests on the interaction of electromagnetic waves with biological systems by Lehmann et al. (1965; 1969); and development of idalized models to determine the intensity of electromagnetic fields in biological tissues by Johnson and Guy (1972), Guy et al. (1974), and Guy (1975).

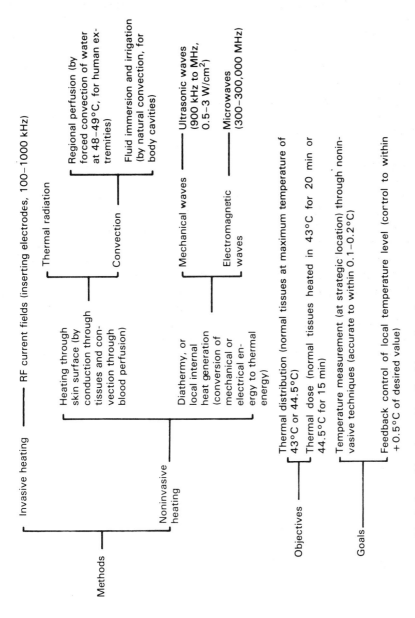

Figure 4.44 Physical considerations of localized hyperthermia.

Microwaves in the frequency range of 27 MHz through 10 GHz have special biological significance since they can be readily transmitted through, absorbed by, and reflected by biological tissues in varying degrees, depending on tissue properties and frequency. The frequencies of 27.12, 40.68, 433, 915, 2450, and 5800 are significant since they are used for industrial, scientific, and medical heating processes. The frequencies of 27.12, 915, and 2450 are used for diathermy purposes in the United States, whereas only 433 MHz is authorized in Europe for these purposes.

The physical model, a multilayered human tissue, consists of skin (denoted by subscript 1), fat (subscript 2), and muscle (subscript 3), as depicted in Fig. 4.45. Plane waves of electromagnetic or acoustic radiation are incident normally on the skin surface. The coordinate system is fixed on the skin surface with y measuring depth. y_1, y_2, and y_3 indicate the depth of each of the triple tissue combinations.

The physical phenomena for heat development in the case of electromagnetic and acoustic radiation are completely analogous (Schwan et al., 1953). All physical, electrical and acoustic prperties of the tissues are considered constant, including density ρ_i, specific heat c_i, thermal conductivity k_i, electrical or acoustic conductivity σ_i, attenuation constant α_i, phase constant β_i, reflection constant r_i, reflection phase coefficient ϕ_i, and velocity of sound u_i. However, the physiological factors such as blood perfusion rate m_{bi}, and metabolic heat generation rate q_{mi}, vary with the tissue temperature T_i, where the subscript b refers to the blood and $i = 1, 2, 3$. It is assumed that the blood enters each tissue at the temperature of location y_i and leaves at the local tissue temperature.

The absorbed power density in each layer of the triple tissue combination can be expressed as (Schwan et al., 1953)

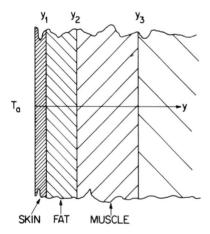

SKIN   FAT   MUSCLE

Figure 4.45 Schematic of skin-fat-muscle combination.

$$q_{ri}''' = \frac{\sigma_i E_{oi}^2}{2} [\exp(-2\alpha_i z_i) + r_i^2 \exp(2\alpha_i z_i) + 2r_i \cos(2\beta_i z_i + \phi_i)] \quad (4.214)$$

E_{oi} signifies the electric field strength at the start of each tissue layer in the electromagnetic case, or excess pressure in the acoustic case. It can be expressed as

$$E_{oi} = E_{oi-1} \exp[\alpha_i(y_i - y_{i-1})] \quad (4.215)$$

For skin, $y_{i-1} = y_0 = 0$, and E_{oi-1} corresponds to the effective incident wave, which is equal to $(2\rho u I)^{1/2}$ in the acoustic case. I denotes the power irradiated on the unit skin surface area while ρ and u are the density and sonic velocity of the coupling medium. z_i is the distance to location y_i measured from the interface with the succeeding tissue layer. The complex reflection coefficient (at the interface with the succeeding layer) due to wave transmitted from one medium to another and thickness greater than depth of penetration can be written as $r_i \exp(j\phi_i)$, where $j = (-1)^{1/2}$. The propagation constant for power transmission through biological tissues can be expressed in terms of $\beta_i - j\alpha_i$, where the wave frequency ω is equal to $\beta_i/2\pi$, and $1/2\alpha$ corresponds to the depth of penetration. Schwan et al. (1953) found that only roughly 2% of the incident energy is reflected at a tissue interface in the acoustic case. Hence, the m_i''' terms in Eq. (4.54) are neglected. The specific conductance is given in the electromagnetic case by

$$\sigma_i = \frac{2\alpha_i \beta_i}{\mu_i \omega} \quad (4.216a)$$

and in the acoustic case by

$$\sigma_i = \frac{2\alpha_i \beta_i}{\rho_i \omega} \quad (4.216b)$$

where μ_i denotes the permeability.

The energy balance equations similar to Eq. (4.54), with $T_b = T_i(y_i, 0)$, are subject to the boundary conditions: continuity of temperatures of heat fluxes. The appropriate initial condition is determined by solving the steady-state equations in the absence of the diathermy heating subject to the boundary conditions [Eq. (205)].

It is known that for vasculated tissues such as skin and muscle, a marked increase in blood flow will occur when the temperature passes through the range of 42–44°C due to vasodilation. As a result of blood cooling, the temperature will drop even though electromagnetic or ultrasonic heating is continued. A semiempirical equation has been developed to simulate changes in the blood flow rate due to vasodilation as the tissue temperature is raised.

Figures 4.46 and 4.47 compare the time history of tissue heated by electromagnetic and ultrasonic diathermy. Tables 4.11–4.13 (Johnson and Guy, 1972; Schwan et al., 1953), with $T_a = 23°C$, $T_b = 36.2°C$, and a skin thickness of 0.2 cm were used in the calculations. It is observed in Fig. 4.46 that when

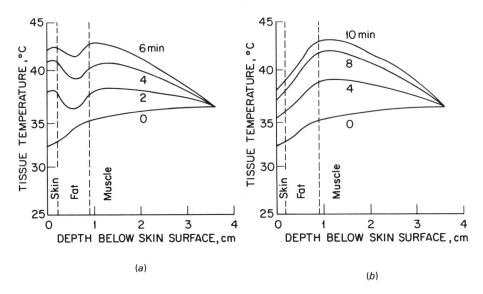

Figure 4.46 Comparison of electromagnetic and ultrasonic diathermy for heating a three-tissue combination with 0.7-cm thick fatty layer. (*a*) Electromagnetic diathermy with P/A = 0.18 mW/cm² at 915 MHz; (*b*) ultrasonic diathermy with P/A = 0.7 W/cm² at 1 MHz.

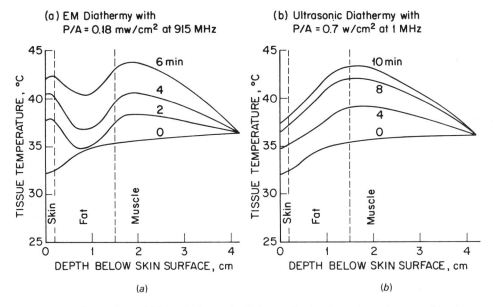

Figure 4.47 Comparison of EM and ultrasonic diathermy for heating a three-tissue combination with fatty layer of 1.3 cm thickness. (*a*) Electromagnetic diathermy with P/A = 0.18 mW/cm² at 915 MHz; (*b*) ultrasonic diathermy with P/A = 0.7 W/cm² at 1 MHz.

Table 4.11 Properties of electromagnetic waves in biological media

Frequency, MHz	Wavelength in air, cm	Relative dielectric constant c I[a]		Conductivity mho/cm I		Wavelength λ, cm I		Depth of penetration ½α, cm I	
		I[a]	II	I	II	I	II	I	II
27.12	1106	113	20	0.612	10.9–43.2	68.1	241	14.3	159
100	300	71.7	7.45	0.889	19.1–75.9	27	106	6.66	60.4
300	100	54	5.7	1.37	31.6–107	11.9	41	3.89	32.1
433	69.3	53	5.6	1.43	37.9–118	8.76	28.8	3.57	26.2
915	32.8	51	5.6	1.60	55.6–147	4.46	13.7	3.04	17.7
2450	12.2	47	5.5	2.21	96.4–213	1.76	5.21	1.70	11.2

Reflection coefficient

Frequency, MHz	Air-I interface r	φ	I-II interface r	φ	Air-II interface r	φ	II-I interface r	ψ
27.12	0.925	+177	0.651	−11.13	0.660	+174	0.651	+169
100	0.881	+175	0.650	−7.96	0.511	+168	0.650	+172
300	0.825	+175	0.592	−8.14	0.438	+169	0.592	+172
433	0.803	+175	0.562	−7.06	0.427	+170	0.562	+173
915	0.772	+177	0.519	−4.32	0.417	+173	0.519	+176
2450	0.754	+177	0.500	−3.88	0.406	+176	0.500	+176

[a] I = skin, muscle, and tissues of high water content; II = fat, bone, and tissues of low water content.
Sources: Johnson & Guy, 1972; Schwan et al., 1953.

Table 4.12 Physical properties of human tissues and blood

	Density ρ, g/cm^3	Specific heat c, mW s/g °C	Thermal conductivity k, mW/cm °C
Skin, muscle	1.07	3471	4.42
Fat	0.937	3258	2.1
Blood	1.06	3889	6.42

Sources: Johnson & Guy, 1972; Schwan et al., 1953.

radiation is applied to a triple tissue combination having a thin fat layer, ultrasonic diathermy produces a maximum temperature at a location deeper from the skin surface than the electromagnetic case. On the contrary, when the fatty layer is thick, then electromagnetic diathermy becomes more effective than ultrasonic diathermy in deep tissue heating, as shown in Fig. 4.47. In both cases, the peak temperature is observed in the musculature near the fat-muscle interface. However, there is a distinct difference in the temperature distributions produced by the two heating techniques. The microwave diathermy is characterized by a minimum temperature in the fat tissue and two peak temperatures, one each in the skin and muscle layers. The differnce between the peak and minimum temperatures grows with an increase in the thickness of the subcutaneous fat tissue. On the other hand, ultrasonic and shortwave heating modalitites produce temperature profiles in the triple tissue configuration with a peak temperature in the musculature and a minimum value at the skin surface.

In conclusion, (1) the temperature distribution from ultrasonic and shortwave heating is characterized by one peak, while that from microwave diathermy has two peaks and one minimum; (2) an increase in power density accelerates the tissue temperature rise; (3) when frequency is increased, the peak temperature in the musculature shifts toward the fat-muscle interface; (4) an

Table 4.13 Physiological properties of human tissues

	Basal blood flow rate, g/s/cm^3 of tissue	Basal metabolic rate, mW/g of tissue
Skin	0.00242	1.0
Fat	0	0
Tissue	0.0005104	0.7

Sources: Johnson & Guy, 1972; Schwan et al., 1953.

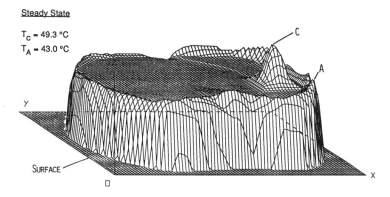

Figure 4.48 Three-dimensional display of temperature distribution with $T_s = 20.0°C$.

increase in the heat transfer coefficient slows down the tissue temperature rise in the skin and part of the fat layer but does not affect the temperature response of deep tissues; (5) ultrasonic diathermy is better than electromagnetic diathermy when the fat layer in a three-layer combination is thin, while electromagnetic diathermy becomes more effective for heating the three-layer tissue with a thick fatty layer. The electromagnetic modality using shortwave produces

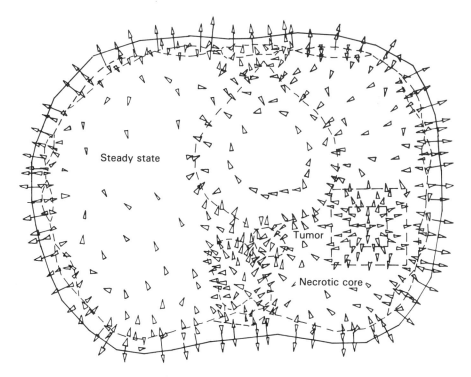

Figure 4.49 Distribution of heat flux vectors with $T_s = 20.0°C$.

very slow but deep tissue heating with a broad temperature plateau in the musculature.

With advances in computational techniques and computer graphics, studies on hyperthermia (e.g., Yang, 1986; Lou et al., 1987; Strohbehn and Roemer, 1984; Paulsen et al., 1985) have been directed toward two-dimensional analysis of complex geometry with a three-dimensional display of temperature distribution. The location of hot spots under radiative heating is thus revealed. For example, Lou et al. (1987) developed a finite element model to analyze the feasibility of the treatment of a deep-seated thoracic tumor by hyperthermia with an annular phased array system. Figure 4.48 is a three-dimensional display of temperature distribution in the cross section of the thorax. C is located in the necrotic center while A is a hot spot in the chest wall. The corresponding distribution of heat flux vectors is depicted in Fig. 4.49.

4.9.3 Thermal Lesion

The pathology of burn trauma and the definition of degrees of burn are briefly summarized in Chao et al. (1977) and Chao (1975). Because of the destruction of the epidermis that encloses the interstitial fluid of the human body at a subatmospheric pressure, the interstitial fluid loses its balance in pressure at the burn site. This results in a substantial increase in outward fluid drainage, as stated in the preceding section. The unburnt skin of burnt patients gives off as much water as it does in a healthy person as normal insensible perspiration (Lamke and Liljedahl, 1971; Lamke and Wedin, 1971). No decrease in cutaneous perspiration is observed. Consequently, the unburnt skin of a burn injury does not compensate for the increased evaporation from the burnt surfaces. These physiological and pathological effects of a burn injury suggest that a change in mass diffusivity can best represent the fluid regulatory function at the burn site of a burn injury. This reasoning enables the mathematical formulation of burnt skin to be carried out in the same manner as normal skin and heat and mass transfer at neighboring burnt and unburnt skins can be calculated independently. Since there is no precise knowledge of the thermal regulatory function inside the tissues underneath the bunrt skin, it is necessary to deduce from knowledge of the regulatory function of burnt skins a picture of the regulatory function of burnt skins at a burn injury.

Following from the preceding argument, it is reasonable to assume that the same mass transfer gradient be maintained through the burnt site as that of the unburnt site, but with a changing mass diffusivity to account for the rehabilitation of burn injury through the formation of eschar, the growth of new tissues and the recovery of local regulatory functions.

4.9.3.1 Heat and mass transfer at partial thickness skin burn.

Superficial partial thickness of first degree burn In this case, the injured skin is approximately one-fifth of the skin thickness. Since there are no heat and mass debts at any instance, the unburnt skin of a burn patient can still maintain its steady-state condition. At the burn site, the mass diffusivity is a function of time, which serves as an indicator of the degree of healing, but there is no heat debt involved. The overall transfer phenomenon is at a quasi-steady state and can be described by steady-state equations instantaneously for a particular value of mass diffusivity D_1, so that the steady state at the unburnt site is not disturbed. Since burn patients are always treated within the temperature range at which there is only insensible perspiration, the governing equations are Eqs. (4.207) and (4.209). With boundary conditions defined at $y_0 = y_1/5$, the burnt skin surface, instead of $y_0 = 0$ in the small forms for normal skin case (see Section 4.7.2.1). $D_1 = D_{11} + D_{12}$, where D_{11} denotes the mass diffusivity of the normal skin and D_{12} is the increase of mass diffusivity induced by burn. M in Eq. (4.210) is replaced by $\dot{m}_1$, while $\dot{m}$ in Eq. (4.211) becomes $\dot{m}_1 + \dot{m}_2$, in which $\dot{m}_2 = D_{12}d\rho_1(y_0)/dy$.

The quantity $\dot{m}_2$ actually represents the exudate which does not take part in cooling the skin because it is required in the regulatory process. The presence of exudate forms a thin film over the burnt surface and becomes a heat dissipating barrier. This results in a higher skin temperature at the burn site. The quantity D_{12} is extrapolated from experimental results and is found to vary with time exponentially in the form of $(A + B) \exp(-ct)$. Like the case of normal skin, $T_1(y_0)$ is obtained through an iterative procedure.

Deep partial thickness burn or second degree burn The formulation for this case is exactly the same as in the preceding case except the boundary conditions at the burn site are moved further inward. The average thickness of burnt skin is three-fifths of the skin thickness. The governing equations in case (1) remain unchanged. The boundary conditions also remain the same except $y_0 = 3y_1/5$.

4.9.3.2 Heat and mass transfer at whole thickness skin burn or third degree burn. In third degree burns, the skin is destroyed completely. Eventually a layer of tissue grows back beyond the subcutaneous tissue. This may not be the same as the original skin tissue physiologically and histologically, but serves the same purpose as the original skin tissue physically. So we will assume that there exists an infinitesimal layer of skin. By doing so, we are allowed to keep the same form of the differential equations which describe the transfer phenomena inside the skin region, except now $y_0 = y_1$.

In Fig. 4.50, the water loss for a 20-day period in normal skin (N_n''), first (N_1''), second (N_2''), and third (N_3'') degree burns are shown and compared with experimental results (Lamke and Liljedahl, 1971). First and second degree burns give off more water than the third degree burns during the first few days. This results from the fact the the local circulation is better established in the former than the latter (Birch et al., 1968). Consequently, the underlying tissue is perfused with a larger amount of fluid. In the third degree burn, the eschar causes certain diffusion resistance. The results for first and second degree burns are mainly for the burnt surface with intact blisters which are only about twice

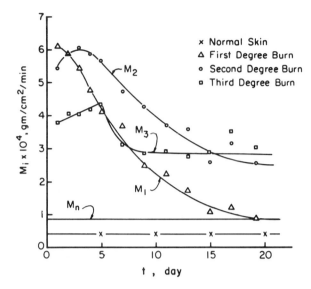

Figure 4.50 Water losses from various degrees of burn and normal skins.

the water loss of normal skin. This does not cause any major concern as long as they remain intact.

Figure 4.51 shows the changes in mass diffusivities at burn sites with reference to that of normal tissue. When the injured tissues resume their proper function, the fluid leakage is reduced during the process of healing. This is manifested clearly by the curve. It is observed that the time it takes for the mass diffusivity to recover to its normal value is less than that for the excessive water loss to be reduced to normal conditions. This is because the surface temperature at the burn site is higher than for the unburnt skin (Fig. 4.52). This results from the insulation of exudate over the burnt surface. When the mass diffusivities return to normal, the excessive fluid loss is terminated, the healing

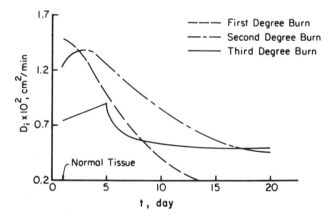

Figure 4.51 Changes in mass diffusivities (D_i) at burn sites.

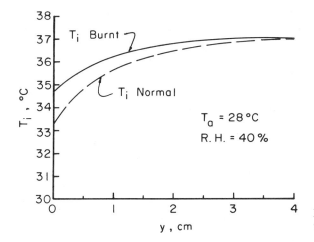

Figure 4.52 Temperature profile in the forearm at burn site.

surface becomes dry, and the temperature begins to drop to normal values. This physiological change causes a time lag in the M_i curves.

The metabolic rates remain at a rather constant value (Davis et al., 1970) since the increase in metabolism cannot significantly compensate for the water loss. Blood flow plays a passive role in burn injury because of the vasoconstriction immediately following burn injury as a natural response to reduce fluid loss. The exact picture of capillary circulation in burnt skin is not known. It is assumed that the local blood flow is maintained at its basal value as soon as the acute period is over.

When a dressing is applied, evaporative water loss is intensively reduced. The amount depends on the physical properties of the dressing and the thickness applied over the burnt surface. The same formulation applies except that the value of h between the skin surface and the environment would have to include the conductive resistance due to the dressing. The only difficulty in obtaining numerical results at this time is the lack of information about the physical properties of the dressings, such as thermal conductivity and mass diffusivity of water in the dressing.

4.10 HEAT TRANSFER EQUIPMENT

The most intimate events occurring to heat and mass transfer researchers are the heat and mass transfer phenomena that take place in their own bodies continuously. For example, the balances of thermal energy and mass have to be maintained in all living organisms. Nevertheless, research in bioheat and mass transfer has not been given adequate attention by most physical scientists and engineers because of its irrelevance to their occupation. The task has then fallen into the hands of medical and biological researchers who, in most cases, have not received proper education and training in heat and mass transfer for both the understanding of basic principles and the application of equipment. Developments in this area are still hampered by two main problems: the lack of satisfactory noninvasive measuring techniques and the lack of accurate information about in vivo thermophysical properties. Both problems are directly related to the lack of heat and mass transfer equipment specifically designed for proper medical applications. Many applications of bioheat and mass transfer are nevertheless emerging. This area is now ripe for new approaches and the introduction of more modern equipment and techniques.

Human beings and other warm-blooded animals have inherent self-regulating mechanisms to maintain their body temperature at a desired level. However, in clinical practice, patients are often and purposely either cooled below or heated above their normal body temperature. The process of cooling or heating a subject is called hypothermia or hyperthermia, respectively. This heating of cooling can be localized or extended to the entire vascular system. The latter case is referred to as systemic hypothermia or hyperthermia.

4.10.1 Hypothermia

Hypothermia has been introduced into surgical practice primarily to protect tissues from the effect of temporary deprivation of their circulation (Cooper and Ross, 1960). A number of important metabolic changes occur when the body temperature of human beings is lowered. These metabolic changes include: (1) acid-base balance; (2) water and electrolytes; (3) changes in renal function; (4) interference with carbohydrate metabolism; and (5) citrate intoxication.

Induction of hypothermia can be achieved by: (1) surface cooling by immersion in cold water at 10°C (then rewarming by blankets perfused with warm water at 40°C back to normal body temperature); (2) bloodstream cooling with a heart-lung circuit or heat exchanger (plastic blood cooling coil in refrigerant); (3) assisted circulation—deep hypothermia achieved with the aid of a heart-lung circuit (shell-and-tube heat exchanger); (4) using a cryoprobe for localized cerebral hypothermia, severe cerebral lesions, and cryosurgery; (5) by a cooling probe or balloon for cavity cooling.

4.10.1.1 Blood heat exchangers. A blood heat exchanger is used to shorten the time normally required to cool a patient before open-heart surgery and to rewarm the patient after surgery. Figure 1.8 shows a schematic oxygenator circuit used in heart and lung bypass surgery. The device can also be used in systemic hyperthermia. This can be accomplished by body surface heating which involves immersing a patient in a heated wax bath. The first commercial blood heat exchanger was developed in 1957. Before the discovery of this invention, the body temperature of a patient was lowered by body surface cooling using either a refrigerated blanket or an ice pack, risky processes that required 1–2 h of anesthesia before the operation could begin. The use of a blood heat exchanger can reduce the time for cooling and rewarming to a matter of minutes and with more precise controlling of the temperature level. Today, it is a standard practice to use the blood heat exchanger for open heart surgery and is employed on all heart-lung machines in the world.

The exchangers are mostly of the shell-and-tube type and are made of various materials such as stainless steel or plastic. They can be either reused or built into disposable blood oxygenators. Furthermore, their use has been expanded to deep hypothermia for special types of surgery and to thermally enhance the anticancer effects of chemotherapy agents used to perfuse isolatable regions of the body.

4.10.1.2 Fluid-cooled garments. This type of clothing functions on the principle of body surface cooling. Its use was well publicized, as see in the body temperature regulation of astronauts in space flight. In clinical practice, fluid-cooled garments have been used for rewarming patients following deep hypothermia (Cooper and Ross, 1960). Their superiority over other means of body surface cooling is in the easy and rapid control of fluid temperature and flow velocity, through which body surface temperature is maintained.

4.10.1.3 Cryoprobes. A cryoprobe, used in cryosurgery or cryotherapy, can be used with excellent results in the specialities of ophthalmology, gynecology, otolaryngology, proctology, podiatry, and veterinary medicine (Pasquale, 1981). It is occasionally used in local hypothermia. Human tissue freezes at $-2.2°C$ with tissue destruction usually occurring between -10 and $-20°C$. The tip temperature of the cryoprobe is between -70 and $-89°C$, depending on the refrigerant being employed. An ice ball is formed at the probe tip following administration of a refrigerant. The outer edge of the ice ball is 0°C and is nonlethal to tissues. The lethal zone is usually 2 to 3.5 mm within the outer edge of the ice ball depending upon the type and location of the target tissue to be frozen.

Refrigerants may be solid (CO_2 snow), liquid (liquid nitrogen or LN_2, freons, N_2O), or compressed gas (CO_2-NO_2 mixtures). They are utilized either in closed probes or sprayed through an open probe (in liquid or snow form) directly to the tissue. A closed probe system consists of a concentric cylinder connected to a probe. The refrigerant enters the probe through the inner tube and exits through the annular space after absorbing heat from the tissue through the probe wall. An open probe is like a hypodermic needle. The more sophisticated and practical probe systems use compressed gases such as a CO_2-CO_2 mixture at a regulated pressure. They utilize the Joule-Thompson effect which induces a temperature drop by means of a rapid expansion of a compressed gas (CO_2-NO_2) through a delivery tube in a hollow expansion chamber. The expanded gas then flows back over the delivery tube and exits into the atmosphere. Defrosting is accomplished by interrupting the flow of exhaust gas and thus preventing gas from expanding in the expansion chamber. The probe tip defrosts and restores its initial temperature within 1–4 s depending on the size of the expansion chamber. This defrost mode facilitates precise control of the cryolesion and allows the removal of the adherent cryoprobe from the tissue.

A needle-like cryoprobe called the cryoseeker is used for the treatment of chronic and postoperative pain and is one of the most promising techniques. The probe is inserted through the skin into the nerve site. The subsequent controlled freezing disrupts the nerve transmission mechanisms and alleviates the pain.

Two factors affect the wide use and rapid growth of cryosurgery in the future: (1) precise control of rapid freezing and a rapid nonelectric defrosting mode, and (2) higher heat removal capability of cryoprobe.

4.10.1.4 Cooling probes and balloons. Cavity cooling can be accomplished using cooling probes for rectal cooling or a thin-walled rubber balloon in the stomach cavity (Cooper and Ross, 1960). They function in a way similar to cryoprobes, differing only in size and cooling temperature.

4.10.2 Hyperthermia

4.10.2.1 Artificial heat sources. A controlled heat source produced by radiofrequency current (0.5 to several MHz) is used in electrosurgery to cut and coagulate living tissue (Caruso et al., 1981). Unmodulated sinusoidal current is used for cutting, while a sequence of exponentially damped sinusoidal pulses is typically employed for coagulation. The device consists of two electrodes of different surface areas which are inserted into tissues. An electric current is passed through the patient's body between the two electrodes. The small-area electrode at the surgical site, called the active or cutting electrode, is where the current is concentrated and produces intense heat within the tissue. The large-area electrode is referred to as the reference electrode, ground plate, or dispersive electrode and is the site through which the current exits. The function of the reference electrode is to divert the applied current over a large surface area in order to prevent high current density from causing tissue damage.

Modern electrosurgical electrodes include the dry metal-plate electrode, the capacitively coupled electrode, a metal plate or foil covered with conductive paste, and the gelled-pad electrode (Geddes et al., 1977). A typical metal-plate electrode is covered with conductive paste. The gelled-pad electrode uses a thick foam pad, impregnated with an electrolytic gel, which is sandwiched between the patient's skin and a thin metallic plate.

Some implanted prosthetic heating devices are powered by internal power sources.

4.10.2.2 Heated probes. Eberhart et al. (1980) presented a brief review on the use of heated probes for measurements of tissue perfusion. Figure 4.53 shows a heated probe developed by Johnson et al. (1979). It consisted of a hollow

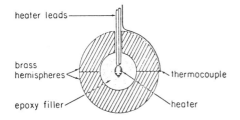

heater leads

brass
hemispheres

thermocouple

epoxy filler

heater

Figure 4.53 Heated probe used for tissue perfusion measurements.

brass sphere (3-mm outer diameter) with a resistant-heated core and a thermocouple bonded to the surface. Transient heat balances were analyzed for the resistance-heated core and the surrounding tissue coupled by heat conduction in the brass sheath. The time-dependent probe surface temperature, predicted from the solution, is matched to the experimental data, obtained with the thermocouple, by a curve-fitting routine performed by a least-squares technique.

4.10.2.3 Convective heating. The most primitive way of heating a patient is through a convective technique of immersing the patient in a liquid bath. This is a very time-consuming process because the heat has to be transferred to the body surface by convection and then to the interior by a heat diffusion process. A heated wax bath was sometimes employed for systemic hyperthermia because it had a relatively higher heat transfer coefficient than that of a water bath.

4.10.2.4 Radiative heating. Both electromagnetic and ultrasonic heating methods transfer the heat from a source to the tissues by radiation and are thus noninvasive. They are suitable for fast and deep tissue heating. The radiation intensity I (or heat flux q''), diminishes exponentially with distance in the tissues as

$$ I = I_0 e^{-ax} \qquad \text{or} \qquad q'' = q_0'' e^{-ax} $$

where I_0 (or q_0'') denotes the intensity (or heat flux) at the body surface, a denotes the absorption coefficient, and x denotes the distance measured from the surface. After being absorbed locally, the electromagnetic or ultrasound waves are converted into heat. In clinical use of diathermy, it is essential that the highest temperatures in the distribution occur at the anatomic site to be treated. Thus, the heating method used, which may be microwave, shortwave, or ultrasonic wave, and the technique of application must be selected according to the specific site of pathology to be treated and to the heating pattern desired.

Recently, hyperthermia has found an important clinical application in cancer treatment because of the destructive effect of heat on malignant cells; this effect is more severe on malignant cells than on normal ones. Two factors, the temperature and duration of heating, which are combined in the form of the thermal death time, provide a crude quantification of the biological effect of heat. An operational range of 41.5–42.0°C seems appropriate for hyperthermia, with an optional temperature of 42°C.

The electromagnetic radiation used in hyperthermia is in the spectral range of 1 MHz to several GHz covering both radiofrequency and microwave ranges. A total body tolerance is considered to be in the range of 10 mW/cm². Microwave generating devices were developed during World War II. Considerable research on the application of electromagnetic waves to heat living tissues has been conducted since the 1960s.

The electromagnetic wave heating device consists of an electromagnetic wave source (diathermy machine) and a horn radiator (applicator or antenna) with an appropriate wave guide. The use of a horn radiator permits simultaneous excitation and thermographic viewing. Some diathermy devices use a direct-contact type applicator instead. The entire arrangement together with the patient must be enclosed in an electromagnetic energy and to reduce spurious reflection. The frequencies of 27.12, 433, 750, 915, and 2450 MHz are most commonly tested or applied clinically (e.g., Guy et al., 1978; Kantor et al., 1978). However, it has been demonstrated that microwave radiation at 915 MHz can penetrate the deepest in the tissues (Yang, 1979a), thus producing the most rigorous therapeutic responses when used clinically to heat deep tissue.

The energy density of plane traveling electromagnetic waves, q''', can be expressed as

$$q''' = \frac{1}{2\sigma E^2}$$

where σ signifies the conductivity of the medium and E is the electric field. This term should be added to the right side of the bioheat transfer equation (4.54) as the radiation absorption term.

The ultrasonic waves commonly used in medical applications range from 0.02 to 1000 MHz or even higher in a few instances. In addition to medical diagnosis by a pulse echo technique, ultrasound is utilized as a therapeutic agent in two distinct ways: for deep-tissue heating when low intensities (about 1 W/cm² or less) are employed and as a selective tissue modifying agent at very high intensities (approximately 1000 W/cm²) in the form of pulse of short duration (about 1 s).

Ultrasonic wave heating devices consist of an ultrasonic wave generator and an applicator. The specific acoustic impedance and intensity of plane traveling waves are $Z_0 = \rho_0 v$ and $I = Z_0^{3/2}$, respectively, where ρ_0 and v are the mean liquid density and phase velocity, respectively. The energy density q''' is, according to Dunn et al., 1969,

$$q''' = \frac{I}{v}$$

As in the case of electromagnetic diathermy, the energy term must be included on the right side of Eq. (4.113) for theoretical analysis.

For this chapter on bioheat transfer, the reader is encouraged to refer to Branemark et al. (1968), Cena and Clark (1978), Chato (1968), Chato and Trezek (1973), Chen et al. (1977), Edwards and Hasall (1971), Hill (1932), Houdas and Ring (1982), Krebs and Kornberg (1957), Lehmann et al. (1965, 1969), Mazur (1960), Monteith and Mount (1974), Morowitz (1960, 1968), Nielsen (1969), Rapatz et al. (1966), Schwan (1957, 1965), Shitzer (1975), Stoll (1967), Valvano et al. (1981), Yang (1979b, 1980), and Yang and Yang (1982).

PROBLEMS

4.1 Derive Eqs. (4.14) through (4.17).

4.2 Consider the case where the left and right surfaces in Fig. 4.4a are cooled by the fluids at T_{a1} (with heat transfer coefficient h_1) and T_{a2} (with h_2), respectively. Note that both T_H and T_L are unknown. Determine the rates of heat removed by the two fluids.

4.3 Derive Eqs. (4.29) and (4.31).

4.4 Consider the case where the inner and outer surfaces in Fig. 4.4b are cooled by the fluids at T_{ai} (with heat transfer coefficient h_i) and T_{ao} (with h_o), respectively. Note that T_i and T_e are unknown. Determine the rates of heat flow into the two fluids.

4.5 Consider a cylindrical muscle of radius $r_0 = 1$ cm and $k = 0.001$ cal/cm s °C with $q''' = 5$ cal/cm³ h, or five times the resting metabolic rate. The surface of each muscle fiber is kept at a temperature $T_s = 37$°C. Determine the largest temperature rise in the fiber (at its center) and the total heat generated per unit length of muscle.

4.6 Consider a long solid cylinder of length l and radius r_0 with a volumetric rate of internal heat generation q'''. The outer surface of the cylinder is cooled by a fluid at T_a with heat transfer coefficient h. Determine the temperature distribution and the rate of heat removal by the fluid.

4.7 Repeat Problem 4.6 for the case where the internal heat generation rate q''' varies as a parabolic function of radius

$$q''' = q_0''' \left[1 - \left(\frac{r}{r_0} \right)^2 \right]$$

4.8 Derive Eqs. (4.32) and (4.33).

4.9 Consider a solid sphere of radius r_0 with a volumetric rate of internal heat generation q'''. The outer surface is cooled by a fluid at T_a with heat transfer coefficient h. Determine the temperature distribution and the rate of cooling by the fluid.

4.10 Derive Eqs. (4.41a), (4.41b), and (4.41c).

4.11 Derive Eqs. (4.50) and (4.51).

4.12 Use the principle of the increase of entropy to prove that heat is dissipated from skin surface at T_s to the environment at T_a only if T_s is higher than T_a.

4.13 Consider a human body to be the cylindrical set of composite layers of core, musculature, functional periphery and skin as shown in Fig. 4.54. Here r_i, k_i, q_{mi}''', and T_i denote the radius, thermal conductivity, volumetric rate of metabolic heat generation, and temperature of each layer, respectively. T_c and T_a represent the body-center temperature (say 37°C) and the ambient temperature (say 25°C), respectively.

(a). Draw an equivalent electrical circuit of heat transfer in the body in the absence of q_{mi}''' (hypothetical case). Also, derive the expression for the overall heat transfer between T_c and T_a.

(b). In the presence of q_{mi}''', as in reality, derive the expression for temperature distribution in each zone.

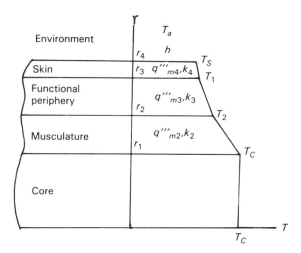

Figure 4.54 Multilayered cylindrical model of human body.

4.14 Repeat problem 4.13(b) for unsteady cases in which (1) T_a undergoes a step change from, say 25 to 0°C (similar to walking into a huge refrigerator), while T_c remains constant; and (2) all q_{mi}''''s undergo a step change, say doubled in strenuous exercise), while T_a and T_c (going up in reality) remain constant. Formulate the problem (write the governing differential equations and appropriate initial and boundary conditions) only.

4.15 Heat transfer in the hand: the hand is vasculated with arterial and venous networks underneath the skin. Arterial blood at T_A carries heat from the body core into the hand and returns to the arm as venous blood at T_V. The heat is then dissipated from the skin surface of area A to the environment at T_a by the combined convection-radiation mechanism with the total heat transfer coefficient $h\,(= h_c + h_r)$. Since the arterial and venous vessels are in direct contact in the hand, the blood temperature comes into equilibrium with the skin temperature T_s at the beginning of the venous vessel. Define the hand as a control volume, as shown in Fig. 4.55. Let $\dot{m}_b$ and c_{pb} be the mass flow rate and specific heat of blood, respectively. If $T_A = 37°C$, $T_a = 20°C$, $A = 0.05\ m^3$, $\dot{m}_b = 40$ ml/min, $T_v = 35°C$, $c_{pb} = 1.0$ kcal/1000 ml °C and $h = 15$ kcal/m² h °C, determine T_s, T_A, and the rate of heat dissipation from the skin to the environment q. In the absence of heat shunt mechanism, namely internal heat exchange, $T_A = T_A'$, and $T_V = T_s$. Determine T_s and q and show the effect of heat shunt mechanism.

Note: Since arterial and venous vessels are situated close to each other, the enthalpy drop in arterial blood is equal to the gain in enthalpy of venous blood, assuming no heat loss to the surroundings. The two flows are countercurrent. This phenomenon is called internal (counter-current) heat exchange or heat shunt mechanism in physiology, which occurs inside the rectangle in Fig. 4.55. In mathematical form, it reads

$$\dot{m}_b C_{pb}(T_A - T_A') = \dot{m}_b C_{pb}(T_V - T_s)$$

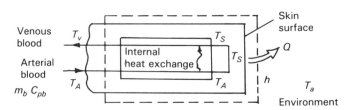

Figure 4.55 Heat transfer in the hand (SSSF process).

or

$$T_A - T'_A = T_V - T_s$$

since c_{pb} differs little in arterial and venous blood. Here, T'_A is the temperature of the precooled arterial blood being supplied to the skin surfaces. In other words, the venous blood returning can cool the arterial blood being supplied to the extremity from T_A to T'_A, while the venous blood in turn is itself being rewarmed from T_s to T_V by the arterial blood.

4.16 The core region that comprises the visceral organs and brain has uniform and constant temperature at 37°c. Since these organs are heat generating, there must be a special means for controlling temperature. There is considerable evidence that the combination of an internal countercurrent heat exchange and an external heat exchange takes place to regulate the temperature of these organs. Cooled venous blood at T_s resulting from the external heat exchange precools arterial blood at T_A through an internal heat exchange process. The precooled arterial blood at T_A then leads to the organ and removes the waste energy from the organ. Now the venous blood leaving the organ at T'_V dissipates heat to the environment at T_a through a complex blood flow network. The venous blood returning at T_s is rewarmed to T_V by the arterial blood during the internal heat exchange process. The dotted line in Fig. 4.56 defines the organ together with the two heat exchangers as the control volume. Determine T_s, T'_A, T'_V, and q in terms of T_A, h, A, T_A, T_V, $\dot{m}_b$, c_{pb}, and q_m, which are defined in problem 4.15.

Note: If there is no metabolic heat generation, i.e., $q_m = 0$, the phenomenon becomes identical with that observed in the hand.

This analysis aids in the development of an understanding of the thermal regulatory mechanism in the body core and in interpreting in vivo test data obtained from such systems.

4.17 During childbirth, measures must be taken to reduce thermal losses from newborn infants to preserve their normal body temperatures. Immediately following delivery, the infant is completely wet and is temporarily exposed to the delivery room environment, resulting in high evaporative and convective heat losses. Determine (1) the approximate Biot number of the infant in the delivery room environment and (2) the body temperature drop in the infant immediately following delivery, using the following data

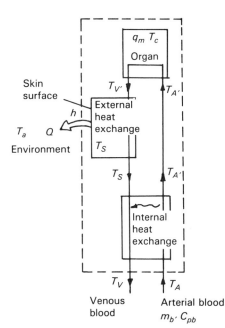

Figure 4.56 Thermal regulation of organs (USUF process).

Infant: weight $m = 2.3$ kg
  surface area $A_s = 0.2$ m^2
  basal metabolic heat production rate $q''' = 2.0$ kcal/kg h
  surface temperature $T_s = 35°C$
  $\rho = 1000$ kg/m^3
  $C_p = 4.18$ kcal/kg K
  $k = 0.606$ W/m K

Delivery room environment: wall temperature $T_w = 22°C$
  air temperature $T_a = 22°C$
  relative humidity = 55%
  partial pressure of water vapor at the temperature of the skin $P_s = 47$ mmHg
  partial pressure of water vapor in the ambient air $P_a = 11$ mmHg
  convective heat transfer coefficient $h_c = 9.8$ kcal/m^2 h °C
  convective surface area $A_c = 0.15$ m^2
  radiative heat transfer coefficient $h_r = 5.0$ kcal/m^2 h °C
  radiative surface area $A_r = 0.15$ m^2
  evaporative heat transfer coefficient $h_v = 11.9$ kcal/m^2 h mmHg
  evaporative surface area $A_v = 0.18$ m^2

Note that the rate of evaporative heat transfer is

$$q_v = h_v A_v (P_s - P_a).$$

4.18 A concentric double-pipe heat exchanger is to be used to regulate the temperature of blood in extracorporeal circulation in open heart surgery. This type of heat exchanging device is also, in principle, found in physiology. Warm water enters the annular space at 50°C and 5 kg/s. Blood at 5 L/min is to be heated in the inner tube from 32 to 37°C which is then returned to the patient's body. If the inner and outer tubes are 1.25 cm ID, 1.40 OD, and 2.5 ID, 2.65 OD, respectively, determine the lengths of the exchanger for the counterflow and parallel-flow cases. Use thermal properties of water for blood.

4.19 Consider a nude human as a 30-cm diameter, 170-cm long vertical cylinder with both ends insulated. The metabolic heat rate of 72 kcal/h is dissipated into the environment by natural convection alone when immersed in water, or by a combination of natural convection and radiative heat transfer (with the surrounding walls at T_w) in air. The effective surface area for radiation is taken to be 0.77 of the total body surface area (including both ends of the cylinder). In two different environments, air and water, convective heat transfer coefficient is h_c and environmental temperature is T_a in air or water that will balance the resting metabolic heat rate. The average skin temperature T_s is 34.2°C and $T_w = 25°C$.

Estimate the radiative heat transfer coefficient h_r using the expression

$$h_r = \sigma \epsilon (T_s + T_w)(T_s^2 + T_w^2)$$

where $\sigma = 5.67 \times 10^{-8}$ w/m^2 K^4 and $\epsilon \cong 1$.

4.20 Construct a plot of the ambient air temperature T_a versus the surrounding wall temperature T_w, indicating a neutral environment for a resting nude subject. The neutral environment means a combination of T_a and T_w for dissipation of the entire metabolic heat produced in the subject. Use the following data: a 30-cm diameter, 170-cm long vertical cylinder with both ends insulated; resting metabolic heat generation rate = 72 kcal/h; effective surface area for radiation = 77% of total body surface area; and mean skin temperature $T_s = 34°c$.

4.21 A nude man is standing in the wind that blow normal to him at a velocity of 3 m/s. Treat him as a 30-cm diameter, 170-cm long vertical cylinder with both ends insulated. Determine the rate of heat loss from him. His mean skin temperature is 34°C and the air temperature is 20°C. Compare this result with the case when the wind stops blowing. Use this as an example to explain the wind-chill chart which is a plot of the equivalent calm-air temperature versus the actual ambient temperature at various wind speeds.

4.22 Consider an adult as a circular cylinder of 30 cm diameter and 170 cm length with both ends insulated (hair on the head and feet on the ground). The skin temperature T_s is 34°C and the ambient air is $T_a = 20$°C. If the person is naked, determine the heat transfer coefficients and the rates of convective heat transfer to the atmosphere for the following cases:

(a) Standing in still air.
(b) Standing in a wind blowing at a velocity u of 10 m/s.
(c) Walking in still air at a velocity of 2 m/s (ignoring the pendulum effect).
(d) Lying down (use one-half of the surface area for heat transfer).
(e) Running at a velocity $v = 4$ m/s against a window blowing at $u = 10$ m/s (ignoring the pendulum effect).

Note: Tables for the physical properties of air are needed.

4.23 In order to produce an infrared thermometer for medical use, the equipment must be adjusted to the characteristic infrared emission of the human body, which has a temperature of 310 K or 37°C. Determine the maximum wavelength λ_{max} and emissive heat flux E_b of the infrared radiation of the human body. Construct the spectral distribution curves ($E_{b\lambda}$ versus λ) of infrared (λ between 0.75 and 1000 μm) from the black body (at $T = 2000$, 1000, 500, 300, and 200 K) and show the locus of λ_{max}.

4.24 Consider the human forearm as a cylinder. Its tissues contain a distributed heat source (e.g., heat produced by tissue metabolism q_m''') and distributed heat sink (e.g., heat removed by blood from tissue q_b''') at each point in the forearm. Bone is treated mathematically the same as soft tissue, and blood flow rate is uniform throughout the forearm. The transport of heat by the circulatory system is given by

$$q_b''' = \dot{m}_b C_{pb}(\beta - 1)(T - T_A)$$

where $0 < \beta < 1$ is a so-called equilibration constant between blood and surrounding tissue. Prove that the tissue and skin surface temperature can be, respectively, given by

$$T = \frac{(T_s - b/a^2)J_0(i\ ar)}{J_0(i\ aR)} + \frac{b}{a^2}$$

$$T_s = \frac{(bk/a^2)[-iJ_1(i\ aR)] + 1.21hJ_0(i\ aR)T_a}{k[-iJ_1(i\ aR)] + 1.21hJ_0(i\ aR)}$$

where $a = [\dot{m}_b C_{pb}(\beta - 1)/k]^{1/2}$
$b = a^2 T_A - q'''/k$

T_s denotes the skin surface temperature at $r = R$, namely $T(R)$. Plot the temperature distribution (T_3 versus r) curves for $\dot{m}_b = 0.00030$, 0.00025, 0.00020, and 0 g/cm³ s at $q_m''' = 0.0001$ cal/cm³ s.

4.25 A primitive (single-layer) model of skin burns considers the layers of the epidermis to be a semi-infinite body with the skin surface being exposed to an extreme heat source: high temperatures sources (e.g., flames, glowing materials, and atom bombs) or low temperatures sources (e.g., wind at 30 kph and -40°C) hitting the bare skin. The epidermis is initially at uniform temperature T_0 and the skin surface is suddenly exposed to an ambient temperature at T_a. Find the temperature time history of the epidermis, assuming the tissue to be infinitely large in extent (valid at small times). The effect of metabolic heating is negligible.

4.26 In unsteady heat transfer analysis, the tissue layer is often treated as a semi-infinite body and initially at a uniform temperature T_0. In order to understand the effects of extreme heat on humans, determine the temperature changes in the tissue layer caused by:

(a) Constant nonpenetrating (surface) heat flux q_s''' absorbed by the skin surface (e.g., a hot radiator) and q_s''' caused by the combined effects of radiation, convection, and evaporation. Plot the temperature-time history at various skin depths ($x = 0$, 1, 2, 3, 4, and 5 mm) for $q_s''' = 0.044$ cal/cm² s, $k = 0.0025$ kcal/m s °C and $\rho C = 1$ cal/cm³ °C.

(**b**) Constant penetrating radiant heat flux I absorbed by the surface and also by the tissue layer with the absorption coefficient α_r (e.g., electromagnetic hyperthermia). The surface heat flux is q_s''' including the effects of radiation, convection, and evaporation. Plot the increase of skin temperature with time for $\alpha r_3 = 00$, 100, 10, and 1 cm^{-1}, $k = 10^{-3}$ cal/cm s °C, $\alpha = 10^{-3}$ cm^2/s and $I = q_s''' = 0.15$ cal/cm^2 s. It should be noted that the absorption coefficient of 10 corresponds to the case of solar radiation.

4.27 A scanning infrared camera is used in the detection of breast cancer through the observation of detailed thermal patterns on the women's thorax. These patterns are related to the tissue temperature and the venous vascular networks spreading near the surface of the breast skin. The venous temperatures are associated with the average tissue temperature in the region perfused by the vein. Determine the tissue temperature distribution by solving the steady, one-dimensional bioheat equation. The tissue thickness is l. The skin surface is exposed to the air at T_a while the interior surface in contact with the core is assumed to be insulated.

4.28 The amount of heat generated by a human infant during a 24-h period is given by a formula Q (in kcal) $= 0.0128 \, lW^{2/3}$, where l denotes the length of the infant in centimeters and W is its weight in kilograms. How much heat is generated by an infant of length 60 cm and weight 3.8 kg? Calculate $\partial Q/\partial l$ and $\partial Q/\partial W$, and interpret these quantities.

4.29 An animal's body loses heat by convection at the rate of $q_c = a(T_s - T_\infty)v^{1/3}$, where T_s and T_∞ denote the temperatures of the animal body surface and of the surrounding air, respectively; v is the wind velocity; and a is a constant. Calculate $\partial q_c/\partial T_s$, $\partial q_c/\partial T_\infty$, and $\partial q_c/\partial v$ and interpret these quantities.

4.30 When an intensity of light I (W/m^2) falls on a plant, the rate of photosynthesis S is found empirically to be given by

$$S = 150I - 25I^2$$

within the range of $0 \leq I \leq 5$. Determine the maximum and minimum values of S when I lies within $1 \leq I \leq 5$.

4.31 Examination of the anatomy of various organs of homeothermic animals discloses the natural close proximity of arteries and veins. The hand is an example. Thus, countercurrent heat exchange between blood vessels occurs as schematically shown in Fig. 4.55. Arterial blood at T_A carries heat from the body core into the hand and returns to the arm as venous blood at T_V. The heat is then dissipated from the skin surface of area A' per unit length to the environment at T_a by the combined convection-radiation mechanism with the overall heat transfer coefficient U. Let x be the distance measured from the arterial flow inlet with a (temperature of T_0) and l be the length of arterial (or venous) flow. Show that the temperature distribution in the arterial and venous flow can be expressed as:

$$\frac{T_{A,V} - T_a}{T_0 - T_a} = \exp\left[\frac{(N_V - N_A)\xi}{2}\right] \frac{\{B \cosh[A(1 - \xi)] \pm \sinh[A(1 - \xi)]\}}{B \cosh A + \sinh A}$$

where the $+$ and $-$ signs are for T_A and T_V, respectively; $\xi = x/l$; $N_k = (UA')_k l/(\dot{m}_b C_{pb})$ (where the index k may be changed by A, V, or the index i); $A = 0.5[(N_A + N_V)(N_A + N_V + 4N_i)]^{1/2}$; $B = [(N_A + N_V + 4N_i)/(N_A + N_V)]^{1/2}$.

Plot the temperature distributions for two special cases: (1) $N_A = N_V = 0$, and (2) $N_A = 0$.

REFERENCES

Abbot, B. C., Hill, A. V., and Howarth, J. V. (1958) *Proc. Roy. Soc. B*, 148:149–187.

Ablin, R. J. (ed.) (1980) *Handbook of Cryosurgery*, Marcel Dekker, New York.

Arrhenius, S. (1915) *Quantitative Laws in Biological Chemistry*, Bell & Sons, London.

Asahina, E. (ed.) (1966) *Proceedings of International Conference on Low Temperature Science*, Sapporo, Japan.

Baeu, D. (1980) Thermal instrumentation. *Ann. N.Y. Acad. Sci.*, 335:481–483.

Balasubramanian, T. A., and Bowman, H. F. (1977) Thermal conductivity and thermal diffusivity of biomaterials: A Simultaneous measurement technique, *J. Biomech. Eng.*, 99:148–154.

Baranski, S., and Czerski, P. (1976) *Biological Effects of Microwaves*, Dowden, Hutchingson & Ross, Stroudsburg, Penn.

Barrett, A. H., and Meyers, P. C. (1975) *Proc. First Euro. Congr. Thermography and Bibl. Radiol.* vol. 6, Karger, Basel, pp. 45–56.

Bergles, A. E. (1978) Enhancement of Heat Transfer. *Heat Transfer—1978*, Hemisphere, Washington, D.C., vol. 6, pp. 89–108.

Birch, J., Branemark, P. I., Nilsson, K., and Lundskog, J. (1968) Vascular reactions in an experimental burn study with infrared thermography and microangiography, *Scand. J. Plast. Reconstr. Surg.*, 2:97.

Boehm, R. F., and Tuft, D. B. (1971) Engineering Radiation Heat Transfer Properties of Human Skin, ASME Paper No. 71-WA/HT-37.

Boltzmann, L. (1884) *Ann. der Physik, Wiedemanns Annalen* 22:291.

Bowman, H. F., Cravalho, E. G., and Woods, M. (1980) Theory, measurement, and application of thermal properties of biomaterials, *Ann. Rev. Biophys. Bioeng.*, 4:43–80.

Branemark, P. I., Breine, U., Joshi, M., and Urbrascheck, B. (1968) Microvascular photophysiology of burned tissues, *Ann. N.Y. Acad. Sci.*, 150:474.

Burton, A. C. (1934) The application of the theory of heat flow to the study of energy metabolism, *J. Nutr.*, 7:497–533.

Burton, A. C. (1953) On the teaching of temperature regulation in a medical physiology course, *J. Appl. Physiol.* 6:65–66.

Carslaw, H. S., and Jaeger, J. C. (1959) *Conduction of Heat in Solids*, Clarendon Press, London, Chaps. 7 and 9.

Caruso, P. M., Pearce, J. A., and DeWitt, D. F. (1981) The Effect of Gelled-Pad Design on Performance of Electrosurgical Dispersive Electrodes. ASME Paper No. 81-WA/HT-22.

Cena, K., and Clark, J. A. (1978) Thermal insulation of animal coats and human clothing, *Phys. in Med. & Biol.*, 23:565–591.

Chao, K. N., (1975) *Heat and Water Losses from Normal and Burnt Skin*, Ph.D. Thesis, University of Michigan, Ann Arbor, Mich.

Chao, K. N., Eisley, J. G., and Yang, W.-J. (1973a) Heat and water migration in regional skins and subcutaneous tissues, *1973 Biomechanics Symposium, Applied Mechanics Division*, vol. 2, ASME, New York, pp. 69–72.

Chao, K. N., Eisley, J. G., and Yang, W.-J. (1973b) Heat and water losses from clothed human skin, in *Recent Advances in Engineering Science* (ed. T. S. Chang), 7:317–324.

Chao, K. N., Eisley, J. G., and Yang, W. J. (1977) Heat and water losses from burnt skin, *Med. Biol. Eng. Comput.* 15:598–603.

Chato, J. (1980) Heat transfer to blood vessels, *J. Biomech. Eng.*, 102:110–118.

Chato, J. C. (ed.) (1968) *Thermal Problems in Biotechnology*, ASME, New York.

Chato, J. C. (1980) Measurement of thermal properties of growing tumors, *Ann. N.Y. Acad. Sci.*, 335:67–85.

Chato, J. C., and Trezek, G. J. (eds.) (1973) *Problems of Heat and Mass Transfer in Biotechnology*, Report on a workshop held at University of Illinois, Champaign, Ill.

Chen, M. M., and Holmes, K. R. (1980) Microvasculature Contributions in Tissue Heat Transfer, in *Thermal Characteristics of Tumors: Applications in Detection and Treatment* (eds. R. K. Jain and P. M. Gullino), Ann. N.Y. Acad. Sci. 335:137–150.

Chen, M. M., and Pantazatos, P. (1980) Tomographical thermography, *Ann. N.Y. Acad. Sci.* 335: 438–442.

Chen, M. M., and Rupinskas, V. (1977) A Simple Method for Measuring and Monitoring Thermal Properties in Tissues, ASME Paper No. 77-WA/HT-42.

Chen, M. M., Pedersen, C. O., and Chato, J. C. (1977) On the feasibility of obtaining three-dimensional information from thermographic measurements, *J. Biomech. Eng.*, 99:58–64.

Connor, W. G., Gerner, E. W., Miller, R. C., and Boone, M. L. M., (1977) *Radiol.*, 123:497–503.

Cooper, K. E., and Ross, D. N. (1960) *Hypothermia in Surgical Practice*, Cassell, London.

Cooper, T. E., (1970) *Bio-Heat Transfer Studies*, Ph.D. Thesis, University of California, Berkeley, CA.

Cravalho, E. G., (1972) *Biomedical Physics and Biomaterials Science* (ed. H. E. Stanley), MIT Press, Cambridge, MA 207–256.

Davis, J. W. L., Liljedahl, S. O., and Birke, G. (1970) *Trans. Third Internat. Congr. on Research in Burns*, Prague.

Davson, H. (1970) *A Textbook of General Physiology*, 4th ed., vol. 1, Churchill, London, pp. 298–367.

Davson, H., and Eggleton, M. G. (eds.) (1962) *Principles of Human Physiology*, Lea & Febiger, Philadelphia, chaps. 32, 33, and 38.

Dunn, F., Edmonds, P. D., and Fry, W. J. (1969) Absorption and Dispersion of Ultrasound in Biological Media, in *Biological Engineering* (ed. H. P. Schwan), McGraw-Hill, New York, chap. 3.

Eberhart, R. C., Shitzer, A., and Hernandez, E. J., (1980) Thermal dilution methods: Estimation of tissue blood flow and metabolism, *Ann. N.Y. Acad. Sci.*, 335:107–132.

Edrich, J., Jobe, W. E., Cacak, R. K., Hendee, W. R., Smyth, C. J., Gautherie, M., Gras, C., Zimmer, R., Robert, J., Thouvenot, P., Escanye, J. M., and Itty, C. (1980) Imaging thermograms at centimeter and millimeter wavelength, *Ann. N.Y. Acad. Sci.*, 335:456–474.

Edwards, M. A., and Hasall, K. A. (1971) *Cellular Biochemistry and Physiology*, McGraw-Hill, London.

Fan, L. T., Hsu, F. T., and Hwang, C. L. (1971) A review on mathematical methods of the human thermal system, *IEEE Trans. Biomed. Eng.* 18:218–234.

Fazzio, R. G., and Jacobs, H. R. (1972) Heat transfer coefficients of blood flow in small tubes and capillaries, *AIChE Symp. Ser.*, 78 (138):233–240.

Fourier, J. B. J. (1822) *Theorie analytique de la chaleur*, Paris; A. Freeman, trans., Dover, New York, 1955.

Geddes, L. A., Silva, L. F., DeWitt, D. P., and Pearce, J. A. (1977) What's new in electrosurgical instrumentation? *Medical Instrumentation*, 2:355–359.

Gibbs, C. G. (1978) Cardiac energetics, *Physiol. Rev.*, 58:174–254.

Giese, A. C. (1968) *Cell Physiology*, 3rd ed. W. B. Saunders, Philadelphia.

Guy, A. W. (1975) *Proceedings of the International Symposium on Cancer Therapy by Hyperthermia and Radiation*, American College of Radiology, Washington, D. C., 179–192.

Guy, A. W., Lehmann, J. F., and Stonebridge, J. B., (1974) Hyperthermia and diathermy, *Proc. IEEE*, 62:55.

Guy, A. W., Lehmann, J. F., Stonebridge, J. B., and Sorensen, C. C. (1978) Development of a 915 MHz direct contact applicator for therapeutic heating of tissues, *IEEE Trans. Microwave Theory Technol.*, MTT-26:550–556.

Hardy, J. D. (ed.), (1963) *Temperature—Its Measurement and Control in Science and Industry*, vol. 3, pt. 3, Reinhold, New York.

Hardy, J. D., Gagge, A. P., and Stolwijk, J. A. J. (eds.) (1970) *Physiological and Behavioral Temperature Regulation*, Charles C Thomas, Springfield, Ill.

Hill, A. V. (1932) A closer analysis of the heat production of nerve, *Proc. Roy. Soc. B* 111:106–164.

Hill, A. V. (1933) The three phases of nerve heat production, *Proc. Roy. Soc. B* 113:345–356.

Holden, H. B. (ed.) (1975) *Practical Cryosurgery*, Pittman Medical Publishing Co., Oxford, U.K.

Houdas, Y., and Ring, E. F. J. (1982) *Human Body Temperature—Its Measurement and Regulation*, Plenum, New York.

Johnson, C. C., and Guy, A. W. (1972) *Proc. IEEE*, 60:692.

Johnson, F. H., Eyring, H., and Polissar, M. (1954) *The Kinetic Basis of Molecular Biology*, Wiley, New York.

Johnson, W. R., Abdelmessih, A. H., and Grayson, J. (1979) Blood perfusion measurements by the analysis of the heated thermocouple probe's temperature transients, *J. Biomech. Eng.*, 99 K:58–65.

Kantor, G., Witters, D. M., Jr., and Greiser, J. W. (1978) The performance of new direct contact application for microwave diathermy, *IEEE Trans. Microwave Theory Technol.*, MTT-26: 563–568.

Krebs, H. A., and Kornberg, H. L. (1957) *Energy Transformations in Living Matter*, Springer, Berlin.

Lamke, L. G., and Liljedahl, S. O. (1971) Evaporative water loss from burns and donor site, *Scand. J. Plast. Reconstr. Surg.*, 5:17.

Lamke, L. O., and Wedin, B. (1971) Water evaporation from normal skin at different climatological conditions, *Acta Dermatovener.* 51:111.

Lehmann, J. F., Brunner, G. D., McMillan, J. A., and Blumberg, J. B., (1959) Comparative study of the efficiency of shortwave, microwave and ultrasonic diathermy in heating the hip joint, *Arch. Phys. Med.*, 40:510–512.

Lehmann, J. F., Johnston, V. C., McMillan, J. A., Silverman, D. R., Brunner, G. D., and Rathbun, L. A. (1965) Comparison of deep heating by microwave at frequencies 2456 and 900 megacycles, *Arch. Phys. Med. Rehab.*, 46:307–314.

Lehmann, J. F., Delateur, B. J., and Stonebridge, J. B. (1969) Selective muscle heating by shortwave diathermy with a helical coil, *Arch. Phys. Med. Rehab.*, 50:117–123.

Lih, M. M. S. (1975) *Transport Phenomena in Medicine and Biology*, Wiley Interscience, New York, chap. 6.

Ling, G. R., and Tien, C. L. (1969) Analysis of Cell Freezing and Dehydration, ASME Paper No. 69-WA/HT-31.

Lou, Z., Yang, W.-J., and Sandhu, T. S. (1988) Numerical analysis of electromagnetic hyperthermia of the human thorax. *Med. Biol. Eng. Comput.* 26:50–56.

Lunardini, V. J. (1981) *Heat Transfer in Cold Climates*, Van Nostrand Reinhold, New York.

Mazur, P. (1960) *Recent Research in Freezing and Drying* (eds. A. S. Parkes and A. J. Smith), Charles C Thomas, Springfield, Ill., p. 65.

Mazur, P. (1963) Kinetics of water loss from cells at subzero temperatures and the likelihood of intracellular freezing, *J. Gen. Physiol.* 47:347–369.

Mazur, P. (1966) Physical and chemical basis of injury in single celled microorganisms subjected to freezing and thawing, *Cryobiology* (ed. H. T. Meryman) Academic Press, New York, chap. 6.

Meryman, H. T. (ed.) (1966) *Cryobiology*, Academic Press, New York.

Miller, R. C., Connor, W. G., Heusinkveld, R. S., and Boone, M. L. M., (1977) Prospects for hyperthermia in human cancer therapy, *Radiology*, 123:489–495.

Monteith, J. L., and Mount, L. E. (eds.) (1974) *Heat Loss from Animals and Man*, Butterworths, London.

Morowitz, H. J. (1960) Some consequences of the application of the second law of thermodynamics to cellular systems, *Biochim. Biophys. ACTA* 40:340–345.

Morowitz, H. J. (1968) *Energy Flow in Biology*, Academic Press, New York.

Myers, P. C., Barrett, A. H., and Sadowsky, N. L. (1980) Microwave thermography of normal and cancerous breast tissue. *Ann. N.Y. Acad. Sci.*, 335:443–455.

Nevins, R. G., and Darwish, M. A. (1970) Heat Transfer Through Subcutaneous Tissue as Heat Generating Porous Material, in *Physiological and Behavioral Temperature Regulation* (eds. J. D. Hardy, A. P. Gagge, and J. A. J. Stolwijk), Charles C Thomas, Springfield, Ill., chap. 21.

Newburgh, L. H. (ed.) (1968) *Physiology of Heat Regulation and the Science of Clothing*, Hafner, New York.

Nielsen, B. (1969) *Thermoregulation in Rest and Exercise*, Aarhuus Stiftsbogtrykkerie A/S, Copenhagen, Denmark.

Pantazatos, P., and Chen, M. M. (1978) Computer-aided tomographic thermography: A numerical simulation, *J. Bioeng.*, 2:397–410.

Pasquale, D. (1981) Cryosurgery: The Present and the Future. ASME Paper No. 81-WA/HT-22, 1981.

Paulsen, K. D., Strohbehn, J. W., and Lynch, D. R. (1985) Comparative theoretical performance for two types of regional hyperthermia systems. *Int. J. Radiation Oncol. Biol. Phys.*, 11: 1659–1671.

Pennes, H. H. (1948) Analysis of tissue and arterial blood temperatures in the resting human forearm. *J. Appl. Physiol.*, 1:93–122.

Rand, R. W., Rinfret, A. D., and von Leden, H. (eds.) (1968) *Cryosurgery*, Charles C Thomas, Springfield, Ill.

Rapatz, G. L., Menz, L. J., and Luyet, B. J. (1966) Anatomy of the Freezing Process in Biological Materials, *Cryobiology* (ed. H. T. Meryman), Academic Press, New York, chap. 3.

Rohsenow, W. M. and Hartnett, J. P. (eds.)(1986) *Handbook of Heat Transfer*, 2nd ed., McGraw-Hill, New York.

Schwan, H. P. (1957) Electrical properties of tissues and cells, *Advan. Biol. Med. Phys.*, 5:147–209.

Schwan, H. P. (1965) Biophysics of Diathermy in Therapeutic Heat and Cold, *Therapeutic Heat and Cold* (ed. S. Licht), Waverly Press, Baltimore.

Schwan, H. P., Carstensen, E. L., and Li, K., (1953) Heating of fat-muscle layers of electromagnetic and ultrasonic diathermy, *Proc. AIEE*, p. 483–488.

Seagrave, R. C. (1971) *Biomedical Applications of Heat and Mass Transfer*, Iowa State University Press, Ames, Iowa.

Shitzer, A. (1975) Studies of Bioheat Transfer in Mammals, in *Topics in Transport Phenomena* (ed. C. Gutfinger), Halstead Press, New York, pp. 211–343.

Shitzer, A., and Eberhart, R. C. (eds.) (1985) *Heat Transfer in Medicine and Biology, Analysis and Applications*, vols. 1 and 2, Plenum, New York.

Smith, A. U. (ed.) (1970) *Current Trends in Cryobiology*, Plenum, New York.

Stefan, J. (1879) sitzungsber. d. Kais. *Akad. d. Wiss., Wien, Math.-Naturwiss. Klasse*, 79:391.

Stoll, A. M. (1967) *Advances in Heat Transfer*, vol. 4, Academic Press, New York, pp. 65–141.

Strohbehn, J. W., and Roemer, R. B. (1984) A survey of computer simulations of hyperthermia treatments, *IEEE Trans. Biomed. Eng.*, BME-31:136–149.

Valvano, J. W., Allen, J. T., and Bowman, H. F. (1981) The Simultaneous Measurement of Thermal Conductivity, Thermal Diffusivity, and Perfusion in Small Volumes of Tissue, ASME Paper No. 81-WA/HT-21.

Von Leden, H., and Cahan, W. G. (eds.) (1971) *Cryogenics in Surgery*, Medical Examination Publishing Co., New York.

Weinbaum, S., and Jiji, L. M. (1985) A new simplified bioheat equation for the effect of blood flow on local average tissue temperature, *J. Biomech. Eng.*, 107:131–139.

White, F. M. (1988) *Heat and Mass Transfer*, Addison-Wesley, Reading, MA.

Wissler, E. G. (1961) Steady-state temperature distribution in man, *J. Appl. Physiol.*, 16:734–740.

Wissler, E. H. (1963) *Temperature—Its Measurement and Control in Science and Industry*, vol. 3, pt. 3 (ed. J. D. Hardy), Reinhold, New York, pp. 603–612.

Wolstenholme, G. E. W., and O'Connor, M. (eds.) (1970) *The Frozen Cells*, Churchill, London.

Yang, W.-J. (1979a) Comparison of Localized Hypothermia by Electromagnetic and Ultrasonic Diathermy, *1979 Advances in Bioengineering*, ASME, New York, pp. 173–176.

Yang, W.-J. (1979b) Heat and Mass Transfer in Skin and Subcutaneous Tissues, in *Applied Physiological Mechanics* (ed. D. N. Ghista), Biomedical Engineering and Computation Mechanics, vol. 1, Harwood Academic, New York, pp. 105–142.

Yang, W.-J. (1980) Human Micro and Macro Heat Transfer—In Vivo and Clinical Applications, in *Perspective in Biomechanics*, vol. 1 (ed. D. N. Ghista), Harwood Academic, New York.

Yang, W.-J., and Wang, J. H. (1977) Shortwave and microwave diathermy for deep tissue heating, *Med. Biol. Eng. Comput.*, 17:518–523.

Yang, W.-J., and Yang, L. H. (1982) Heat and Mass Transfer Equipment with Medical Applications, ASME Paper No. 82-WA/HT-87.

Yang, W.-J. (1986) High heat and its effects on the body. *Mech Eng.*, 108:82–85.

Zacarian, S. A. (1968) *Cryosurgery of Skin Cancer*, Charles C Thomas, Springfield, Ill.

FLOW VISUALIZATION TECHNIQUES IN MEDICAL AND BIOLOGICAL APPLICATIONS

5.1 INTRODUCTION

Flow visualization methods have contributed significantly to investigations of normal and morbid anatomy as well as to explorations of normal and pathological physiology. In clinical applications, they serve as a diagnostic guide in surgery, a means to determine the activity of bloodborne pharmacological agents in normal and malignant cells, to aid in the diagnosis and localization of tumors, and to establish the in vivo topography of the vascularization (angiogenesis) of a growing malignant tumor.

The flow systems in medicine and biology include vascular vessels and excretory passages ranging from regular to microscopic size. To facilitate recognition of the morphology, function, and necrosis of the flow systems, contrast media are needed for image intensification. They are nothing but dyes, which include stains, fluorescein, and radiopaque and radioactive agents. These contrast media may be in gaseous form but more commonly are liquids or fine particulates suspended in solution. The image is intensified through tinting, staining, fluorescence, interaction between x-rays and radioactive contrast media, and radioisotope labeling. Observations are made with the unaided eye or through an optical device. The quality of visualization can be enhanced by means of appropriate vision magnification or image intensification. Photography and radiology (when interactions between x-rays and radioactive contrast media are involved) are commonly employed in either a single or sequential recording.

Here we first discuss contrast media and then visualization aids. Applications of flow visualization techniques are classified into three categories: medicine, pharmacology, and botany. These methods are of course applicable in zoology as well.

Because of potential hazards to living organisms as a result of long exposure to radioactive contrast media, other diagnostic methods are being constantly sought and developed in an effort to replace radiology in medicine. Two such alternative means of visualization—thermography and ultrasound—are presented in Section 5.7.

5.2 CONTRAST MEDIA

Contrast media in medicine and biology may be in the liquid or gas state, or as fine particulates suspended in solution. Contrast medium is used in visualization to intensify the image in contrast to its surrounding tissues (1) through tinting, smearing or staining (dyes), (2) as the direct result of absorption of radiation of different wavelength (fluorescein), (3) through interactions with x-rays (radioactive contrast media), and (4) by tagging a radioactive tracer to immunoglobulins (radioisotope labeling).

5.2.1 Dyes

Dyes were conventionally employed for visualizing blood vessels in tissue sections. For example, the staining technique has facilitated recognition of small blood vessels in various sections. Stains (i.e., dyes) regularly used for blood are of the so-called Leishman (1901) type. They are essentially mixtures of red, blue, and purple stains made by mixing a solution of eosin (a yellowish red acid dye) with methylene blue, methyl violet, methylene azure, and other basic analine dyestuffs to make a polychrome stain. Numerous other compounds such as eosinates, are also available. When blood is stained by one of these variants of the original Leishman mixture, red cells turn pink, and white cells, containing nuclei and granules, become distinctly colorful. Nuclei exhibit various shades from royal purple to bright magenta, while granules show red, rose, blue, or purple in pale blue cell bodies. Platelets are pink and purple. Through the microscope, these colors appear bright and dark by transmitted light. The Groat-Jenner stain (Groat, 1936) is a modification of the Leishman stain and can be used for photographing blood.

Luxol fast-blue stain is an alcohol-soluble amine salt of sulfonated copper phthalocyanine developed by Kluver and Barrera (1953) as a myelin stain. It gives an intense bright blue coloring of myelin, which is resistant to light, heat, acids, and alkalies. Margolis and Pickett (1956) used the stain in combination with the periodic acid-Schiff (PAS) reaction and hematoxylin, thus obtaining maximum contrast, to study the nervous system in humans and experimental animals. The PAS-positive vascular endothelium gave good contrast with blue-stained erythrocytes and hematoxylin-stained nuclei. Tannock and Steel (1969) used luxol fast-blue stain, while Endrich et al. (1979) injected hematoxylin and eosin.

Stained smears are recorded photomicrographically in natural color or by transference of these colors into black and white halftone contrasts. Details of black and white photographic technique are given in Zimmer (1947). All conventional histological techniques of using dyes to show blood vessels in tissue sections have some important shortcomings: they cannot provide information about the speed of blood flow and, furthermore, they often give rather poor contrast between blood vessels and other tissue elements.

5.2.2 Fluorescein

Fluorescein is a dye that produces fluorescence under radiation of different wavelengths. Fluorescence, a kind of luminescence, is an emission of radiation energy resulting from absorption of the incident radiation. When incident photons give energy to an acceptor molecule A some of the electrons within the molecule are raised to a higher energy level. The electrons may then return to their ground level with the emission of photons equal to the energy level difference. The fluorescence is usually less energetic than the incident photons. Where the acceptor A is nonfluorescent, the transferred energy is not reemitted; this process is referred to as quenching:

$$h\nu_1 + A \longrightarrow A^*$$

$$A^* \longrightarrow A + h\nu_2 \quad \text{(fluorescence)}$$

or

$$A^* \longrightarrow A + \text{thermal energy} \quad \text{(quenching)}$$

Here h is the Planck constant; ν is the frequency of the radiation emitted or absorbed by the molecule; and A^* is A in the excited state. According to Stokes' law, the fluorescent light is always of longer wavelength than that of the exciting radiation. Radiation throughout the range from the near infrared to gamma rays is capable of exciting fluorescence in suitable substances. However, attention is confined to visible fluorescent light excited by ultraviolet or visible light, since the majority of medical and biochemical applications fall within this category.

Fluorescence is applied in medicine in (1) fluorescent roentgen-ray screens (fluorscopy) and roentgen-ray photography; (2) clinical diagnosis and therapy by ultraviolet fluorescence (through fluorescence of body substances); (3) fluorescence and photosensitization; (4) fluorescence to trace drugs, hormones, enzymes, and vitamins; (5) fluorophotometry or fluorimetry; (6) the fluorescence microscope; (7) the phosphorescence microscope.

5.2.3 Radiographic Contrast Media

A certain part of the body, for example, the stomach, may attenuate radiation to no greater or lesser extent than the tissues surrounding it. Then the radiograph yields no separate visualization of the organ. To overcome this difficulty, the body cavity (kidney, stomach, or blood vessel) may be filled with a contrast medium which consists of a material with an atomic number and density different from the surrounding tissues.

The contrast media currently used in clinical diagnostic radiology are of two basic types: highly radiolucent and densely radiopaque. They serve to enhance radiographic contrast by virtue of their differential absorption of x-rays. Enhancement of contrast improves diagnostic visualization of the size, shape, and structure of organs, of spaces in and surrounding organs, and the course and patency of certain vessels and ducts.

The radiolucent-type media have atomic numbers and densities less than the surrounding structures. Air, oxygen, and carbon dioxide are in this category. They produce a clear outline of the gastrointestinal tract, of cavities such as the ventricles and cisternae of the brain and the subarachnoid spaces of the spinal cord. Gaseous media may also be injected into joint cavities and into retroperitoneal connective tissue spaces.

There is a greater number of the radioopaque type compounds with higher atomic numbers and densities than body tissues. Except for barium sulfate, they contain iodine. The increased attenuation of these media is due largely to increased photoelectric attenuation. Some are soluble in water and others are used as a solid or suspended in an aqueous or oily vehicle. The contrast agent may be administered directly into a lumina or cavity; examples include instillations into the tracheobronchial tree (bronchography), common bile duct (cholangiography), excretory tracts of the urinary system (retrograde ureteropyelography, cystography, and urethrography), ductal tracts of the salivary glands and seminal vesicles (sialography and seminal vesiculography), genital system (ureterosalpinography), and various sinus tracts and fistulas. They may be injected into the spinal subarachnoid space (myelography) and selectively into certain sections of the circulatory system for direct opacification of arteries, veins, cardiac chambers, and lymphatics (arteriography, venography, angiocardiography, and lymphangiography). The lumen of the gastrointestinal tract can be outlined by using both barium sulfate suspensions and the iodine compounds.

Some radiopaque media are introduced indirectly to the organ to be examined. A material of high atomic number is combined with another substance so as to produce a compound which the organ will absorb from the bloodstream. Following intravascular or oral administration, the medium is transported via the bloodstream to any organ that selectively concentrates and excretes the medium, thus imparting radiopacity to the organ and its excretory passages, for example, by the hepatobiliary system (cholecystrography and cholangiography) and by the renal system (excretory urography). Sodium tetraiodophenolphthalein is the contrast medium often used. After oral administration, it is secreted by the liver and is then concentrated by the gall bladder.

The absorption of x-rays is governed by the formula

$$A = C\lambda^3 Z^4 + 0.2$$

Absorption A thus varies with the third power of the wave length λ and the fourth power of the atomic number Z. C denotes a constant while 0.2 is the factor to account for scattering effects, depending on the wavelength at which the observation is made.

Besides having a different mass attenuation coefficient, the contrast media must have other essential features: low systemic and local tissue toxicity, minimal pharmacodynamic action, selective localization, ease of administration, and prompt and complete elimination.

5.3 VISUALIZATION AIDS

The morphology and function of vascular and excretory systems may be observed by the unaided eye or with the aid of an optical instrument. Appropriate vision magnification or image intensification may be required to enhance the quality of visualization. Commonly, photography and radiology are used in either a single or sequential recording.

5.3.1 Magnification of Vision and Image Intensification

Major portions of the vascular system such as fine arteries, arterioles, and capillaries are too small to be visualized by standard angiographic technique. Some form of magnification is required to delineate these structures. Three methods are available to magnify vision or to produce a magnified image: photographic or optical magnification, direct geometric magnification, and electronic magnification (Greenspan, 1971). The microscope is a basic instrument used to magnify the range of vision. Microphotography and microradiography are the processes of recording magnified images.

The process involved in making the fluoroscopic image brighter is known as image intensification or image amplification. Three types are in use. (1) The image intensifier is a vacuum type which converts the light image from a fluoroscopic type of screen into an electron image. The electrons thus produced are then accelerated, focused, and made to form a smaller and much brighter light image. (2) In the electrooptical image intensifier, the large fluoroscopic screen image is transferred by the mirror optical system to the input phosphor of the image-intensifier tube. Another optical system transfers the output phosphor image of the intensifier tube, without a change in size, to a cine camera. (3) In solid-state image intensification, an alternating voltage is applied between two outer electrodes. Due to this alternating voltage, emission of light from the electroluminescent layer increases when the photoconductor is irradiated.

5.3.2 Radiology

Radiology makes use of the interactions between x-rays and matter; the transmitted beam emerging from the subject contains information. The x-ray beam has a more or less uniform intensity before the attenuation processes (including unmodified scattering, photoelectric absorption, Compton effect, and pair effect) take place. When it leaves the subject, the beam contains an image made up of regions of different intensities.

Boyd (1955), Rogers (1967), and Fisher and Werner (1971) provide useful surveys on the application of autoradiographic techniques in medicine and biology, including discussions of the basic principles of autoradiography and photographic processes, and of practical problems related to them.

5.4 APPLICATIONS IN MEDICINE

5.4.1 Angiography

Angio is a prefix meaning blood vessel (i.e., vascular). Angiography is the radiographic visualization of blood vessels in living subjects by roentgen contrast methods. With the improvement in opaque media and high-speed filming equipment, it has become possible to obtain quantitative information on blood flow and shunt volumes, to estimate the elasticity of vessels, and to demonstrate in vivo the functional influence of pharmacological agents of the vascular system.

5.4.1.1 Historical development of angiography (Abrams, 1971a). Only a month after the announcement of Roentgen's discovery of x-rays in December 1895, visualization of human blood vessels was achieved by Haschek and Lindenthal (1896). They injected Teichman's mixture into the blood vessels of an amputated hand (Fig. 5.1). A photograph of their roentgenogram clearly showed the potential of the method for visualizing the vascular bed. Kassabian (1907) studied the blood vessels of infants and adults by injecting a concentrated emulsion of bismuth subnitrate, a strong solution of litharge (red oxide of lead) or metallic mercury. He also studied the blood vessels of the kidney, heart, brain, spleen, liver, and stomach in cadavers. In 1910, Franck and Alwens (1910) introduced a suspension of bismuth and oil into the heart of dogs and rabbits and observed the passage of the oily droplets from the heart into the lungs. Sicard and Forestier (1923) used Lipiodol to study the bronchial tree, the spinal subarachnoid space, and the cardiovascular system, first in dogs and later in human subjects. The first arteriograms and venograms in humans were obtained with 20% strontium bromide by Berberich and Hirsch (1923). Brooks (1924) observed the vessels of the human lower extremity by intraarterial injection of sodium iodide.

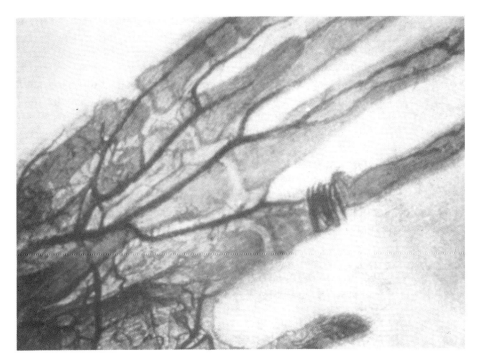

Figure 5.1 Roentgenogram made by Haschek and Lindenthal after injection of Teichman's mixture into blood vessels of amputated hand (Abrams, 1971a).

In 1928, Moniz et al. (1928) applied the technique of carotid angiography to the study of cerebral lesions. Forssmann (1931) developed the technique of cardiac catheterization to visualize the right heart and the pulmonary vessels in dogs. Moniz et al. (1931) used sodium iodide to visualize the pulmonary vessels in a variety of conditions. Osborne et al. (1923) described the visualization of the collecting system of the kidney by the intravenous and oral administration of large doses of sodium iodide. After the use of an organic iodide (Selectan) by Swick (1929) to opacify the collecting structures of the kidney during its excretion after intravenous injection, several opaque organic iodides were discovered to have the ability to be administered intraveneously to delineate specific segments of the human vascular bed.

Ameuille et al. (1936) succeeded in opacifying the heart using the catheter method. While Castellanos et al. (1938) opacified only the right heart chambers, Robb and Steinberg (1938) obtained sequential opacification of the right and left sides of the heart. In the case of thoracic aortography, Nuvoli (1936) used the direct puncture technique to study the human thoracic aorta, showing aneurysm, tortuosity, and other conditions. It was followed by countercurrent or retrograde brachial aortography of Castellanos and Pereiras (1939) and catheter aortography of Radner (1948). In the study of the abdominal aorta, Farinas (1941) used the retrograde passage of a catheter from the femoral arery for aortography, which has been extensively adopted for study of the visceral branches of the abdominal aorta.

Angiography is now applied to all of viscera for diagnosis of diseases, such as tumors. It is a definitive method of demonstrating congenital vascular malformations in all parts of the human body. Arteriograms can reveal the wear and tear on arteries caused by atheromatous change and can also delineate the site of thrombosis.

The most important technical elements in angiographic studies consist of the radiographic equipment and the opaque agent together with injectors and catheters. A major portion of the vascular system (fine arteries, arterioles, and capillaries) cannot be visualized, and thus some form of magnification of an x-ray image is required.

5.4.1.2 Radiopaque media. Radiopaque media for angiography are water soluble or water miscible and are used for contrast in x-ray filming. They can be classified into two groups: iodinated organic contrast media and inorganic contrast media (Strain and Rogoff, 1971). These media have permitted the visualization of not only the heart and blood vessels but also of other viscera which are nonopaque to x-rays. The media used for angiographic procedures throughout the past three decades include Diodrast, Neo-Iopax, Urokon, Miokon, Hypaque, Renografin, Ditriokon, Conray, Angi-Conray, and Isopaque (Abrams, 1971b). None of the opaque media currently in use is entirely satisfactory. In general, the diatrizoate drugs (Hypaque and Renografin) have been widely accepted and are the media of choice in most situations. Table 5.1 shows the recommended dosage of media for different angiographic procedures (Abrams, 1971b) for an adult weighing 64 kg.

5.4.1.3 Radiographic equipment. Equipment for high-speed filming of rapidly changing x-ray images is needed for explorating normal and pathological physiology and for dynamic, manipulative roentgenology. Rapid large-film changers (Amplatz, 1971) and cinefluorographic equipment (Abrams, 1971c) are widely used.

Rapid film changers permit the filming of several radiographs per second. Their emergence represented a major advance in the diagnostic application of angiography. Basically two systems are in use: (1) the rapid transport and mechanical change of radiographic cassettes and (2) rapid transport of film with intermittent closure of intensifying screens. The major advantages of rapid large-film systems over cinefluorography include (1) significantly better quality, (2) adequate field size, and (3) easy film processing. Disadvantages are (1) the limited speed of exposure (2) the absence of monitoring while filming, and (3) the inability to check the catheter position during injection.

The major advantage of fluoroscopy over roentgenography is that it permits the study of motion. Motion picture photography of the fluoroscopic screen during angiography provides a permanent recording of events under observation. Cinefluorography has direct and indirect types. Direct cinefluorography is the motion picture recording of events as they are produced on a conventional fluoroscopic screen by the exposure of the subject to x-radiation, as shown in Fig. 5.2*a*. Many films produced in this way are of high quality. Indirect cinefluorography (or cineradiography) is the motion picture photography of the output phosphor of an image intensifier (Fig. 5.2*b*). The light of the fluoroscopic image is increased by means of the image intensifier tube. The cine camera photographs the output phosphor of the intensifier. A solid-state amplifying screen may be used to replace the image-intensifier tube. A photoconductive layer combined with an electroluminescent phosphor furnishes the light-amplifying element. The cine camera photographs the image on the screen (Fig. 5.2*c*).

In kinescopy, the television monitor rather than the output phosphor of the image intensifier is employed to make the motion pictures (Fig. 5.2*d*). Television has the advantages of greater ease in viewing, simultaneous viewing by an unlimited number of individuals, an additional increase in image intensification, and being transmitted remotely or recorded on videotape.

The major advantages of cinefluorography include its capacity to study motion, high frame rates, possibility of employing relatively less contrast agents, monitoring the intensifier screen or the television screen throughout the examination, and analysis at variable film rates. The major disadvantages are its poor resolving power, no standardization of cine techniques, limited field size, and more complicated film processing and viewing as compared to conventional large film.

5.4.2 Microangiography

Microangiography was first introduced to study fine histological sections by Goby in 1913.

5.4.2.1 Microcirculation in normal tissues and organs. Application of microangiography to visualize vascular changes at the level of small arterioles and capillaries did not take place until 1935 by Grenchishkin and Prives (1935). Barclay (1947) refined the method to make the renal vascular bed visible to the human eye. Tirman et al. (1951) applied microradiography to illustrate the microcirculation of the kidneys, intestines, gallbladder, liver, spleen, heart, and lung. Most angiographic studies were performed following the sacrifice of animals injected prior to death; Bellman (1953) recorded the capillary circulation in living animals. This was limited to thin sections of the anatomy. Circulation within the rabbit ear in response to cold, heat, burns, and other agents was visualized. Saunders et al. (1957) introduced a photographic plate beneath the muscle of a living animal and observed vasodilatation and vasoconstriction. Carlson et al. (1960) conducted microangiographic studies of radiation-induced

Table 5.1 Choice and volume of media in angiography*

Procedure	Site of injection	Method of injection	Medium of choice	Concen- tration, %	Volume, cc	Duration of injection†
Angiocardiography						
Intracardiac	Variable	Catheter	Renografin	76	40–45	1.5–2
Intravenous	Antecubital vein	Robb-Steinberg needle	Renografin	76	50	1.5–2
Arteriography						
Adrenal	Adrenal arteries	Catheter	Renografin	60	3–5	0.5
Aortic arch	Aorta	Catheter	Renografin	76	45	2
Aorta, abdominal	Aorta	Catheter	Renografin	76	35	2
Brachial	Brachial artery	Needle	Renografin	60	20	0.6
Aorta, thoracic	Aorta	Catheter	Renografin	76	45	2–3
Carotid	Carotid artery	Needle or catheter	Renografin	60	4–12	1
Celiac	Celiac artery	Catheter	Renografin	76	40	4
Coronary						
Selective	Coronary artery	Catheter	Renografin	76	3–5	0.5
Nonselective	Aortic root	Catheter	Renografin	76	50	2
Femoral	Femoral artery	Needle or catheter	Renografin	60	15	2
Hepatic	Hepatic artery	Catheter	Renografin	76	20	1–2
Pancreatic	Celiac artery	Catheter	Renografin	76	30–40	1–2

Pelvic	Common iliac artery	Catheter	Renografin	76	30	2
Pulmonary						
Bilateral	Main pulmonary artery	Catheter	Renografin	76	40	1–2
Selective	Right or left pulmonary artery	Catheter	Renografin	76	25	1
Renal						
Selective	Renal artery	Catheter	Renografin	60	5–7	1
Bilateral	Aorta	Catheter	Renografin	76	35	2
Subclavian	Subclavian artery	Catheter	Renografin	60	15	0.5–1
Superior mesenteric	Superior mesenteric artery	Catheter	Renografin	76	30	2–3
Vertebral	Vertebral artery	Catheter	Renografin	60	6	0.5
Lymphangiography	Variable	Needle	Ethiodol	–	10	1 h
Venography						
Adrenal	Adrenal vein	Catheter	Renografin	76	6–8	2
Inferior vena cava	Both femoral veins	Catheter	Renografin	76	40	3
Peripheral veins	Variable	Needle	Renografin	60	35	3–5
Splenoportography	Spleen	Needle or catheter	Renografin	76	30	4
Superior vena cava	Superior vena cava	Catheter	Renografin	76	40	3

*The figures are those for an adult weighing 64 kg.
†In seconds, except where noted.
Source: Abrams (1971b).

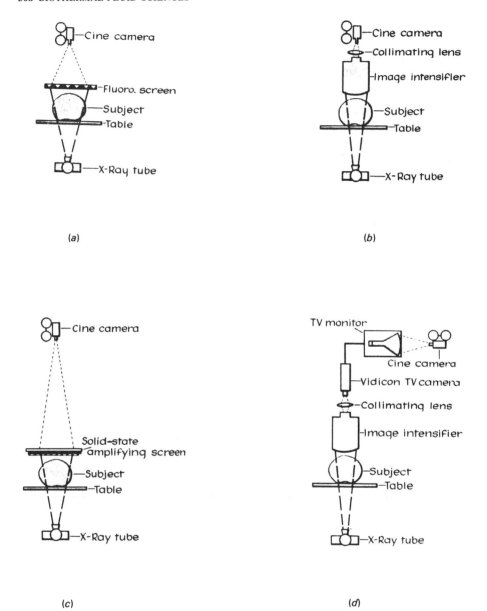

Figure 5.2 Motion picture photography of (*a*) the conventional fluoroscopic screen, (*b*) the output phosphor of an image intensifier, (*c*) the image on a solid-state amplifying screen, and (*d*) a television monitor, image intensifier-vidicon system (Abrams, 1971a).

vascular changes in bone following exposure of the hind limb to large doses. Chang and Trembly (1961) studied the effect of irradiation on the renal and brain circulation.

In the angiographic study of living animals, the animal is first anesthetized and then has its blood replaced by a gelatinous contrast medium. This is not close to a normal physiological state. Rubin et al. (1964) displaced the blood by catheterizing both the arterial and venous side. As the contrast medium (lead oxide particulate suspended in gelatin) is injected into one catheter, the blood flows out into the other. The heart continues to beat after all the blood is replaced and virtually all tissues are perfused with the radiopaque medium. Figure 5.3 shows microangiograms of the kidney.

5.4.2.2 Microcirculation of tumors. The vascular morphology, or in more precise terms, the degree of blood supply in vital and nonvital regions of natural or transplanted tumors, has long been of interest in tumor biology in its various aspects. Many tumors contain substantial regions which cannot readily be reached by bloodborne substances. These regions, which contain living cells capable of starting tumors when transplanted into new hosts, can easily be mapped out by changing the color of the systemic blood with a harmless dye that also colors the interstitial fluid without entering the living cells. In some tumors in rats and mice, the only region presenting an open connection with the systemic circulation is a thin peripheral zone varying in thickness from a few millimeters to a tenth of a millimeter or less (Goldacre and Sylven, 1962). Moore (1953) reviewed the use of fluorescent and radioactive-tracer methods for diagnosis and localization of brain tumors.

The activity of bloodborne canceroidal drugs and tumor antisera depends on the mechanism through which these agents are brought to the malignant cells. Therefore, the information on blood circulation in the terminal vascular bed of neoplastic tissue is of vital importance.

Neovascularization occurs in various human and animal tumors. This is particularly evident in experimentally transplanted tumors in which initially there are no cells near the neoplastic cells. However, after several days, the tumors become more and more vascularized, new cells are developed, and the implanted tissue piece grows rapidly thereafter. A visualization study is performed to determine (1) the gross histological features of new blood vessel formation in tumors and (2) blood circulation in the terminal vascular bed of neoplastic tissue.

All visualization techniques can be classified into two categories: in vivo and in vitro.

In vivo techniques Development of transparent chambers in the rabbit ear, in the superficial cheek pouch of hamsters, and in the dorsal skin flap of mice and rats allows direct microscopic visualization of the vascular arborization in time and permits functional studies under direct observation. This methodology is

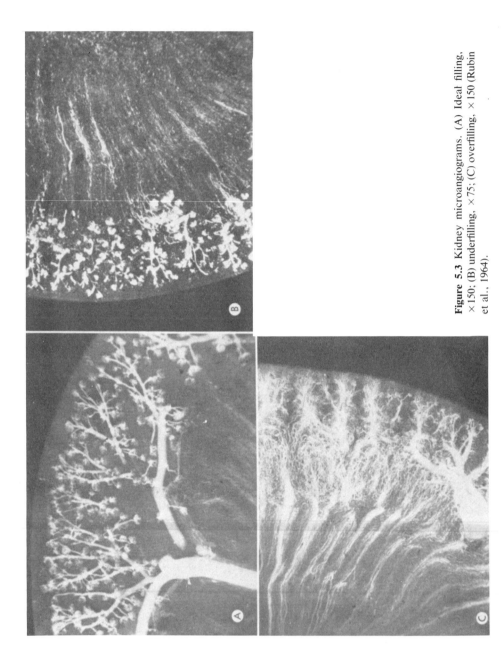

Figure 5.3 Kidney microangiograms. (A) Ideal filling, ×150; (B) underfilling, ×75; (C) overfilling, ×150 (Rubin et al., 1964).

used in conjunction with photomicrography, video technique, or radiography to visualize the hemodynamics of tumor tissue in vivo. The main restriction is that the tumor must be located in the subcutaneous tissues and that it can only grow in a lateral direction to a limited size.

Endrich et al. (1979) conducted in vivo experiments on WAG inbred Rijswijk rats fitted with Algire window chambers. Observations were started 14 days after implantation of a tiny tissue piece of BA 1112 sarcoma. With translumination by a tungsten-halogen projector lamp, blood flows in the microcirculation of the tumor were observed microscopically. The image of the microcirculatory bed was televised and recorded on videotape (Fig. 5.4). The blood cell velocity can be determined from video recordings of RBC flow and an on-line correlation. Figure 5.5 shows a typical photograph obtained 24 days after implantation.

In vitro techniques Injection of dyes, plastics, and radiopaque media permits the study of unrestricted tumor growth in a variety of sites. Limitations of this approach are that animals must be sacrificed and a large number of technical factors need careful standardization to achieve reproducibility. Correlative histological studies are possible but functional studies are limited.

Rubin and Casarett (1966) list all the techniques that have been employed to visualize microcirculation of tumors. They include the methods of India ink injection, transparent chambers in rabbit ears, transparent chambers in skin flaps in mice, staining of red blood cells, vinyl cases of blood vessels, angiography in vivo, direct photographic observations in vivo and microangiography. Jee and Arnold (1960) and Tiboldi et al. (1968) used a mixture of India ink and a solution of calfskin gelatin. Angulo et al. (1958) injected a homogeneous suspension of fine particulate carbon in a gelatin solution, while Rubin and Casarett (1966) used a mixture of lead oxide particulates (0.5–3.5 μm size) in a gelatin solution as a radiopaque contrast medium. Tannock and Steel (1969) employed ^{51}Cr labeling of red cells to visualize the functional state of blood vessels in tumors by autoradiography. Rassmussen-Taxdal et al. (1955) used fluorescence by hematoporphyrin to demonstrate and delimit lymphatic and cancer tissues in humans.

5.5 PHARMACOLOGICAL APPLICATIONS

In clinical pharmacology, both naturally occurring and artificial radionuclide are used as radiation sources in therapy. They are also used as radioactive tracers or labels for drug evaluations to gain information on drug distribution in metabolic and physiological function studies. Visualization of drug movement and distribution in the vascular system and tissue sections, within cells and in the terminal vascular bed of neoplastic tissue, is very useful (1) for eliciting a therapeutic response at the target site with a relatively slight effect on the rest of the body and (2) for determining the mechanism through which bloodborne

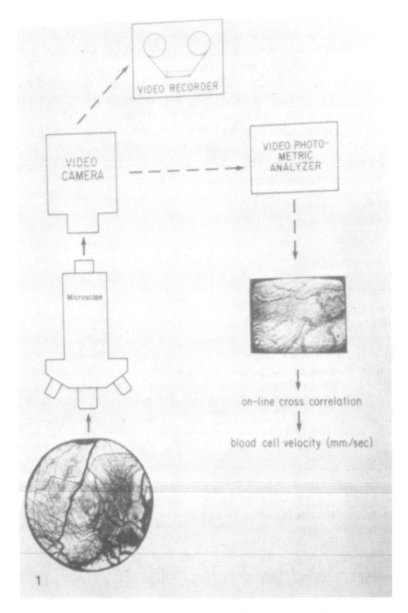

Figure 5.4 Diagram of experimental apparatus (Endrich et al., 1979).

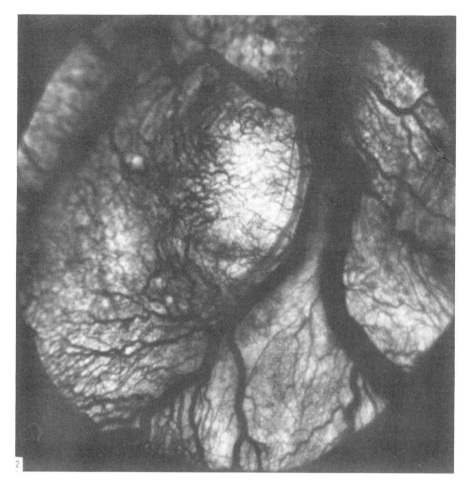

Figure 5.5 Microscopic visualization of vascular arborization of implanted tumor (Endrich et al., 1979).

drugs and antisera are brought to the cell level (normal and malignant cells). The use of a radionuclide in therapy is of no interest here, since the emphasis of flow visualization is on radionuclide distribution at the target site.

5.5.1 Radioactive Tracers or Labels

Movement of certain drugs and antibodies in the vascular system, in tissues, or within cells can be visualized by means of radioisotope labeling (also known as tapping) under ultraviolet light illumination. This method is used for visualization, identification, and localization of various antigens, protein hormones and many other substances (Coons, 1958; Nairn, 1962). Fluorescein-labeled antiserum (i.e., a fluorescein tagged to an antibody), which retains its immunological reactivity, is used as a marker. The most generally used fluorescent material is flurorescein isothiocyanate. The use of a labeling technique to visualize antigen in tissue sections was first perfected by Coons (1958) in the 1950s. At present, immunofluorescent antibody producers are used in one of three basic variations: (1) direct, (2) indirect or "sandwich," and (3) complement fluorescence (Friedman, 1967).

5.5.2 Thermographic detection of drug effects in peripheral circulation. Thermography, the visualization of a thermal image, is one of the three most important methods for evaluating many pharmocological peripherally acting drugs (Barnes and Gershon-Gohen, 1963; Gershon-Cohen et al., 1965; Winsor, 1967). In analyzing the effects of drugs on the peripheral circulation, one must take into account total circulation. Thermal patterns of arms, legs, and head often show clearly the division of blood from one vascular bed to another. This device is ideally suited for measuring temperature distribution resulting from drug administration. Figure 5.6 is a thermograph showing a division of blood from nose to cheek. Before the drug is administered (Fig. 5.6A), shows a warm (white) nose as compared with a cool (dark) cheek. Twenty minutes after administration of 50 mg of β-pyridylcarbinol, the nose is cool and the cheek warm (Fig. 5.6B). See Section 5.7 for more information on thermography.

5.6 BOTANICAL APPLICATIONS

Two kinds of conducting tissues are represented in the vascular system of the vascular plant: xylem, concerned with conduction of water and substances dissolved in it, and phloem, which serves as a conduit for organic food materials. The vascular tissue forms a continuous unit extending from near the tips of the roots to the vein endings in the leaves, as well as to the floral parts, fruits,

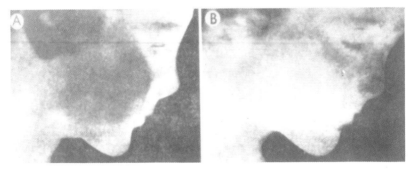

Figure 5.6 Thermograph showing diversion of blood from nose to cheek. (A) Nose is warm (white) and cheek is cool (dark). (B) Twenty minutes after administration of 50 mg of β-pyridylcarbinol, the nose is cool and the cheek is warm (Winsor, 1967).

and seeds if these are present. The phenomenon of transport of water and solutes by plants, called translocation, is as fundamental to the life of plants as is the circulation of blood to animal life. The speed of translocation can be measured or observed by introducing a foreign substance and following its spread.

Visualization studies of long-distance transport in phloem and xylem and very slow transport (called transfer) between cells have been successfully performed. Fluorescent dyes and radioactive substances are commonly used in these studies.

5.6.1 Fluorescent Dyes

Schumacher (1933) applied fluorescein to the sieve tube (which constitutes the channels of transport of the phloem) of leaves and observed its movement in the parietal cytoplasm. Later, Bauer (1949), Eschrich (1953), and others also observed fluorescent dyes moving in the sieve tubes. The use of fluorescein, a less toxic fluorescent dye, as an indicator for the translocation of foods in plants continued into the 1950s (Canny, 1973; Crafts, 1961). Bauer (1949) listed several other fluorescent dyes he found to move in sieve tubes: berberin sulfate, primulin, rhodamine B, rhodamine 6G and chinin chloride. However, the conditions for translocation and for good microscopic images are directly antithetical; for the former, intact plants, undamaged phloem, and a minimum of handling are required; for the latter, the thinnest layer of cells possible, separated from its neighboring tissues, is needed. The compromises that have been achieved have not been satisfactory both as translocation systems and as microscopic images.

5.6.2 Radioactive Tracers

The production of radioactive isotopes ushered in a new era in physiological research on translocation. They provided the long-sought tracers that could be employed in whole, intact plants, and since their introduction, more real knowledge of normal transport processes has been acquired than in all the preceding years.

Use of these tracers in animal physiology followed their discovery while progress in botanical applications of the technique was slow, probably due to the different nature of plant cells versus animal cells. Plant cells with a prominent cell wall and a large central sap vacuole create different preparative problems from animal cells. In earlier applications in plant work, migration of the tracer was detected by a Geiger-Müller (GM) counter (see, e.g., Stout and Hoagland, 1939; Biddulph, 1940). However, the radiograph of ^{35}S and ^{32}P translocation in the phloem by Biddulph (1956) led the way for widespread use of microradiography in plant physiology, mostly related to translocation problems. Many microautoradiographic studies of long-distance transport in both phloem and xylem have been successfully performed. Problems of short-distance transport and cell compartmentation have also been investigated using the same technique. Table 14.1 of Canny (1973) collects the available data on translocational speed measurement by a variety of radioactive tracer methods. All these experiments measure directly the speed of an object observed to move. It was concluded that the translocational speed fell in the range of 1–150 cm/h. In this speed range, cineradiography is unnecessary as static autoradiographic images can be interpreted in relation to the dynamic event of translocation.

5.6.3 Microradiography

Resolution in microautoradiography decreases substantially with increasing range of penetration of the radiation emitted by the radioisotope. Therefore, isotopes with low-energy radiation have been most favored in biological specimens. A selection of radioisotopes is listed in Table 5.2 (Lüttge, 1972). The principal elements of organic compounds (i.e., H and C) have radioisotopes ^{3}H and ^{14}C emitting a radiable component appropriate for microautoradiography. ^{35}S and ^{33}P are two other suitable isotopes of elements that are important in many organic substances. Organic molecules occurring in plant cells are composed mainly of H, C, N, O, P, and S. There are no radioactive isotopes from N and O with a half-life sufficiently long for use as tracers. As monoesters, $^{32}PO_4^{3-}$ and $^{35}SO_4^{2-}$ can be precipitated as lead and barium salts, respectively. ^{3}H- and ^{14}C-labeled compounds are used for the localization of labeled water-soluble organic compounds.

Canny (1973) conducted a survey of the literature on tracer methods of measuring translocational speed in plants. These methods allow the visualization of movement of the tracer. Figure 5.7a illustrates the plant of *Phaseolus* which assimilated $^{14}CO_2$ at a patch on a single lateral leaflet of the first trifoliate leaf and was allowed to translocate the products for 4 h before it was oven dried. The autoradiograph formed is shown in Fig. 5.7b. Note the absence of radiolabeled translocation from the mature and half-grown leaves, and the presence in the young buds and the roots. Figure 5.8 is a microautoradiograph of a series of three (a, b, c) subsequent longitudinal 1-μm sections the phloem.

Other photometric techniques, such as optical micrographs, scanning electron microscope photographs, etc. have also been employed. Lüttge (1972) emphasized the application of microautoradiography to plant physiology, particularly the transport of water-soluble inorganic ions, water-soluble organic compounds, and organic compounds insoluble in a wide range of polar and nonpolar solvents (DNA, RNA, protein, polysaccharide, etc).

5.7 THERMOGRAPHY AND ULTRASOUND IN MEDICINE

Any use of radiation (electromagnetic wave phenomenon) for visualization, both fluoroscopy and radiography, is accompanied by hazards, particularly when the time of exposure is prolonged. For this reason, diagnostic methods

Table 5.2 Nuclides used in microautoradiography of water-soluble inorganic ions

Compound nuclide used in	Radioisotope	β* %†	β* Energy, MeV	γ %†	γ Energy, MeV	X-rays, %†	Half-life (t₁/₂)	Specific activity available‡
Organic compounds	³H	100	0.018	—			12.3 yr	
	¹⁴C	100	0.156	—			5730 yr	
	³⁵S	100	0.167	—			87 days	$^{35}SO_4^{2-}$: carrier free
	³³P	100	0.248	—			25.2 days	$^{33}PO_4$: carrier free
Inorganic anions	³²P	100	1.71	—			14.3 days	$^{32}PO_4$: 50–140 Ci/mg P
	³⁶Cl	98.3	0.71	—		1.7	3×10^5 yr	$^{36}Cl^-$: > 3 mCi/g Cl
	⁴⁵Ca	100	0.256	—			165 days	$^{45}Ca^{2+}$: 10–25 mCi/mg Ca
	⁵⁹Fe	46	0.27	56	1.10		45 days	$^{59}Fe^{3+}$: 3–30 mCi/mg Fe
		53	0.46	44	1.29			
Inorganic cations		plus 2 others		plus 3 others				
	⁸⁶Rb	91	1.77	9	1.08		18.7 days	$^{86}Rb^+$: 2–10 mCi/mg Rb
		9	0.69					
Misc. limited use	²²Na	89.7	0.54	99.94	1.28	10.2	2.62 yr	$^{22}Na^+$: 100 mCi/mg Na
		0.06	1.83					
	⁴²K	82	3.54		largely		12.4 h	$^{42}K^+$: 250 mCi/g K
		18	1.98	18	1.52			

*β⁻ particles except for ²²Na where β⁺ particles are emitted.

†Percentages are based on the number of disintegrations in the parent radionuclide = 100%. Different γ-rays may be emitted in coincidence so that percentages > 100 are possible.

‡Data from suppliers catalogues.

Source: Luttge (1972).

317

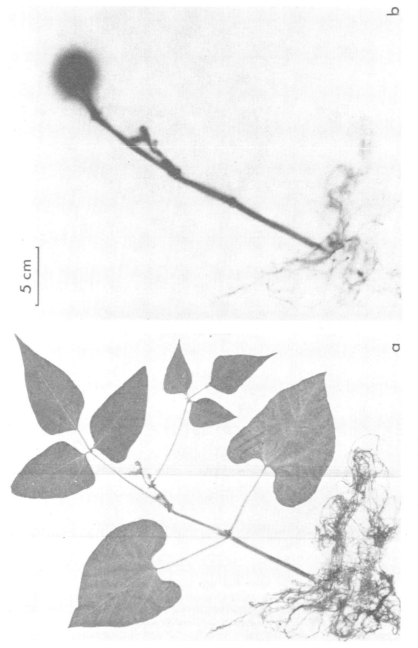

5 cm

b

a

Figure 5.7 (a) Phaseolus plant assimilating $^{14}CO_2$ at a patch on a single lateral leaflet of the first trifoliate leaf and being allowed to translocate the products for 4 h before being oven dried. (b) Autoradiograph formed by exposing the plant to x-ray film for 2 days (Canny, 1973).

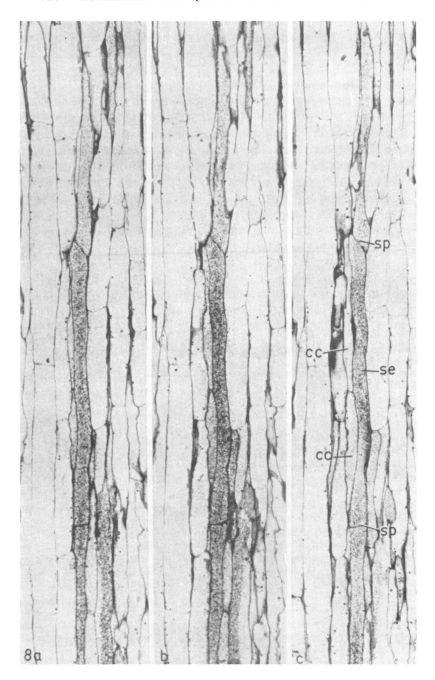

Figure 5.8 Microautoradiographs of a series of three subsequent longitudinal 1-μm sections of the phloem. Stained with gentian violet; cc, companion cell; se, sieve element; sp, sieve plate 255 $\times$ (Lüttge, 1972).

are being constantly sought and developed in an effort to replace radiology in medicine. Total replacement seems unlikely in the near future but efforts are being directed toward two means of visualization: thermography and ultrasonics.

Thermography is a method that gives a photographic display of point-to-point temperature differences of the surfaces of the body. The device consists of an infrared detector maximally sensitive to the 10-μm heat waves from the body, which produces voltage variations that are proportional to temperature. The voltage variations activate an electric light bulb which is mounted on a scanner. The light falls on a Polaroid film to produce a picture. Using a standard thermal gray scale, which consists of a series of heat standards, the temperature of the part under study can be determined (Barnes and Gershon-Cohen, 1963; Gershon-Cohen et al., 1965). A temperature difference of 0.1°C can be detected. Quantitative results can be obtained by placing the photographic film in a reader and evaluating the amount of reflected light from various points on the picture as compared on the gray scale.

As an aid in visualization, the skin may be photographed after being treated with a phosphor, since upon irradiation with ultraviolet light, it emits visible light at different intensities depending on temperature. Many pathological conditions, such as tumors, produce local changes of temperature. It is also used in the detection of drug effects on the peripheral circulation, as indicated in Section 5.5. In botanical applications, thermography records the thermal image of plant leaves as a guide in the study of heat and mass transfer between the leaves and the ambient air.

Ultrasounds refer to mechanical vibrations of frequencies greater than those normally audible to the human ear. A detector records the intensity of sound waves that are reflected from region to region of the area scanned. The intensity depends on the density of the structure from which the waves are reflected. Ultrasonography, a medical diagnostic procedure, determines any abnormalities.

In summary, flow visualization is as important as flow measurement in experimental thermal sciences. The following additional references are recommended to the reader: Eschrich and Steiner (1967), Fritz and Eschrich (1970), Geiger et al. (1969), Heyser et al. (1969), Mortimer (1965), Perkins et al. (1959), Schmitz and Willenbrink (1967, 1968), Trip and Gorham (1967), Veret (1987), Webb and Gorham (1964), and Yang (1985, 1988).

REFERENCES

Abrams, H. L. (1971a) Introduction and Historical Notes, in *Angiography* (ed. H. L. Abrams), 2d ed., vol. 1, chap. 1, Little, Brown, Boston.

Abrams, H. L. (1971b) The Opaque Media, in *Angiography* (ed. H. L. Abrams), 2d ed., vol. 1, chap. 2, Little, Brown, Boston.

Abrams, H. L. (1971c) Cinefluorographic Equipment, in *Angiography* (ed. H. L. Abrams), 2d ed., vol. 1, chap. 6, Little, Brown & Co., Boston.

Ameuille, P., Ronneaux, G., Hinault, V., Desgrez, J. and Lemoine, J. M. (1936) Remarques sur quelques cas d'arteriographie pulmonaire chez l'homme vivant, *Bull. Mem. Soc. Med. Hosp. Paris,* 52:729.

Amplatz, K. (1971) Rapid Film Changers, in *Angiography* (ed. H. L. Abrams), 2d ed., vol. 1, chap. 5, Little, Brown, Boston.

Angulo, A. W., Hessert, E. C., Jr., and Kownacki, V. P. (1958) A carbon-gelatin injection mass for minute vascular and respiratory passages, *Stain Technol,* 33:63.

Barclay, A. A. (1947) Micro-arteriography, *Br. J. Radiol.,* 20:394.

Barnes, R. B., and Gershon-Cohen, J. (1963) Clinical thermography, *J. Am. Med. Assoc.,* 185: 949.

Bauer, L. (1949) Über den Wanderungsweg fluoreszierender Farbstoffe in den Siebröhren, *Planta,* 37:221.

Bellman, S. (1953) Microangiography, *Acta Radiol.,* Suppl. 102.

Berberich, J., and Hirsch, S. (1923) Die röntgenographische Darstellung der Arterien und Venen am Lebenden, *München. Klin. Wochenschr.,* 49:2226.

Biddulph, O. (1940) Absorption and movement of radiophosphorus in bean seedlings, *Plant Physiol.,* 15:131.

Biddulph, S. F. (1956) Visual indication of ^{35}S and ^{32}P translocation in the phloem, *Am. J. Bot.,* 43:143.

Boyd, G. A. (1955). *Autoradiography in Biology and Medicine,* Academic Press, New York.

Brooks, B. (1924) Intraarterial injection of sodium iodide, *J. Am. Med. Asso.,* 82:1016.

Canny, M. J. (1973) *Phloem Translocation,* Cambridge University Press, London.

Carlson, H. C., Williams, M. M. D., Childs, D. S., Jr., Dockerty, M. B., and Janes, J. M. (1960) Microangiography of bone in study of radiation changes, *Radiology,* 74:113.

Castellanos, A., and Pereiras, R. (1939) Counter-current aortography, *Rev. Cubana Cardiol.,* 2: 187.

Castellanos, A., Pereiras, R., and Vasquez-Paussa, A. (1938) On a special automatic device for angiocardiography, *Bol. Soc. Cubana Pediat.,* 10:209.

Chang, C. H., and Trembly, B. (1961) Microangiographic studies of kidney and brain in animals: Technical aspects. *Yale J. Biol. Med.,* 33:451.

Coons, A. H. (1958) Fluorescent Antibody Methods, in *General Cytochemical Methods* (ed. J. F. Danielli), vol. 1, Academic Press, New York, 400 pp.

Crafts, A. S. (1961) *Translocation in Plants,* Holt, Rinehard & Winston, New York.

Endrich, B., Reinhold, H. S., Gross, J. F., and Intaglietta, M. (1979) Tissue perfusion inhomogeneneity during early tumor growth in rats, *J. Natl. Cancer Inst.,* 62:387.

Eschrich, W. (1953). Beiträge zur Kenntnis der Wundsiebröhrenentwicklung bei Impatiens holsti, *Planta,* 43:37.

Eschrich, W., and Steiner, M. (1967) Autoradiographische Untersuchungen zum Stofftransport bei Polytrichum commune, *Planta,* 74:330.

Farinas, P. L. (1941) A new technique for the arteriographic examination of the abdominal aorta and its branches, *Am. J. Roentgenol.,* 46:641.

Fisher, H. A., and Werner, G. (1971) *Autoradiographie,* Walter de Gruyter, Berlin.

Forssmann, W. (1931) Über Kontrastdarstellung der Höhlen des levenden rechten Herzens und der Lungenschlagader, *Munchen. Med. Wschr.,* 78:489.

Franck, O., and Alwens, W. (1910) Kreislaufstudien am Röntgenschirm, *München. Med. Wschr.,* 51:950.

Friedman, H. (1967) Immunologic Techniques with Pharmocologic Application, in *Animal and Clinical Pharmacologic Techniques in Drug Evaluation,* (ed. P. E. Siegler and J. H. Moyer), III, vol. 2, chap. 50, Year Book Publishers, Chicago.

Fritz, E., and Eschrich, W. (1970) ^{14}C-Mikroautoradiographie Wasserlöslicher Substanzen im Phloem, *Planta* 92:267.

Geiger, D. R., Saunders, M. A., and Cataldo, D. A. (1969) Translocation and accumulation of translocate in the sugar beet petiole, *Plant Physiol.* 44:1657.

Gershon-Cohen, J., Haberman-Brueschke, J. A. D., and Brueschke, E. E. (1965) Medical thermography: A summary of current status, *Radiol. Clin. North Am.*, 3:403.

Goby, P. (1913) New application of x-ray: Microradiography, *J. Roy. Microsc. Soc.*, 4:373.

Goldacre, R. J., and Sylven, B. (1962) The access of bloodborne dyes to various tumor regions, *Br. J. Cancer*, 16:306.

Greenspan, R. H. (1971) Magnification Angiography in *Angiography* (ed. H. L. Abrams), 2d ed., vol. 1, chap. 8, Little, Brown, Boston.

Grenchishkin, S. V., and Prives, M. G. (1935) Weiche Röntgenstrahlen in Medizin und Embryologie, *Vesn. Rentgenol. Radiol.* 14:201.

Groat, W. A. (1936) A general-purpose polychrome blood stain, *J. Lab. Clin. Med.*, 21:978.

Haschek, E., and Lindenthal, O. T. (1896) A contribution to the practical use of the photography according to Roentgen, *Wien. Klin. Wochenschr.*, 9:63.

Heyser, W., Eschrich, W., and Evert, R. F. (1969) Translocation in perennial monocotyledons, *Science*, 164:572.

Jee, W. S. S., and Arnold, J. S. (1960) India ink-gelatin vascular injection of skeletal tissues, *Stain Technol.*, 35:59.

Kassabian, M. K. (1907) *Roentgen Rays and Electrotherapeutics with Chapters on Radium and Phototherapy*, J. B. Lippincott, Philadelphia.

Kluver, H., and Barrera, E. (1953) A method for the combined staining of cells and fibers in the nervous system, *J. Neuropathol. Exp. Neurol.*, 12:400.

Leishman, W. B. (1901) A simple and rapid method of producing Romanowski staining in malaria and other blood films, *Br. Med. J.*, 2:757.

Lüttge, U. (ed.) (1972) *Microautoradiography and Electron Probe Analysis—Their Application to Plant Physiology*, Springer-Verlag, Berlin.

Margolis, G., and Pickett, J. P. (1956) New applications of luxol fast blue myelin stain: Myeloangio-cytoarchitectonic method; myelin-neuroglia method; myelin-fat method; myelin-axis cylinder method, *Lab. Invest.*, 5:459.

Moniz, E., Diaz, A., and Lima, A. (1928) La radioarteriographie et la topographie cranioencephalique, *J. Radiol. Electrol.*, 12:72.

Moniz, E., de Carvalho, L., and Lima, A. (1931) Angiopneumographie, *Presse Med.*, 53:996.

Moore, G. E. (1953) *Diagnosis and Localization of Brain Tumors*, Charles C Thomas Publisher, Springfield, Ill.

Mortimer, D. C. (1965) Translocation of the products of photosynthesis in sugar beet petiole, *Can. J. Bot.*, 43:269.

Nairn, R. C. (ed.) (1962) *Fluorescent Protein Tracing*, Livingstone, London.

Nuvoli, I. (1936) Arteriografia dell'aorta toracica mediante punctura dell'aorta ascendente o del ventricolos, *Policlin.* (Pract.), 43:227.

Osborne, E. D., Southerland, C. G., Scholl, A. J., and Roundtree, L. T. (1923) Roentgenography of the urinary tract during excretion of sodium iodide, *J. Am. Med. Assoc.*, 80:368.

Perkins, H. J., Nelson, C. D., and Gorham, P. R. (1959) A tissue-autoradiographic study of the translocation of ^{14}C-labeled sugars in the stems of young soybean plants, *Can. J. Bot.*, 37: 871.

Radner, S. (1948) Thoracic aortography by catheterization from the radial artery, *Acta Radiol.* (*Stockholm*), 29:178.

Rassmussen-Taxdal, D. S., Ward, G. E., and Figge, F. H. J. (1955) Fluorescence of human lymphatic and cancer tissues following high doses of intravenous hematoporphyrin, *Cancer*, 8:78.

Robb, G. P., and Steinberg, I. (1938) A practical method of visualization of chambers of the heart, the pulmonary circulation, and the great blood vessels in man, *J. Clin. Invest.*, 17:507.

Rogers, A. W. (1967) *Techniques of Autoradiography*, Elsevier, Amsterdam.

Rubin, P., and Casarett, G. (1966) Microcirculation of tumors, Part I: Anatomy, function, and necrosis. *Clin. Radiol.* 17:220.

Rubin, P., Casarett, G. W., Kurohara, S. S., and Fujii, M. (1964) Microangiography as a technique—radiation effect versus artifact, *Am. J. Roentgenol.,* 92:378.

Saunders, R. L. de C. H., Lawrence, J., and Maciver, D. A. (1957) Microradiographic Studies of Vascular Patterns in Muscle and Skin, in *X-Ray Microscopy and Microradiography* (ed. V. E. Cosslett, A. Engstrom, and H. H. Pattee), Academic Press, New York.

Schmitz, K., and Willenbrink, J. (1967) Histoautoradiographischer Nachweis ^{14}C-markierter Assimilate im Phloem, *Z. Pflanzenphysiol.,* 58:97.

Schmitz, K., and Willenbrink, J. (1968) Zum Nachweis tritiierter Assimilate in den Siebröhren von Cucubita, *Planta,* 83:111.

Schumacher, W. (1933) Untersuchunger über die Wanderung des Fluoreszeins in den Siebrohren, *Jahrb. Wiss. Bot.,* 77:685.

Sicard, J. A., and Forestier, G. (1923) Injections intravascularies d'huile iodée sous controle radiologique, *C. R. Soc. Biol.,* 88:1200.

Stout, P. R., and Hoaglund, D. R. (1939) Upward and lateral movement of salt in certain plants as indicated by radioactive isotopes of potassium, sodium and phosphorus absorbed by roots, *Am. J. Bot.,* 26:320.

Strain, W. H., and Rogoff, S. M. (1971) The Radiopaque Media: Nomenclature and Chemical Formulas, in *Angiography* (ed. H. L. Abrams), 2d ed., vol. 1, chap. 3, Little, Brown, Boston.

Swick, N. (1929) Darstellung der Niere und Harnwege in Röntgenbild durch intravenöse Einbringung eines neuen Kontraststoffes, des Uroselectans, *Klin. Wochenschr.,* 8:2087.

Tannock, I. F., and Steel, G. G. (1969) Techniques for study of the anatomy and function of small blood vessels in tumors. *J. Natl. Cancer Inst.,* 42:771.

Tiboldi, T., Kurcz, M., and Kovacs, K. (1968) Examination of the blood supply of oestrogen hormone induced pituitary tumors in rats with India ink method, *Neoplasma,* 15:259.

Tirman, W. S., Caylor, C. E., Banker, H. W., and Caylor, T. E. (1951) Microradiography: Its application to study of vascular anatomy of certain organs of rabbits, *Radiology* 57:70.

Trip, P., and Gorham, P. R. (1967) Autographic study of the pathway of translocation, *Can. J. Bot.,* 45:1567.

Veret, C. (1987) *Flow Visualization IV,* Hemisphere, Washington, D.C.

Webb, J. A., and Gorham, P. R. (1964) Translocation of photosynthetically assimilated C[14] in straight-necked squash. *Plant Physiol,* 39:663.

Winsor, T. (1967) Clinical Techniques for Evaluating Drug Effects on Peripheral Circulation, in *Animal and Clinical Pharmacologic Techniques in Drug Evaluation* (ed. P. E. Siegler and J. H. Moyer, III), vol. 2, chap. 40, Year Book Medical Publishers, Chicago.

Yang, W.-J. (ed.) (1985) *Flow Visualization III,* Hemisphere, Washington, D.C.

Yang, W.-J. (ed.) (1988) *Handbook of Flow Visualization,* Hemisphere, Washington, D.C.

Zimmer, S. M. (1947) Photomicrography of Blood, in *Medical Physics* (ed. O. Glasser), vol. 1, Year Book Publishers, Chicago, p. 1013.

MASS TRANSFER

The most typical physiological problems lie in the field of mass transfer. For example, the transport of oxygen, carbon dioxide, nutrients, end-products and various ions across the cell membrane is vital to the survival of cells. Mass transfer is the process in which a component or species in a mixture travels from a region of high concentration to one of low concentration. If the mixture is stagnant, the transfer occurs by molecular diffusion. If there is bulk mixing of the component in the mixture from mechanical stirring or because of a density gradient, mass transfer occurs primarily by the mechanism of forced or natural convection. These mechanisms are analogous to heat transfer by conduction and by convection; however, there is no counterpart in mass transfer for thermal radiation. The analogy is so straightforward that equations developed for heat transfer are often found to apply to mass transfer by a mere change in the meaning of the symbols. Very often, heat and mass transfer processes take place simultaneously, for example, with heat and water losses from the skin surface into the environment. Chemical reactions may also take place in the traveling species under the mass transfer situation.

6.1 COMPOSITIONS, FLUXES, AND PHASE EQUILIBRIUM

The first step needed in mass transfer studies is to define the compositions in a mixture, the velocities, and the fluxes (rate of transfer per unit area). Table 6.1 summarizes the composition definitions and their relations. The average velocity v, the species (i) velocity v_i, and the diffusion velocity v_{Di}, are all defined in reference to a certain fixed frame. The velocity measured with a

Table 6.1 Definitions of a species i and their relations

Quantity	Symbol	Definition	Relation
Mass	m_i		$m = \Sigma\, m_i$
Mole	n_i		$n = \Sigma\, n_i$
Mass concentration	ρ_i	m_i/V	$\rho = \Sigma\, \rho_i$
Molar concentration	C_i	n_i/V	$C = \Sigma\, C_i$
Mass fraction	$(mf)_i$	m_i/m	$\Sigma\, (mf)_i = 1$
Mole fraction (liquid)	x_i	n_i/n	$\Sigma\, x_i = 1$
Volume	V_i		$V = \Sigma\, V_i$
Volume fraction	$(vf)_i$	V_i/V	$\Sigma\, (vf)_i = 1$
Partial pressure	P_i		$P = \Sigma\, P_i$
Mole fraction (gas)	y_i	P_i/P	$\Sigma\, y_i = 1$

probe is the mass average velocity or the hydrodynamic velocity. The species velocity is the velocity of an individual species, while v is the weighted average of all the species velocities:

$$v = \Sigma\, (mf)_i v_i \qquad (6.1)$$

Their difference is defined as the diffusion velocity:

$$v_{Di} = v_i - v \qquad (6.2)$$

The molar average velocity $\bar{v}$ is defined by

$$\bar{v} = \Sigma\, x_i v_i \qquad (6.3)$$

and its corresponding diffusion velocity $\bar{v}_{Di}$ is

$$\bar{v}_{Di} = v_i - \bar{v} \qquad (6.4)$$

The flux of species i relative to a fixed frame consists of two components: diffusional and convective fluxes. The latter is induced by the bulk motion of a species v_i, whereas the former is relative to the bulk average velocity $(v_i - v)$. Each flux can be expressed on a mass or molar basis. Accordingly, one writes Fick's first law of diffusion as

$$N_i'' = \rho_i(v_i - v) + (mf)_i \Sigma\, \rho_i v_i \quad \text{(mass basis)} \qquad (6.5a)$$

$$= C_i(v_i - v) + x_i \Sigma\, C_i v_i \quad \text{(molar basis)} \qquad (6.5b)$$

$$\underset{\text{diffusional}}{} \qquad \underset{\text{convective}}{}$$

In a binary mixture consisting of species A and B, the molar flux is related to the concentration gradient by

$$N_{Az}'' = -CD_{AB}\frac{\partial x_A}{\partial z} + x_A(N_{Az}'' + N_{Bz}'') \qquad (6.5c)$$

where N_{Az}'' is the z-component of the vector N''_A and D_{AB} denotes the mass diffusivity.

For dilute solutions where x_i is very low, the convective contribution becomes negligible. The mass flux is reduced to

$$N_i'' = -D \frac{dC_i}{dz} = -DC \frac{dx_i}{dz}$$

In general, the total flux is a summation of individual fluxes

$$N'' = \Sigma N_i'' \qquad (6.6)$$

Figure 6.1 illustrates partial pressures P_i and compositions x_i or y_i at equilibrium between an ideal gas and an ideal liquid solution for a binary-component mixture under constant pressure P. For both species, the partial pressure is represented by a straight line with a slope equal to the vapor pressure:

$$P_i = x_i p_i \qquad (6.7)$$

where p_i denotes the vapor pressure of species i. The relation [Eq. (6.7)] is called Raoult's law which is often used to determine P_i. The general form of Henry's law reads

$$P_i = x_i H_i \qquad (6.8)$$

where H_i represents the Henry constant which is usually determined empirically. Usually, Raoult's law applies to the high-concentration range and Henry's law applies to low concentrations.

6.2 MASS DIFFUSION

Diffusion is defined as the random motion of all molecules and ions in all directions. It results from the kinetic energy stored in all these particles, making them move continually, bouncing against each other, each particle weaving its way among all the others and continuing to go wherever it is not limited by some solid structure.

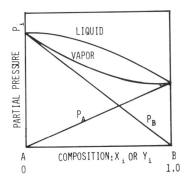

Figure 6.1 Equilibrium between ideal gas and ideal liquid solution.

The diffusional flux is described by Fick's first law as

$$N'' = -D\frac{dC}{dz} \tag{6.9}$$

or

$$N'' = -D\frac{d\rho}{dz} \tag{6.10}$$

in the z direction where D is the mass diffusivity and dC/dz and $d\rho/dz$ denote the concentration gradient.

Theoretical determination of concentration variations is demonstrated in Sections 6.2.1 and 6.2.2 for a few simple cases.

6.2.1 Steady One-dimensional Diffusion

Consider oxygen diffusing through a plane layer of fluid of thickness δ such as a membrane (Fig. 6.2a). With the origin being fixed on the left surface, Eq. (6.9) can be integrated to yield

$$N = DA\frac{\Delta C}{\delta} \tag{6.11}$$

In the gas phase, the equation can be written as

$$N = \frac{DA}{RT}\frac{\Delta P}{\delta} \tag{6.12}$$

where R denotes the gas constant.

In binary mixtures, this transfer phenomenon is also known as an equimolar counter diffusion since the rates of interdiffusion of both species are equal (i.e., $N''_A = -N''_B$). It requires that both components are of essentially equal densities. The total concentration C or ρ is constant.

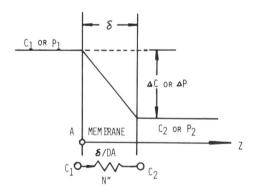

Figure 6.2 (a) Steady, one-dimensional diffusion through a plane layer.

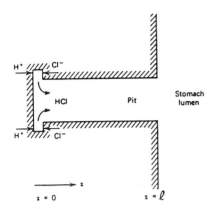

Figure 6.2 (*Cont.*) (*b*) Diffusion model for gastric HCl secretion.

Example 6.1 HCl/water interdiffusion in stomach pits (Rehm et al., 1953)
During gastric secretion, pure HCl is secreted from the base of a cylindrical pit and migrates through the pit into the stomach lumen, as shown in Fig. 6.2*b*. Water follows the HCl to maintain osmotic equilibrium with the blood flowing through the surrounding tissue. The water is regulated by the osmotic permeability of secretory tissues to water and the rate of secretion of osmotically active substances. The HCl secretion rate is treated as known and the stomach itself is filled with pure water.

The system in binary, with the subscripts A and B denoting HCl and water respectively. Equation (6.5c) is employed:

$$N''_{A0} = -CD_{AB} \frac{dx_A}{dz} + x_A(N''_{A0} + N''_{B0})$$

Here, the subscript 0 denotes the state at $z = 0$. The boundary conditions are

$$x_A(0) = x_{A0}; \qquad x_A(l) = 0$$

in which x_{A0} is the mole fraction of acid required to produce a solution isotonic with blood. The solution is obtained as

$$\frac{N''_{A0}}{N''_{A0} - x_{A0}(N''_{A0} + N''_{B0})} = \exp \frac{(N''_{A0} + N''_{B0})l}{CD_{AB}}$$

Since the acid solution secreted is mostly water, one can approximate

$$N''_{A0} + N''_{B0} \cong N''_{B0} = C_B v_B \cong Cv$$

and

$$\frac{N''_{A0}}{N''_{A0} - C_{A0}v} = \exp \frac{vl}{D_{AB}}$$

Here, v and v_B denote the mass average velocity and velocity of water, respectively. The result indicates that there is both a strong convective contribution to acid transport and substantial water migration under the specified conditions. Here water movement is passive and dependent on osmotic gradient. Should the osmolality of the stomach contents become sufficiently greater than that of the blood flowing through the surrounding tissue, water movement is reversed.

6.2.2 Unsteady One-dimensional Diffusion

Consider the gall bladder as an infinitely flat plate, since the serosa thickness δ is very small compared with the radius. The initial concentration of a salt such as NaBr in the serosa is C_0. Then the surface s concentration is suddenly raised to C_s, resulting in the diffusion of the salt in the serosa with diffusivity D. It is postulated that it can not diffuse through the epithelial cells. With the origin fixed at the interface between the serosa and the epithelial cells, the governing equation reads (Lih, 1975)

$$\frac{\partial C}{\partial t} = D \frac{\partial^2 C}{\partial z^2} \tag{6.13}$$

It is subjected to the initial and boundary conditions

$$C(0, z) = C_0 \qquad \frac{\partial C(t, 0)}{\partial x} = 0 \qquad C(t, \delta) = C_s \tag{6.14}$$

The problem may be solved by the method of separation of variables or Laplace transformation. The time history of salt concentration can be obtained as

$$\frac{C_s - C(t, z)}{C_s - C_0} = 2 \sum_{n=0}^{\infty} \frac{(-1)^n}{\lambda_n \delta} e^{-(n+1/2)\pi^2 \tau} \cos \lambda_n z \tag{6.15}$$

where

$$\tau = \frac{Dt}{S^2} \qquad \lambda_n = \frac{(2n+1)\pi}{2\delta} \qquad n = 0, 1, 2, \ldots \tag{6.16}$$

The volume-average salt concentration in the serosa C_a is

$$\frac{C_s - C_a(t)}{C_s - C_0} = \frac{2}{\pi^2} \sum_{n=0}^{\infty} \frac{(-1)^n}{(n+\frac{1}{2})^2} \exp\left[-(n+\tfrac{1}{2})^2 \pi^2 \tau\right] \tag{6.17}$$

The results are useful in determining the mechanism of solute transport by the gall bladder.

A great variety of problems in heat conduction have been solved in the literature. According to the heat and mass transfer analogy, the solutions of most of these problems can be utilized to find the solutions of the corresponding mass diffusion problems. Lightfoot (1974) also treats some general problems of transport phenomena in living organisms.

6.3 CONVECTIVE MASS TRANSFER

The rate of mass transfer from a phase boundary (a solid or fluid interface) to a fluid stream is given by the expression

$$N = h_D A (C_S - C_\infty) \qquad (6.18)$$

Here, h_D denotes the convective mass transfer coefficient and $(C_S - C_\infty)$ is the concentration difference between the interface and the bulk fluid.

6.3.1 Interface Transfer

For interphase transfer (Fig. 6.3) the mass flux across the interface can be expressed as

$$N'' = h_{DI}(C_I - C_{IS}) = h_{DII}(C_{IIS} - C_{II}) \qquad (6.19)$$

The equilibrium relation at the interface gives

$$C_{IS} = m C_{IIS} \qquad (6.20)$$

in which m is the equilibrium curve slope.

Due to difficulty in measuring the interfacial compositions C_{IS} and C_{IIS}, it is common to use overall coefficients U_D as in the heat transfer case. Let the equilibrium concentrations of C_I and C_{II} be C_I^* and C_{II}^*, respectively. Then,

$$C_I^* = m C_{II} \qquad C_{II}^* = \frac{C_I}{m} \qquad (6.21)$$

One may use $C_I - C_I^*$ and $C_{II}^* - C_{II}$ as the overall driving force for mass transfer between the two phases such that

$$N'' = U_{DI}(C_I - C_I^*) = U_{DII}(C_{II}^* - C_{II}) \qquad (6.22)$$

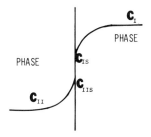

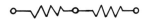

Figure 6.3 Interfacial transfer.

Then the overall coefficients must be

$$\frac{1}{U_{DI}} = \frac{1}{h_{DI}} + \frac{m}{h_{DII}} \quad \text{and} \quad \frac{1}{U_{II}} = \frac{1}{h_{DII}} + \frac{1}{mh_{DI}} \tag{6.23}$$

The equations signify that there are two resistances to the overall mass transfer between the two phases as illustrated in the electrical circuit in Fig. 6.3.

6.3.2 Membrane Transfer

Consider the transfer of a solute from one phase to another (Fig. 6.4). The two liquids are separated by a membrane that is permeable to the solute. The hemodialysis membrane is an example of this. The mass flux can be expressed as

$$N'' = h_{DI}(C_I - C_{IS})$$
$$= K^*(C_{IS}^* - C_{IIS}^*) \tag{6.24}$$
$$= h_{DII}(C_{IIS} - C_{II})$$

where K^* denotes the permeability (i.e., mass transfer coefficient) of the membrane defined as the diffusion coefficient of the solute in membrane D^*, divided by the membrane thickness δ,

$$K^* = \frac{D^*}{\delta} \tag{6.25}$$

C_{IS}^* and C_{IIS}^* are the concentrations in the membrane at the interface. They are related to the interfacial concentrations C_{IS} and C_{IIS} by

$$\frac{C_{IS}^*}{C_{IS}} = \frac{C_{IIS}^*}{C_{IIS}} = \zeta \tag{6.26}$$

where ζ is the distribution coefficient which depends on the solubility of the solute in the membrane. It is convenient to define an overall mass transfer

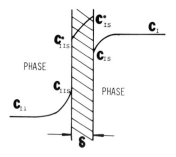

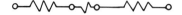

Figure 6.4 Membrane transfer.

Table 6.2 Heat and mass transfer analogy

Transport mechanism	Heat	Mass
Driving potential	ΔT	ΔC or ΔP
Diffusion rate	$q_K = -kA\, \partial T/\partial n$	$N = -DA\, \partial C/\partial n$
Biot number	$h_c L/k$	
Diffusivity ratio	Prandtl no. (Pr) $= \nu/\alpha$	Schmidt no. (Sc) $= \nu/D$
Convective transfer rate	$q_c = h_c A\, \Delta T$	$N = h_D A\, \Delta C$ or $h_D A\, \Delta P$
Dimensionless transfer coefficient	Nusselt no. (Nu) $= h_c L/k$	Sherwood no. (Sh) $= h_D L/D$
	Local $\mathrm{Nu}_x = h_{cx} x/k$	Local $\mathrm{Sh}_x = h_{Dx} x/D$
Forced convection	$\mathrm{Nu} = f(\mathrm{Re},\, \mathrm{Pr})$	$\mathrm{Sh} = f(\mathrm{Re},\, \mathrm{Sc})$
Natural convection	$\mathrm{Nu} = f(\mathrm{Gr},\, \mathrm{Pr})$	$\mathrm{Sh} = f(\mathrm{Gr}_m \mathrm{Sc})$
Grashof number	$\mathrm{Gr} = \rho^2 g\beta\, \Delta T L^3/\mu^2$	$\mathrm{Gr}_m = \rho^2 g\beta_m\, \Delta C L^3/\mu^2$
Coefficient of volumetric expansion	$\beta = $ thermal	$\beta_m = $ concentration

coefficient U_D and to express the mass flux in terms of the bulk solution concentrations.

$$\frac{1}{U_D} = \frac{1}{h_{\mathrm{DI}}} + \frac{1}{\zeta K^*} + \frac{1}{h_{\mathrm{DII}}} \tag{6.27}$$

The equation indicates the existence of three resistances between two phases separated by a membrane as shown in Fig. 6.4.

6.3.3 Formulas

Table 6.2 summarizes the analogous physical quantities in heat and mass transfer. The Sherwood number may be correlated in terms of basic nondimensional quantities in a manner similar to the Nusselt number.

6.3.3.1 Forced convective mass transfer.

External flows Laminar flow over a flat plate ($\mathrm{Re} < 5 \times 10^5$):

$$\mathrm{Sh}_x = 0.332\, \mathrm{Re}_x^{1/2} \mathrm{Sc}^{1/3} \quad \text{(local)} \tag{6.28}$$

$$\mathrm{Sh} = 0.664\, \mathrm{Re}^{1/2} \mathrm{Sc}^{1/3} \quad \text{(average)} \tag{6.29}$$

Turbulent flow over flat plate ($\mathrm{Re} > 5 \times 10^5$):

$$\mathrm{Sh}_x = 0.0296\, \mathrm{Re}_x^{4/5} \mathrm{Sc}^{1/3} \quad \text{(local)} \tag{6.30}$$

$$\mathrm{Sh} = 0.037\, \mathrm{Re}^{4/5} \mathrm{Sc}^{1/3} \quad \text{(average)} \tag{6.31}$$

Cross flow circular cylinders (characteristic length $L = $ diameter):

$$\mathrm{Sh} = C\, \mathrm{Re}^m \mathrm{Sc}^{1/3} \tag{6.32}$$

The *C-m*-Re relationship follows that for transfer flow over a sphere (L = diameter):

$$Sh = 2 + 0.6 \, Re^{1/2}Sc^{1/3} \tag{6.33}$$

Internal flows ($L = D_h$) Fully developed laminar flows (Re < 2100):

$$Sh = 1.86 \left(ReSc \, \frac{D_h}{l} \right)^{1/3} \tag{6.34}$$

where Sh = 4.36 for a constant surface mass flux ($l \rightarrow \infty$)
Sh = 3.66 for constant surface concentration ($l \rightarrow \infty$)

Fully developed turbulent flow (Re $\geq$ 10,000):

$$Sh = 0.023 \, Re^{4/5}Sc^{1/3} \tag{6.35}$$

6.3.3.2 Natural convective mass transfer. The following expressions are for a gross approximation only.

Laminar flow on a vertical plate or cylinder (Ra < 10^9):

$$Sh = 0.59 \, Ra^{1/4} \tag{6.36}$$

Turbulent flow on a vertical plate or cylinder (Ra > 10^9):

$$Sh = 0.1 \, Ra^{1/3} \tag{6.37}$$

A long horizontal cylinder (L = diameter):

$$Sh = C \, Ra^n \tag{6.38}$$

The *C-n*-Ra relationship follows that in heat transfer for spheres (L = diameter):

$$Sh = 2 + 0.43 \, Ra^{1/4} \qquad 1 < Ra < 10^5 \tag{6.39}$$

6.4 MASS TRANSFER IN MICROCIRCULATION

In microcirculation, capillaries have a basic function to supply oxygen and nutrient to the cells of the tissues and organs and to carry away carbon dioxide and other waste products. Various mass transfer processes may take place simultaneously across the capillary membranes and ultimately across the cell membranes in the tissues. Such processes include

1. Concentration-driven diffusion resulting from a concentration gradient,
2. Convective transport of solute (where the bulk flow is due to a hydrostatic pressure differential),

3. Osmosis as the result of an osmotic pressure gradient, and
4. Active transport against a concentration gradient at the expenditure of metabolic energy.

For each membrane, it is convenient to sum up the hydrostatic and osmotic pressure differentials into the overall pressure differential ΔP, which drives the net overall convective flow v. ΔP and v are related as

$$K_v \Delta P = v \qquad (6.40)$$

in which K_v denotes the flow permeability of the capillary or cell membrane in moles per area per time per mmHg. Then, the total mass flux across a membrane can be expressed as

$$N_i'' = C_i K_v \Delta P + D_i \frac{\Delta C_i}{\Delta z} \qquad (6.41)$$

Equation (6.41) includes two driving potentials, the concentration gradient $\Delta C_i / \Delta z$, and the overall pressure difference ΔP. $D_i / \Delta z$ is known as the diffusional permeability. With the application of either Raoult's or Henry's law, ΔP may be replaced by ΔC_i and the above equation can be rewritten as

$$N_i = KA \Delta C_i = KA \frac{\Delta P}{R_g T} \qquad (6.42)$$

Here, K denotes the overall permeability of the membrane which consists of the flow permeability K_v, and the chemical permeability $D_i / \Delta z$. R_g is the gas constant. The mass transfer resistance is

$$R = \frac{1}{KA} \quad \text{or} \quad R = \frac{R_g T}{KA} \qquad (6.43)$$

Mass transport processes of foods and products of metabolism take place in two different ways: internal and external respiration. We can use oxygen as

Table 6.3 Partial pressure in the systemic capillary-tissue cells system

Location	Partial pressure, mmHg	
	O_2	CO_2
Alveolar air	104	40
Arterial blood	96	40
Venous blood	40	46
Tissue cells (metabolism)	35	55

Table 6.4 Pulmonary gas exchanges at resting conditions

Parameter	Rate, ml/min		
Circulation system			
Alveolar ventilation	5250		
Pulmonary blood flow	5000		
O_2 metabolism	250		
CO_2 metabolism	200		
Pulmonary system			
Partial pressure in alveoli	mmHg	100	40
Partial pressure in arterial blood	mmHg	96	40
Partial pressure in venous blood	mmHg	40	46

Source: Matthews (1967).

an example. The internal respiration occurs in the circulation system (as opposed to pulmonary) capillaries where oxygen is transferred from the cells, through plasma, across the capillary wall, and eventually into the tissue. The process is reversed for external respiration. In pulmonary capillaries, oxygen is transported from the alveolar sac, through plasma, across the walls and membranes, and finally into the red cells. Some basic information on internal and external respiration is presented in Tables 6.3 and 6.4, respectively.

6.4.1 Internal Respiration and Nutritional Exchanges

In systemic capillaries, reactants for the cells, such as oxygen, transfer out from an arterial capillary through the interstitial fluid into tissue cells, while products of cellular reactions, such as carbon dioxide and urea, migrate out from tissue cells through the interstitial fluid into a venous capillary. This mass transfer process of oxygen and carbon dioxide in the tissues (between the capillaries and the tissue cells) is called internal respiration. The mechanism of mass transport can also be applied to the nutritional (e.g., molecules and wastes) exchange in the tissues.

The mass transfer problem can be treated using the lumped and distributed parameter analyses. In the former approach, the transfer process is considered to consist of three principal resistances in series, as shown in Fig. 6.5: R_1, R_2, and R_3 are resistances due to the flowing plasma, the capillary wall, and the tissue fluid layer, respectively. There is good evidence that both the membranes and the intercellular fluids in the tissue and red blood cells present negligible resistance. The overall resistance to mass transfer in the capillary-tissue system is, therefore, the summation of these three resistances:

$$R_T = \sum_{I=1}^{3} R_I \qquad (6.44)$$

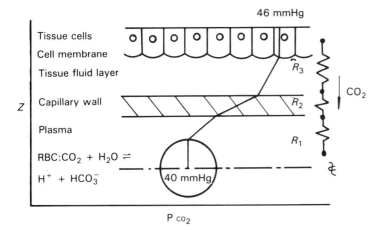

Figure 6.5 Carbon dioxide transfer in the tissue-venous capillary (metabolic) system.

The rate of mass transfer at steady state is

$$N = \frac{(\Delta P)_{\text{overall}}}{R_T} \tag{6.45}$$

in which $\Delta P_{\text{overall}}$ is the overall partial-pressure difference between tissue cells and red corpuscles.

Figure 6.6 is Krogh's cylindrical model for oxygen transport in the capillary tissue system. The tissue is a hollow cylinder composed of the inner and outer radii, R_1 and R_2, respectively. It is assumed that oxygen transport is convection controlled in the tissue. At steady state, mass balance relations yield

Tissue:
$$\frac{D_r}{r}\frac{\partial}{\partial r}\left(r\frac{\partial C_t}{\partial r}\right) + D_z\frac{\partial^2 C_t}{\partial z^2} = \frac{A^*C_t}{B^* + C_t} \tag{6.46}$$

Capillary:
$$A_c v_z \frac{dC_c}{dz} = -h_D P^*[C_c(z) - C_t(R_1, z)] \tag{6.47}$$

Here D_r and D_z denote the oxygen diffusivities in the radial and axial directions, respectively. V_z is the blood velocity in the capillary while A_c and P^* are, respectively, the cross-sectional area and perimeter of the capillary. A^* is the maximum rate of metabolic oxygen consumption while B^* is the Michaelis-Menten constant.

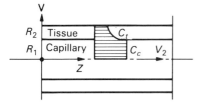

Figure 6.6 Krogh's model for oxygen transport in the capillary-tissue system.

The appropriate boundary conditions read

$$-D_r \frac{\partial C_t(R_1, z)}{\partial r} = h_D[C_c(z) - C_t(R_1, z)] \tag{6.48}$$

$$\frac{\partial C_t(R_2, z)}{\partial r} = 0 \tag{6.49}$$

$$\frac{\partial C_t(r, 0)}{\partial z} = 0 \qquad C_t(r, 0) = C_c(0) = C_O \tag{6.50}$$

$$\frac{\partial C_t(r, 1)}{\partial z} = 0 \tag{6.51}$$

where C_O is the arterial oxygen concentration

Hypoxia is a special case in which the tissue is low in oxygen concentration, $C_t \ll B$. The reaction becomes first order to $k_1 C_t$ where $k_1 = A/B$. With $D_z = 0$, Eq. (6.46) is reduced to

$$\frac{1}{r} \frac{d}{dr}\left(r \frac{dC_t}{dr}\right) = \frac{k_1}{D_r} C_t \tag{6.52}$$

The solutions are obtained as

$$C_c(z) = C_O \exp\left[-\frac{2h_D z}{(1 + b)v_z R_1}\right] \tag{6.53}$$

$$C_t(r) = \left(\frac{C_c b}{1 + b}\right) \frac{K_1(aR_2)I_0(ar) + I_1(aR_2)K_0(ar)}{K_1(aR_2)I_0(aR_1) + I_1(aR_2)K_0(aR_1)} \tag{6.54}$$

where

$$a = \frac{k_1}{D_r} \qquad b = \frac{K_1(aR_2)I_1(aR_1) + I_1(aR_2)K_1(aR_1)}{K_1(aR_2)I_0(aR_1) + I_1(aR_2)K_0(aR_1)} \tag{6.55}$$

I_0, I_1 and K_0, K_1 are the modified Bessel functions of the first and second kinds, respectively.

6.4.2 External Respiration

In the case of pulmonary capillaries, the lumped parameter model consists of five major resistances in series between the alveolar space and red corpuscles in a capillary: the alveolar gas film, the alveolar membrane, the interstitial fluid layer, the pulmonary capillary wall, and the flowing plasma (Fig. 6.7). The overall resistance between the alveolar space and the red blood cells is

$$R_T = \sum_{I=1}^{5} R_I \tag{6.56}$$

The equation for N in Section 6.4.1 is used to determine the rate of mass transfer under $(\Delta P)_{overall}$ between the alveolar space and red blood cells.

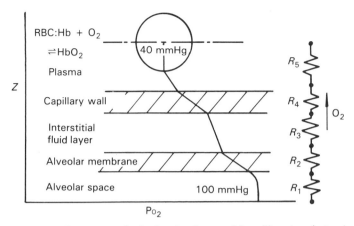

Figure 6.7 Oxygen transfer in the alveolar-arterial capillary (respiratory) system.

Oxygen transport in pulmonary capillaries takes place between blood and alveolar space in which mass transfer is by diffusion only. Warner and Seagrave (1970) treated the lung as a lumped parameter system. Various mammals can breathe with oxygen-rich liquids, such as Ringer solution and oxygenated fluorocarbon emulsion, in their lungs. Kylstra et al., (1966) introduced Ringer's solution with known oxygen and carbon dioxide concentrations into the dog lungs *by gravity* at frequencies between 6 and 21 breaths per minute and measured the concentration changes with respect to time and position in the lobule.

Consider a lobule as a spherical chamber of radius R as depicted in Fig. 6.8. The unsteady mass transfer equation for O_2 or CO_2 reads

$$\frac{\partial C}{\partial t} = \frac{D}{r^2} \frac{\partial}{\partial r} \left(r^2 \frac{\partial C}{\partial r} \right) \tag{6.57}$$

subject to

$$C(r, 0) = C_0 \qquad \frac{\partial C(0, t)}{\partial r} = 0 \tag{6.58}$$

$$C(R, t) = C^* \text{ for } CO_2 \qquad -D \frac{\partial C(R, t)}{\partial r} = N'' \text{ for } O_2$$

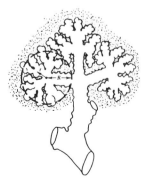

Figure 6.8 The primary lobule as a hypothetical gas exchange unit.

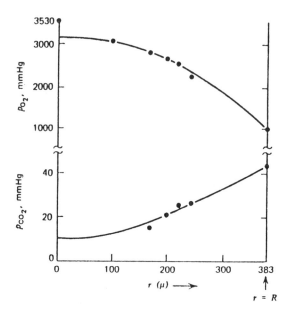

Figure 6.9 Agreement between experimental data (dots) and theoretically calculated oxygen and carbon dioxide tension profiles (curves) in a spherical lobule.

Here C_0 is the initial concentration. At the lobule wall, the oxygen flux from the liquid to the pulmonary capillaries, N'', is considered constant, while the CO_2 concentration is taken to be constant at C^* from the pulmonary capillaries.

The solutions for both O_2 and CO_2 can be obtained by the method of separation of variables. Figure 6.9 compares the solutions (solid lines) and experimental results (dots) in terms of the partial pressure (Guyton, 1954). The diffusion of O_2 and CO_2 in water and $R = 420$ μm were used in theory.

6.4.3 Removal of Waste Products in the Nephron

In the lungs, venous blood capillaries are brought into close proximity to the alveoli (micropores) into which CO_2 is released. In a similar manner, blood capillaries are brought close to urinary tubules (microtubes) into which waste products carried by the blood are drained. Moreoover, like the lungs which are comprised of alveolar-capillary units, each kidney consists of tubular-capillary units, called nephrons. In each kidney, are approximately 1.25 million tubules being perfused by about 500 ml of blood per minute.

Figure 6.10 illustrates the tubular-capillary unit. Mass transfer processes in each nephron may be divided into six steps:

1. Glomerulus: filtration;
2. Proximal tubule: reabsorption (diffusion) of permeating solute (glomerular filtrate) coupled with osmosis of water out of the tubule;

3. Descending loop of Henle: same as step 2;
4. Ascending loop of Henle: active transport;
5. Distal tubule: same as step 2; and
6. Collecting duct: osmosis, diffusion, and active transport.

The transfer structure of the tubular-capillary unit is similar to that of the alveolar-capillary unit shown in Fig. 6.11. For example, the transfer of filtrate from the tubule channel to blood in the capillary encounters five major resistances: filtrate film (R_1), proximal tubule cells (R_2), peritubule fluid layer (R_3), capillary wall (R_4), and blood flow film (R_5), as shown in Fig. 6.11. Therefore, the rate of transfer is

$$N = \frac{(\Delta C)_{\text{overall}}}{R_T} \qquad R_T = \sum_{I=1}^{5} R_I \qquad K = \frac{1}{R_T}$$

where $(\Delta C)_{\text{overall}}$ is the concentration difference between the tubule and the capillary. The equation applies to all kinds of mass transfer mechanisms. K, the reciprocal of the overall resistance R_T, is called permeability. The value of K for the tubular-capillary unit is about 1×10^{-4} to 4×10^{-4} ml/min/mosm, where mosm is milliosmoles of osmolarity.

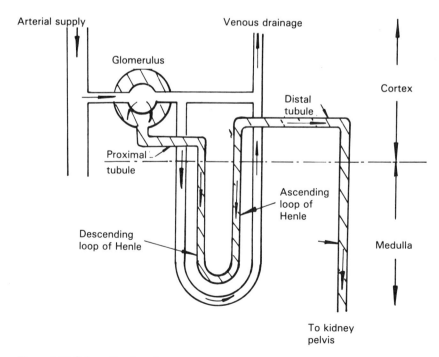

Figure 6.10 Schematic of nephron.

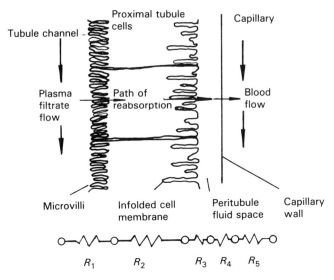

Figure 6.11 Tubular-capillary unit.

6.5 BLOOD–GAS INTERACTION IN ERYTHROCYTES

When oxygen reaches its destination, the intracellular space, it is used for chemical reactions with the nutrients to produce energy. As a result of metabolism, carbon dioxide is produced in the intracellular space and must be disposed into the atmosphere.

The following chemical equations summarize the mechanisms of blood carrying oxygen and carbon dioxide (Matthews, 1967).

Oxygen:

$$RBC: Hb^- + H^+ \rightleftharpoons HHb \text{ (a chemical buffer)}$$

$$H^+Hb + O_2 \rightleftharpoons HbO_2 + H^+$$

Carbon dioxide:

$$Plasma: CO_2 + H_2O \rightleftharpoons H_2CO_3 \rightleftharpoons H^+ + HCO_3 \ (5\%)$$

$$RBC: CO_2 + H_2O \rightleftharpoons H_2CO_3 \rightleftharpoons H^+ + HCO_3^- \ (65\%)$$

$$CO_2 + HbNH_2 \rightleftharpoons HbNHCOOH \ (30\%)$$

The solubility of oxygen in plasma is extremely low (Roughton, 1959). With the erythrocytes as an oxygen carrier, blood can carry approximately 20 ml $O_2/100$ ml blood. The oxygen molecule is attached to or associated with the hemoglobin Hb to form oxygenated HbO_2, called oxyhemoglobin. The percent saturation of Hb is defined as

$$y_{O_2} = \frac{[HbO_2]}{[HbO_2] + [Hb]} \qquad (6.59)$$

It is dependent on the partial pressure (or tension) of oxygen P_O as

$$y_{O_2} = \frac{K(P_{O_2})^n}{K(P_{O_2})^n + 1} \qquad (6.60)$$

where K is a constant, dependent on the pH and temperature of the hemoglobin solution, and n is an empirical constant between 2.5 and 2.6. A plot of y_{O_2} versus P_{O_2} (Fig. 6.12), is called the oxygen saturation (or dissociation) curve.

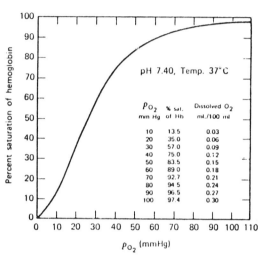

P_{O_2} mm Hg	% sat. of Hb	Dissolved O_2 ml./100 ml
10	13.5	0.03
20	35.0	0.06
30	57.0	0.09
40	75.0	0.12
50	83.5	0.15
60	89.0	0.18
70	92.7	0.21
80	94.5	0.24
90	96.5	0.27
100	97.4	0.30

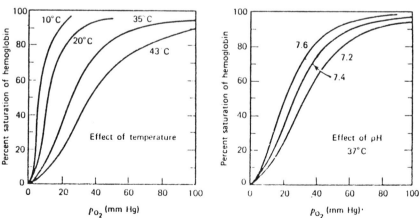

Figure 6.12 Oxygen curves.

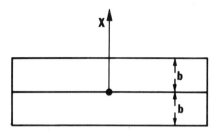

Figure 6.13 The disk model of the red blood cell.

As shown in Fig. 6.12, at a constant P_{O_2}, y_{O_2} decreases with pH value. In both, internal reaction of oxygen with hemoglobin takes place.

Figure 6.13 shows the thin disk model of the red blood cell. It consists of a slab with thickness $2b$ and contains hemoglobin at initial oxygen concentration C_0. Oxygen diffuses from both disk surfaces at O_2 with a concentration of C_s. The origin of the x axis is fixed at the disk center. During the initial stage following the introduction of a red blood cell into plasma or a solution, an insufficient amount of oxyhemoglobin has built up to cause its dissociation. The O_2 transport equation is

$$\frac{\partial C}{\partial t} = D \frac{\partial^2 C}{\partial x^2} - kC \tag{6.61}$$

where D is the diffusivity of O_2 in hemoglobin and k denotes a reaction constant. Neglecting resistance by the red blood cell membrane, it is subject to

$$C(x, 0) = C_0 \qquad \frac{\partial C(0, t)}{\mu x} = 0 \qquad C(b, t) = C_s \tag{6.62}$$

The solution can be obtained by the Laplace transform method as

$$\frac{C(x, t) - C_0}{C_s - C_0} = \frac{\cosh Bx}{\cosh B} - 2 \sum_{n=0}^{\infty} \frac{(-1)^n (n + 1/2)\pi}{\lambda_n^2} e^{-\lambda_n^2 \tau} \cos\left(n + \frac{1}{2}\right)\pi x \tag{6.63}$$

in which

$$X = \frac{x}{b} \qquad \tau = \frac{Dt}{b^2} \qquad \lambda_n^2 = (n + \tfrac{1}{2})^2\pi^2 + B^2 \qquad B^2 = b^2 \frac{k}{D} \tag{6.64}$$

6.6 MODELING OF MASS TRANSFER SYSTEMS

6.6.1 Respiratory System

The transport of O_2 and CO_2 in the respiratory system has been studied by compartment models (Fig. 6.14). The simplest is a two-compartment model consisting of alveolar space and tissue space, while a more elaborate model adds pulmonary space and arterial space to form a four-compartment model. In the latter case, part of the blood from the tissue space is shunted to the arterial space for remixing. The governing equations are dervied for both models. Note that an expression for an equilibrium relation is needed in the four-compartment model. The transport lags between the compartments are neglected.

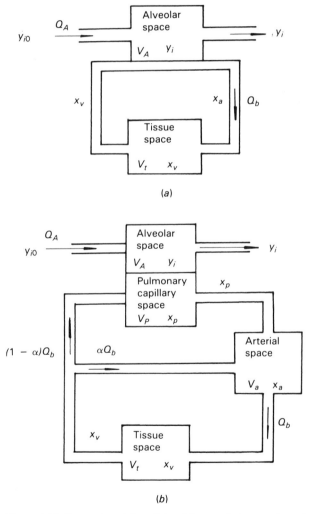

Figure 6.14 Compartmental models of the respiratory system. (*a*) Two-compartment model; (*b*) four-compartment model.

6.6.1.1 Two compartment model. The mass balance equations for species i read

Alveolar space:
$$V_A \frac{dy_i}{dt} = Q_A(y_{i0} - y_i) + Q_b(x_v - x_a) \qquad (6.65)$$

Tissue space:
$$V_t \frac{dx_v}{dt.} = Q_b(x_a - x_v) + r_m(RQ) \qquad (6.66)$$

Here, V_A and V_t denote the volumes of the lungs and the equivalent tissue space, respectively; y_i, the mole fraction of species i in the gas phase; Q_A, the alveolar ventilation rate; Q_b, the blood flow rate; x_v and x_a, the volume fraction of i in the venous and alveolar spaces, respectively; r_m, the metabolic production rate in milliliters of O_2 consumed per minute; and RQ, the respiratory quotient. x_a and y_i are related by the equilibrium relation

$$x_a = f(y_i) \qquad (6.67)$$

The defect of this simple model is the failure of the $x_a - y_i$ equilibrium relationship when the mass transport in the lungs is hindered.

6.6.1.2 Four compartment model. The mass balance equations in the four chambers read

Alveolar space:
$$V_A \frac{dy_i}{dt} = Q_A(y_{i0} - y_i) - U_{DA} A_A(y_i - y^*) \qquad (6.68)$$

Pulmonary space:
$$V_P \frac{dx_P}{dt} = Q_b(\alpha x_v - x_P) + U_{DA} A_A(y_i - y^*) \qquad (6.69)$$

Arterial blood space:
$$V_b \frac{dx_a}{dt} = Q_b[\alpha x_v + (1 - \alpha)x_P - x_a] \qquad (6.70)$$

Tissue space:
$$V_t \frac{dx_v}{dt} = Q_b(x_a - x_v) + r_m(RQ) \qquad (6.71)$$

Equilibrium relation:
$$y^* = f(x_P, x_v) \qquad (6.72)$$

where U_{DA} signifies the overall mass transfer coefficient between the alveolar space and the capillary blood; A_A, the mass transfer area in the alveolar space; y^*, the mole fraction of the inspired gas in equilibrium with volume fractions in the venous and arterial blood; and α, the fraction of blood shunted.

6.6.2 Circulatory System

While the respiratory system deals with mass transfer of gases, the circulatory system involves all substances that are associated with metabolism. Mass transfer processes for metabolism and their automatic regulation take place in both the systemic and pulmonary circulation. These tasks are summarized in the compartmental model shown in Fig. 6.15. The model consists of six chambers:

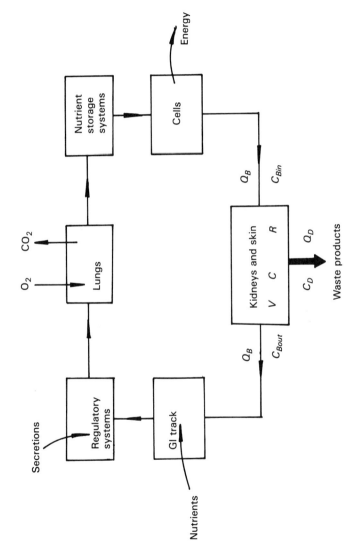

Figure 6.15 Compartmental model for mass transfer in the circulatory system.

(1) Lungs (O_2 and CO_2 exchange); (2) nutrient storage systems (in all cells especially the liver and muscle cells); (3) cells (metabolism to produce energy and waste products); (4 and 5) kidneys and skin (removal of waste products); and (6) gastrointestinal track (nutrient absorption).

This model differs from the previous (lumped parameter) model in that each compartment has more than one input and/or output (i.e., a multiple input-output compartment). Therefore, the concentration of species i in the compartment C differs from the outgoing concentrations in the bloodstream $C_{B\,out}$ and in other streams C_D. The mass balance for species i in the compartment j, for example, the kidneys, reads

$$V \frac{dC}{dt} = Q_B(C_{B\,in} - C_{B\,out}) + r \pm Q_D C_D \qquad (6.73)$$

where Q_B and Q_D denote the volumetric rates of the bloodstream and other streams, respectively; r, the production rate of species i; $+$ denotes carrying into and $-$ denotes carrying out of the compartment by other flows Q_D. The rate of mass transfer of species i to the bloodstream in the chamber is

$$NA = Q_B(C_{B\,in} - C_{B\,out}) = KA(C - C_{B\,ave}) \qquad (6.74)$$

where A represents the area of transfer of species i from compartment to bloodstream; K, the overall mass transfer coefficient for i in j; and $C_{B\,ave}$, the average concentration of i in the bloodstream in j, say $(C_{B\,in} + C_{B\,out})/2$.

Each compartment has a balance equation and a rate equation. These equations for the six chambers are coupled and must be solved simultaneously. This model is useful in establishing some quantitative relationships for the amount of various species present in the metabolic system. The relative importance of mass transfer processes, the control scheme, and their effects on the overall metabolic process may be elucidated from the solution of these equations.

6.7 MASS TRANSFER IN SOME ORGANS

6.7.1 Placenta

The placenta is the only pathway for transporting oxygen, carbon dioxide, nutrients, and metabolic wastes between the fetus and the maternal body. Therefore, it is the first vital lifeline of mammals including human beings. Figure 6.16a is a diagram of the placenta circulation in rabbits and sheep (Middleman, 1972). In these animals, mass transfer between the maternal and fetal capillaries is a counterflow type (Fig. 6.16b) analogous to heat transfer in the counterflow heat exchanger. In the human placenta, there are no capillaries on the maternal side (Guyton, 1954). The fetal capillaries are immersed in a "pool" of maternal blood, as depicted in Fig. 6.17. The mass transfer operation is then analogous to the heat transfer process in the shell-and-tube heat exchanger (a unit con-

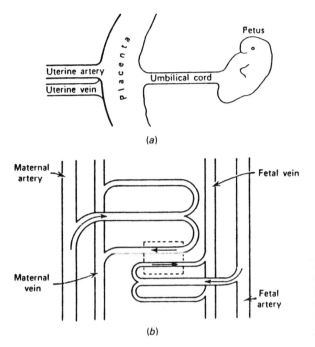

Figure 6.16 The countercurrent model of animal placental circulation and mass exchange: (*a*) fetal-uterine interface; (*b*) model with countercurrent unit shown in the dashed rectangle.

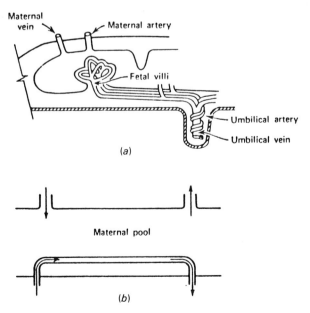

Figure 6.17 Diagram of human placental circulation: (*a*) capillary arrangement in the human placenta; (*b*) model of "pool."

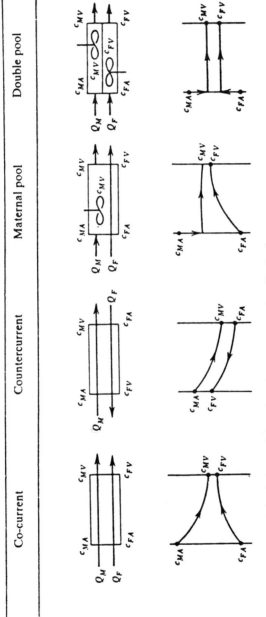

Figure 6.18 Placental circulation models and pertinent concentration distributions.

350

sisting of multiple small tubes placed in a large diameter tube) (Faber, 1969; Meschia et al., 1967). There are other mammals with other kinds of placental circulation such as a parallel flow and double pool type. Figure 6.18 illustrates four placental circulation models and pertinent concentration distributions.

Figure 6.19 illustrates a schematic diagram of the maternal pool model of the human placenta. The maternal blood is considered well mixed in the pool. The concentration of the transported substance (oxygen, nutrients, carbon dioxide, or wastes), C_M, abruptly drops from C_{MA} in the arterial end to C_{MV} in the venous end, which is equal to C_M. Mass balance on the fetal bloodstream yields

$$Q_F \frac{dC_F}{dx} = \frac{P}{l} (C_{MV} - C_F) \tag{6.75}$$

where P denotes the permeability in cubic centimeters per second (cm³/s) of the substance in the membrane between the maternal and fetal capillaries; l is the length for mass transfer. The boundary condition is

$$C_F(0) = C_{FA} \tag{6.76}$$

The concentration distribution is found to be

$$\frac{C_F(x) - C_{MV}}{C_{FA} - C_{MV}} = \exp\left(-\frac{P_x}{Q_F l}\right) \tag{6.77}$$

At the exit of Q_F where $C_F(l) = C_{FV}$, one obtains

$$\frac{C_{MV} - C_{FV}}{C_{MV} - C_{FA}} = \exp\left(-\frac{P}{Q_F}\right) \tag{6.78}$$

The fetal and maternal transport functions, y_F and y_M, respectively, mea-

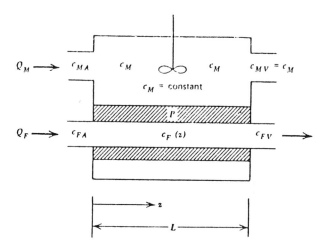

Figure 6.19 Detailed scheme of the maternal pool model of the human placenta.

sure the efficiencies of mass transfer on both sides in terms of the fraction of the initial difference between the concentrations in the two entering flows. They are defined as

$$y_F = \frac{C_{FV} - C_{FA}}{C_{MA} - C_{FA}} \qquad y_m = \frac{C_{MA} - C_{MV}}{C_{MA} - C_{FA}} \qquad (6.79)$$

The diffusion clearance or clearance is the mass transfer rate per unit initial concentration difference between the two entering flows:

$$D_C = \frac{N}{C_{MA} - C_{FA}} \qquad (6.80)$$

where N is the rate of mass transfer and can be expressed as

$$N = Q_M(C_{MA} - C_{MV}) = Q_F(C_{FV} - C_{FA}) \qquad (6.81)$$

Therefore,

$$y_F = \frac{1 - \exp{(-P/Q_F)}}{1 + (Q_F/Q_M)[1 - \exp{(-P/Q_F)}]} \qquad (6.82)$$

$$D_C = \frac{Q_F[1 - \exp{(-P/Q_F)}]}{1 + (Q_F/Q_M)[1 - \exp{(-P/Q_F)}]} \qquad (6.83)$$

The placental mass transfer process is a series operation consisting of the maternal side convective mass transfer, membrane diffusion, and the fetal side convective mass transfer. It is analogous to the heat transfer process between hot and cold fluids flowing across the separating tube wall in a double pipe heat exchanger. The overall mass transfer coefficient U_D (in cm/sec) is defined as

$$\frac{1}{U_D A} = \left(\frac{1}{h_D A}\right)_M + \frac{1}{P} + \left(\frac{1}{h_D A}\right)_F \qquad (6.84)$$

in which h_D represents the convective mass transfer coefficient between the membrane and the bloodstream in meters per second. $U_D A$ can be based on either the maternal side or the fetal side.

Depending on the relative magnitudes among the three terms on the right-hand side of Eq. (6.84), there are two limiting cases in the mass transfer process: diffusion- and convection-limited or controlled. When the flow rates are much larger than the permeability, the transport is diffusion controlled. Then Eq. (6.84) is reduced to

$$U_D A = P \qquad (6.85)$$

When the permeability is large compared with the flow rates, the process is convection controlled. Equation (6.84) can be approximated as

$$\frac{1}{U_D A} = \left(\frac{1}{h_D A}\right)_M + \left(\frac{1}{h_D A}\right)_F \qquad (6.86)$$

6.7.2 Percolation Problems

The following unsteady transfer phenomena in a single-pipe mass exchanger is called a percolation problem. This is also applicable to a similar mass transfer system.

The simplest mass exchange device consists of an insulated pipe (with ID $= 2r_i$ and OD $= 2r_o$) through which a fluid flows at a velocity v, as shown in Fig. 6.20. Let C_f and C_w be the fluid and tube-wall concentrations of species A, respectively. Initially, both concentrations are the same at a reference value, say 0. Then, the inlet fluid concentration of species A undergoes a sudden change to C_{f0}. The resulting changes in both $C_f(x, t)$ and $C_w(x, t)$ are functions of the axial location x and the time t. This is the so-called percolation problem which is commonly observed in transport phenomena of heat and mass.

By applying the mass conservation law on the differential elements, dx, in both the fluid and tube wall, one gets

$$\frac{\partial C_f}{\partial t} + v \frac{\partial C_f}{\partial x} = a(C_w - C_f) \tag{6.87}$$

$$\frac{\partial C_w}{\partial t} = b(C_f - C_w) \tag{6.88}$$

where

$$a = \frac{2h_D}{r_i} \qquad b = \frac{2h_D r_i}{r_0^2 - r_i^2} \tag{6.89}$$

h_D is the convective mass transfer coefficient between the fluid and the inside tube wall. The appropriate initial and boundary conditions are

$$C_f(x, 0) = C_w(x, 0) = 0 \tag{6.90a}$$

$$C_f(0, t) = C_{f0} \tag{6.90b}$$

The Laplace transformation of Eqs. (6.87) and (6.88) with respect to t, subject to Eq. (6.90a) yields

$$s\overline{C}_f + v \frac{d\overline{C}_f}{dx} = a(\overline{C}_w - \overline{C}_f) \tag{6.91}$$

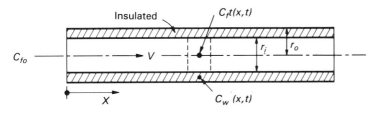

Figure 6.20 Single-pipe mass exchanger.

$$s\bar{C}_w = b(\bar{C}_f - \bar{C}_w) \tag{6.92}$$

where s is the Laplace variable and $\bar{C}_f(x, s)$ and $\bar{C}_w(x, s)$ are, respectively, the Laplace transformed functions of $C_f(x, t)$ and $C_w(x, t)$. Solving for $\bar{C}_w$ from Eq. (6.92) and inserting the result into Eq. (6.91) yields

$$\frac{d\bar{C}_f}{dx} + \frac{s}{v}\left(1 + \frac{a}{b+s}\right)\bar{C}_f = 0 \tag{6.93}$$

The Laplace transformed form of Eq. (6.90b) is

$$\bar{C}_f(0, s) = \frac{C_{f0}}{s} \tag{6.94}$$

The solution of Eq. (6.93) that satisfies Eq. (6.94) is

$$\frac{\bar{C}_f(x, 0)}{C_{f0}} = \frac{1}{s}e^{-(sx/v)[1+a/(b+s)]} \tag{6.95}$$

Introducing Eq. (6.95) into Eq. (6.92) and rearranging gives

$$\frac{\bar{C}_w(x, s)}{C_{f0}} = \frac{b}{s(b+x)}e^{-(sx/v)[1+a/(b+s)]} \tag{6.96}$$

The inverse Laplace transformation of Eq. (6.95) and (6.96) produces, when $t < x/v$,

$$C_f(x, t) = C_w(x, t) = 0$$

When $t \geq x/v$,

$$\frac{C_f(x, t)}{T_{f0}} = e^{-ax/v}\left[e^{-bt^*}I_0\left(2\frac{abxt^*}{v}\right) + b\int_0^{t^*}e^{-b\zeta}I_0\left(2\frac{abx\zeta}{v}\right)d\zeta\right] \tag{6.97}$$

$$\frac{C_w(x, t)}{T_{f0}} = be^{-ax/v}\int_0^{t^*}e^{-b\zeta}I_0\left(2\frac{abx\zeta}{v}\right)d\zeta \tag{6.98}$$

where I_0 is the modified Bessel function of the first kind of order zero and $t^* = t - x/v$. The function x/v is known as the transport lag or dead time.

6.7.3 Transient Mass Transfer in the Liver

Gorsky (1963) measured the sinusoidal and extravascular volumes of the liver using the indicator method. Figure 6.21 shows a model of the liver at the sinusoidal and extravascular compartments of volumes V_s and V_e, respectively. Let C_s and C_e be the concentrations of the injected indicator in these compartments. The mass balance on these compartments yield

$$\frac{\partial C_s}{\partial t} + v\frac{\partial C_s}{\partial x} = \frac{h_D A(C_e - C_s)}{V_s} \tag{6.99}$$

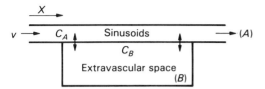

Figure 6.21 Model for the liver.

$$\frac{\partial C_e}{\partial t} = \frac{h_D A}{V_e} (C_s - C_e) \tag{6.100}$$

where v is the blood flow velocity in the sinusoidal compartment; h_D, the mass transfer coefficient; and A, the interfacial area between the two compartments. The appropriate initial and boundary conditions are

$$C_s(x, 0) = C_e(x, 0) = 0 \tag{6.101}$$
$$C_s(0, t) = C_{s0}$$

It is obvious that there exists a similarity in the governing equations and the appropriate initial and boundary conditions between these two transport phenomena.

Example 6.2 Counterflow plate type dialyzer In a dialyzer, circulating blood is put in contact with a buffered electrolyte across a cellulosic membrane which permits the passage of water and small solute by diffusion. Normally, water is ultrafiltered from the blood, which is too slow to have an appreciable effect on solute mass transfer. Consideration is then given to convective mass transfer between fluids separated by a semipermeable membrane, at net mass transfer rates too small to affect velocity profiles.

Consider a plate type dialyzer (Fig. 6.22), with semipermeable membranes separating the blood and the dialyzate in counterflow. The origin is fixed at the blood entrance with z measuring the distance in the direction of blood flow. Let U_D be the overall mass transfer coefficient between the two fluids and B be the width of the membrane sheets. Q_b and Q_d are the volumetric flow rates of the blood and the dialyzate, respectively. Note that Q_d is negative for counterflow. The solute concentrations in the blood and the dialyzate are denoted by C and C^*, respectively. One obtains the following governing equations:

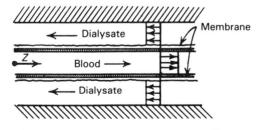

$Z = 0$

$Z = L$ **Figure 6.22** Flow model of dialyzer.

$$Q_b \frac{dC}{dz} = 2U_D B(C^* - C)$$

$$Q_d \frac{dC^*}{dz} = 2U_D B(C - C^*)$$

subject to the boundary conditions

$$C(0) = C_0 \qquad C^*(l) = C_0^*$$

Here, C_0 and C_0^* represent the solute concentrations at the inlets of the blood and dialyzate, respectively and l is the length of the dialyzer. The solutions are

$$\frac{C(z) - C_0}{C_0^* - C_0} = \frac{1 - \exp[-(\tau + \tau^*)z]}{1 + (\tau^*/\tau) \exp[-(\tau + \tau^*)l]}$$

$$\frac{C^*(z) - C_0}{C_0^* - C_0} = \frac{1 + (\tau^*/\tau) \exp[-(\tau + \tau^*)z]}{1 + (\tau^*/\tau) \exp[-(\tau + \tau^*)l]}$$

where $\tau = 2U_D B/Q_b$ and $\tau^* = 2U_D B/Q_d$.

6.8 ARTIFICIAL ORGANS FOR MASS TRANSFER

6.8.1 Oxygenators (Artificial Lungs)

Artificial devices for the extracorporeal oxygenation of blood have their primary application in cardiac surgery, where it is advantageous to bypass blood not only around the heart but around the lungs as well. An artificial heart-lung machine consists of a blood pump to substitute for the heart's function and a gas exchange unit to replace the natural lungs, as illustrated in Fig. 1.8.

The first operational oxygenator was developed by J. H. Gibbon, Jr., in 1939 (Galletti and Brecher, 1962); however, developmental work was delayed by World War II. It was not until 1953 that Gibbon succeeded in using a screen type oxygenator for total cardiopulmonary bypass for operating on a patient with an arterial septal defect. The first stage of the development was capped by a technically sophisticated film type device known as the Gibbon-Mayo oxygenator.

The second generation of apparatus was the simple, disposable bubble oxygenators, constructed by DeWall and Rygg-Kyrsgaard. This type of device was haunted by the ever-mounting evidence of serious blood damage due to large blood-gas interfaces. It prompted the invention of membrane oxygenators, which represent the third generation and present era in artificial generation. Pioneer work on membranes was led by Kolff (1976) and Clowes, Pierce, and Galletti (Galletti and Brecher, 1962). The design of membrane oxygenators became strictly an engineering feat.

The ideal oxygenator should be efficient in gas exchange and gentle to blood. The mass transfer rate N can be determined by the equation

$$N = h_D A(P_G - P_B) \tag{6.102}$$

where h_D denotes the mass transfer coefficient; A the blood-gas contact area; P_G, the partial pressure of oxygen in the gas phase; and P_B, the partial pressure of oxygen in the blood.

To generate A to enhance N, various methods can be utilized, for example, by dispersing the gas phase in a pool of blood; by dispersing blood into a gas phase via foaming or spraying; by spreading blood in a thin film on a stationary or moving solid plate, screen, etc.; and by allowing blood and gas to flow on opposite sides of a large semipermeable membrane. h_D can be augmented through blood-side mixing by mechanical means, by turbulent gas flow, or by moving or rotating the solid surfaces with blood films. The amount of oxygen transferred is equal to Nt where t is the blood-gas contact time; this can be prolonged by continually recycling some blood from the outlet back to the inlet of the device.

The three most successful types of artificial lungs are the bubble, film, and membrane oxygenators (Fig. 6.23) (Galletti and Brecher, 1962). In bubble oxygenators, the gas phase is broken into a large number of small bubbles in order to gain a large blood-gas interfacial area. The DeWall oxygenator is a popular bubble type unit in which the gas phase is bubbled through a column

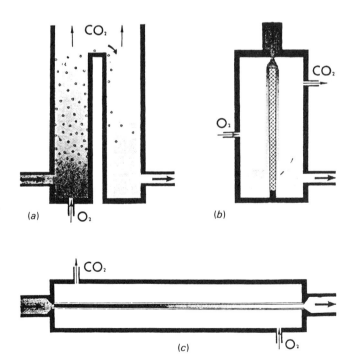

Figure 6.23 Three basic classes of artificial lungs: (*a*) bubble, (*b*) film, and (*c*) membrane.

of venous blood. After flowing across a defoaming surface (coated with silicone compounds), the blood runs down a long helical settling chamber from which any remaining bubbles are removed by buoyant rising. Such oxygenators suffer from excessive hemolysis.

Film oxygenators can be classified into stationary and moving types. The former consists of a solid vertical stationary support over which venous blood flows down as a thin film into a settling chamber. In order to improve mixing, the support can be replaced by corrugated materials, screen or porous materials. The moving type has moving supports such as spinning cylinders or rotating discs.

Both the bubble and film oxygenators suffer from producing blood trauma such as froth formation, fibrin deposition, protein denaturation, and hemolysis. In membrane oxygenators, blood and gas are separated by membranes (silicone rubber or Teflon) which are essentially permeable to gases but not liquids. However, the presence of a membrane between the blood and gas adds a mass transfer resistance to the system. In general, CO_2 transfer in membrane units is dependent on the membrane barrier, while O_2 transfer is controlled by the blood film resistance. Membrane oxygenators enjoy the advantages of low blood trauma, no gas or fibrin emboli, and simplicity in blood-volume control. However, they are bulky and hard to assemble.

Detailed reviews and surveys on oxygenators are available in most books on artificial organs such as Cooney (1976), Flower (1968), and Galletti and Brecher (1962).

6.8.2 Artificial Kidneys and Dialyzers

Numerous techniques have been developed to artificially remove metabolic wastes and excess electrolytes from the body when natural kidney function fails. Practically all of the procedures for waste removal involve contacting a body fluid with a dialysate solution across a semipermeable membrane. The membrane is permeable to all species except proteins and blood cells. The dialysate is an aqueous solution that is similar in composition to normal body fluids excluding wastes. On contact, the body fluid loses wastes such as urea, creatinine, uric acid, sulfates, and phenols to the dialysate by diffusion across the membrane. The mass transfer process is called dialysis and the device is known as a dialyzer.

There are various designs of artificial kidneys; some require blood pumps, while others provide blood flow due to arterial-venous pressure differences. Some operate on a counterflow or parallel-flow of blood and dialysate while others use crossflow. Some have dialysate on a once-through operation, while others recycle it. Membranes may be aligned in parallel flat sheets, flattened tubing rolled into coils, or parallel cylindrical semipermeable hollow fibers. Artificial kidneys may be classified according to construction, for example, flat-plate membrane types, coiled membrane types, and hollow fiber or capillary types (Flower, 1968; Cooney, 1976; Babb et al., 1967). They are most commonly used in clinical practice.

The practical dialyzer in widespread use is the coil type unit developed in 1943 by Kolff (1976) and is called the Kolff twin coil. The membrane consists of two cellophane tubes which are flattened, placed on a nylon open-mesh spacer material, and rolled into a coil. The coil is then held in a cartridge that is open at each end. The cartridge is put in a large dialysate bath where the dialysate is pumped upward through the coil, thus forming a crossflow device. A brief historical development of coil type devices is available in Cooney (1976).

The flat-plate type was first developed in the 1940s. Among a large variety of designs, the most widely used is the Kiil dialyzer. The patient's blood is passed between two flat sheets of regenerated cellulose of rectangular plan. The membranes are clamped between parallel epoxy boards having longitudinal grooves and carved-out longitudinal channels. The dialysate flows in counter-current with the blood.

The hollow fiber or capillary dialyzer is a more recent type. It resembles a shell-and-tube heat exchanger and consists of a large number (more than 10,000) of thin-walled capillaries made of regenerated cellulose (see Fig. 6.24). No blood pump is needed. These units are efficient, disposable, and preste-

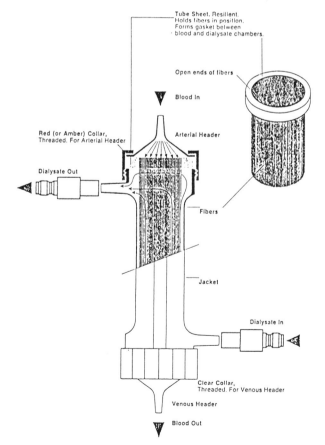

Figure 6.24 Capillary artificial kidney.

rilized and require low blood priming volumes. These advantages indicate that this unit will grow in popularity.

6.8.3 Artificial Livers

Prosthetic livers have been experimentally used as a substitute for whole livers in the extracorporeal support of comatose patients. Temporary liver support using extracorporeal circuitry may serve two purposes (Eiseman and Soyer, 1972): (1) maintenance of a recipient until a donor organ becomes available and (2) treatment of a potentially regenerative ability from liver failure (patients experience failure due to acute viral or chemical toxins).

In addition to earlier models of prosthetic livers in the 1950s, the following three devices were developed in the 1970s by Eisenman and Soyer (1972). Mark I consisted of a cylinder containing nine layers of thinly sliced fresh liver. The perfusate entered the bottom and exited the top of the perfusion chamber. Slices of liver rested on a sling of wide mesh Dacron sewn peripherally to the solid plastic ring. The stack of nine solid plastic rings each with its contained Dacron sling filled the cylindrical perfusion unit. The slices of liver floated between but were trapped between layers of the wide mesh Dacron slings. The perfusate consisted of equal parts of frozen homologous swine serum, freed of lipoprotein, and of McCoy's tissue culture. The second device, Mark II, had the liver slices resting on plastic disks rather than on a Dacron mesh. Perforations, which permitted passage of perfusate, covered only half of the disk area while the other half remained solid. By alternately positioning the perforations of the lower disk opposite the unperforated portion of the disk above, the perfusate was forced into a zigzag pathway that ensured maximum contact with the liver. The third unit, Mark III, consisted of a slender cylinder containing single liver cells in a plasma suspension. Oxygenated plasma entered the bottom of the cylinder and exited at the top. Cells were held within the chamber by a 2-μm filter backed by a metal screen that fitted into the top of the cylinder. The perfusate flow was reversed when the filter began to become clogged.

A great variety of these systems can be used to obtain contact with plasma from a patient in liver failure with functioning liver cells or subcellular elements. Liver cells can be impregnated into a membrane over which blood or plasma can be passed, as in a membrane oxygenator.

6.8.4 Artificial Pancreas

The pancreas is composed of two major tissues: (1) the acini which secrete digestive juices called the pancreatic juice into the duodenum and (2) the islets of Langerhans which secrete two hormones called insulin and glucagon directly into the blood. The main functions of insulin include regulating glucose, lipid, and protein metabolism. An insufficient secretion of insulin results in an increase in the blood glucose concentration from a normal value. Conversely, a

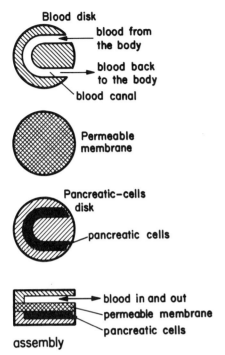

Blood disk
blood from
the body
blood back
to the body
blood canal

Permeable
membrane

Pancreatic−cells
disk
pancreatic cells

blood in and out
permeable membrane
pancreatic cells

assembly

Figure 6.25 Artificial pancreas.

great amount of insulin can cause a fall in the blood glucose concentration. A prolonged lack of insulin may result in the disease called diabetes (or diabetes mellitus). Though much more rare than diabetes, increased insulin production known as hyperinsulinism does occasionally occur.

One treatment in diabetes is to artificially inject insulin as needed or through an automated procedure. An artificial pancreas has recently been developed. As shown in Fig. 6.25, it consists of a disk with a U-shaped blood canal, a semipermeable membrane, and a disk with a U-shaped trough for placing healthy pancreatic cells. The blood disk is placed on top of the pancreatic-cells disk with the canal lined up with the trough. The membrane is sandwiched between the disks so that the blood can be in contact with the pancreatic cells. This is a hybrid device since it contains both pancreatic cells and blood.

This is a special case of percolation problems in which C_w (corresponding to T_w in Section 6.7.2) is constant or $b = 0$.

PROBLEMS

6.1 Consider water vapor (species A) diffusing at a constant rate from the liquid surface up through a layer of stagnant air (B) of thickness δ. Derive the expression for the rate of water vapor diffusion in terms of partial pressure.

6.2 Derive the expressions for the concentration distributions of the transported substance $C_M(x)$

and $C_F(x)$ on the maternal and fetal sides, respectively, for both the counterflow and the parallel-flow models.

6.3 Derive the expressions for the fetal and maternal transport fractions (y_F, y_M) and the diffusion clearance D_c for both the counterflow and parallel-flow models.

6.4 The counterflow and parallel-flow models are two extreme cases with the single-pool and double-pool cases. Construct the y_M versus y_F diagrams with P/Q_M and Q_F/Q_M as the parameters. For the same values of P/Q_M and Q_F/Q_M, is the human placenta more efficient than the placental circulation of rabbits and sheep? If not explain why.

6.5 Diamond (1962) used the plate model for solute transport by the gall bladder. A salt such as NaBr at a surface concentration C_s diffuses in a layer of the serosa with a thickness of l. The diffusion of the salt in the serosa is D. Let x be the distance measured from the interface with epithelial cells that do not contain salt. If the initial salt concentration in the serosa is C_0, the salt buildup in the serosa can be described by

$$\frac{\partial C}{\partial t} = D \frac{\partial^2 C}{\partial x^2}$$

subject to

$$C(x, 0) = C_0 \qquad \frac{\partial C(0, t)}{\partial x} = 0 \qquad C(l, t) = C_s$$

Determine the salt concentration as functions of x and t.

6.6 Hemolysis of cells (Fig. 6.26) in hypotonic saline or water is usually considered as a swelling of the cells to the critical hemolytic volume, because of the difference in osmotic pressure (or chemical activity of water) across the membrane. Since the surface area cannot be increased, it gives way and hemoglobin escapes. Derive the expression for the time history of hemoglobin concentration in the ghost (hemolytic cell in a spherical form with diameter D). The initial hemoglobin concentration in the ghost is C_i and K is the membrane permeability. (*Hint:* use the compartmental analysis.)

6.7 When a substance is injected into a vein, the concentration of the substance in the vein varies with time t and the distance z from the point of injection. Under certain conditions, the concentration can be expressed as

$$C(z, t) = \frac{a}{t^{1/2}} \exp\left(-\frac{z^2}{Dt}\right)$$

in which a is a constant and D denotes the mass diffusivity of the substance. Show that $C(z, t)$ satisfies the mass diffusion equation

$$\frac{\partial C}{\partial t} = \frac{D}{4} \frac{\partial^2 C}{\partial z^2}$$

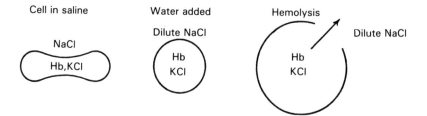

Figure 6.26 Diagram illustrating hemolysis.

Show that the function $C(z, t) = a \exp(bz + Db^2t/4)$ also satisfies the mass diffusion equation.

6.8 A substance is injected into a vein. When the drift due to the motion of the blood and the diffusion of the substance through the bloodstream are taken into consideration, the concentration can be described by a function of the form

$$C(z, t) = \frac{a}{t^{1/2}} \exp\left[-\frac{(z - vt)^2}{Dt}\right]$$

where v is the velocity of the blood. Show that $C(z, t)$ satisfies the following equation:

$$\frac{\partial C}{\partial t} + v\,\frac{\partial C}{\partial z} = \frac{D}{4}\,\frac{\partial^2 C}{\partial z^2}$$

6.9 During the process of metabolism, a bacterium absorbs a chemical substance at the rate of M and distributes it throughout the volume V. The rate can be expressed by $M = aA/V$, where A is the surface area of the bacterium and a is a constant. For a cylindrical bacterium of radius R and length l, determine $\partial M/\partial R$ and $\partial M/\partial l$ and find how an increase in radius or in length would affect the rate of metabolism.

REFERENCES

Babb, A. L., Grimsrud, L. Bell, R. L., and Layno, S. B. (1967) Engineering Aspects of Artificial Kidney Systems, in *Chemical Engineering in Medicine and Biology* (ed., D. Hershey), Plenum, New York, pp. 289–332.

Cooney, D. O. (1976) *Biomedical Engineering Principles,* Marcel Dekker, New York.

Dawids, S. G., and Engell, H. C. (1976) *Physiological and Clinical Aspects of Oxygenator Design,* Elsevier, Amsterdam.

Diamond, J. M. (1962) The mechanism of solute transport by the gallbladder, *J. Physiol.,* 161:474–502.

Eiseman, B., and Soyer, T. (1972) Prosthetics in Hepatic Assistance, in *Artificial Organs and Cardiopulmonary Support Systems* (ed. F. T. Rapaport and J. P. Merrill), Grune & Stratton, New York, pp. 125–130.

Faber, J. (1969) Application of the theory of heat exchangers to the transfer of inert materials in placentas, *J. Circ. Res.,* 24:221–234.

Flower, J. R. (1968) *Chemical Engineering in Medicine—The Artificial Kidney and Lung Machines,* Institute of Chemical Engineers, London.

Galletti, P. M., and Brecher, G. A. (1962) *Heart Lung By-pass,* Grune & Stratton, New York.

Gorsky, C. A. (1963) A linear method for determining liver sinusoidal and extravascular volumes, *Amer. J. Physiol.,* 204:626–640.

Guyton, A. C. (1954) *Textbook of Medical Physiology,* 3d. ed., W. B. Saunders, Philadelphia, p. 1154.

Kolff, W. J. (1976) *Artificial Organs,* Halsted Press, New York.

Kylstra, J. A., Paganelli, C. V., and Lanphier, E. H. (1966) Pulmonary gas exchange in dogs ventilated with hyperbarically oxygenated liquid, *J. Appl. Physiol.,* 21:177–184.

Lightfoot, E. N., Jr. (1974) *Transport Phenomena in Living Systems,* Wiley Interscience, New York.

Lih, M. M. S. (1975) *Transport Phenomena in Medicine and Biology,* Wiley Interscience, New York, pp. 343–354.

Matthews, B. F. (1967) *Chemical Exchanges in Man,* Oliver & Boyd, Edinburgh.

Meschia, G., Battaglia, F. C., and Burns, P. D. (1967) Theoretical and experimental study of transplacental diffusion, *J. Appl. Physiol.,* 22:1171–1178.

Middleman, S. (1972) *Transport Phenomena in the Cardiovascular System,* Wiley Interscience, New York.

Rapaport, F. T., and Merrill, J. P. (1972) *Artificial Organs and Cardiopulmonary Support Systems,* Grune & Stratton, New York.

Rehm, W. S., Schlesinger, H. and Dennis, W. H. (1953) Effect of osmotic gradients on water transport, hydrogen ion and chloride ion production in the resting and secreting stomach. *Amer. J. Physiol.,* vol. 175, pp. 473–486.

Roughton, F. J. W. (1959) Diffusion and simultaneous chemical reaction velocity in haemoglobin solutions and red cell suspensions, *Prog. Biophys. Biophysic. Chem.* 9:55–104.

Warner, H. R., and Seagrave, R. C. (1970) *Mass Transfer in Biological Systems* (ed. A. L. Shrier and T. C. Kaufman), AIChE, New York, p. 12.

NOMENCLATURE

A	transport area, m^2
A_c	flow cross-sectional area, m^2
a	activity, kJ/kg; or cell radius, μ
Bi	Biot number
C	mass concentration, kg/m^3; or electrical capacitance, F
C_f	drag coefficient
C_p, C_v	specific heats, kJ/kg °C
c	pulse wave velocity, m/s
D	mass diffusivity, m^2/s; tube diameter, m
D_h	hydraulic diameter, m
D^*	diffusion coefficient of solute in membrane
E	Young's modulus, dyn/cm^2; energy of the system, kJ; emissive power, kW/m^2
E_k	elasticity modulus, dyn/cm^2
e	specific energy of the system $= E/m$, kJ/kg; specific enthalpy, kJ/kg
F	Faraday constant, coulombs/equiv; shape factor
F_D	drag force, N
F_0	Fourier number
f	friction factor (fanning); fugacity, N
G	shear modulus, dyn/cm^2; Gibbs free energy, kJ; irradiation heat flux, kW/m^2
Gr	Grashof number
Gr_m	Grashof number for mass transfer

Gz	Graetz number ($= \mathrm{RePr}D_h/l$)
g	gravitation acceleration ($= 9.8$ m/s^2)
g_c	conversion factor
H	enthalpy, kJ; Henry's constant, mmHg
h	heat transfer coefficient, kW/m^2 °C
h_D	mass transfer coefficient, m/s
h_D^*	permeability, m/s
h_{fg}	latent heat of vaporization, kJ/kg
I	radiation intensity, kW/m^2
I_s	insulation ($= \delta/k$), m^2 K/W
i	electric current, amp
J	radiosity, kW/m^2
K	thermodynamic equilibrium constant; overall permeability of membrane, m/s
K_c	Casson's viscosity
K_v	flow permeability of capillary or cell membrane, mol/m^2 s mmHg
k	thermal conductivity, kW/m °C
L	characteristic length, m; electrical inductance, Henry
LW	lost work, kJ
l	length, m
M	molecular weight, g
m	mass, kg
$\dot{m}$	mass flow rate, kg/s
m'''	volumetric rate of internal mass generation, kg/s m^3
mf	mass fraction
N	rate of mass transfer, kg/s
N''	mass flux, kg/s m^2
Nu	Nusselt number
n	exponent; number of moles
P	pressure, N; permeability of membrane, m^3/s
Pr	Prandtl number
p	vapor pressure, N
Q	heat, kJ; or volumetric flow rate, m^3/s
q	rate of heat transfer, W
q''	heat flux, W/m^2
q'''	volumetric rate of internal heat generation, W/m^3
R	inner radius of tube, m; electrical resistance, ohm; thermal resistance, s/W; gas constant
R_D	mass transfer resistance, s/m^3
Ra	Rayleigh number
Re	Reynolds number
r	radial distance, m
(r, θ, z)	cylindrical coordinates
(r, θ, ϕ)	spherical coordinates

S	entropy, kJ/K; cell-to-cell spacing, cm
Sc	Schmidt number
Sh	Sherwood number
s	specific entropy ($= s/m$), kJ/kg K; Laplace variable, s^{-1}
T	temperature, °C
t	time, s
U	velocity of erythrocyte, cm/s; internal energy, kJ; overall heat transfer coefficient, W/s m^2
U_D	overall mass transfer coefficient, m/s
u	specific internal energy, U/m, kJ/kg
V	volume, m^3
$\dot{V}$	volumetric flow rate, m^3/s
v	velocity (subscripts showing direction), m/s
v_i	species velocity, m/s
v_{Di}	diffusion velocity of species i, m/s
$\bar{v}$	mean velocity of suspension, m/s
v^*	specific volume ($= V/m$), m^3/kg
vf	volume fraction
W	work, kJ
$\dot{w}_b$	blood perfusion rate per unit volume of tissue, kg/s m^3
x	distance, m; mole fraction (liquid phase)
(x, y, z)	Cartesian coordinates
y	mole fraction (gas phase)
Z	elevation, m
z	distance, m
α	α parameter; thermal diffusivity, m^2/s; absorptivity
β	bulk modulus of elasticity of blood, dyn/cm^2; coefficient of thermal expansion, K^{-1}
β_m	coefficient of volumetric expansion (concentration), m^3/kg
γ	correction factor; shear strain
δ	thickness of wall, tissue, layer, coat, or membrane, cm
δ_e	effective wall thicknes, cm
ϵ	oxidation-reduction potential; emissivity
ζ	distribution coefficient
ζ_r	radial displacement of wall, cm
ζ_z	longitudinal displacement of wall, cm
η	apparent viscosity or shear viscosity, poise or g/s cm
η_k	viscosity modulus, dyn s/cm^2
θ	angle; angle of inclination
λ	wavelength, μm
λ_1	stress relaxation time, s
λ_1	strain relaxation time, s, poise or g/s cm
μ	absolute viscosity; μ_P, power-law model; μ_B, Bingham-plastic model
ν	kinematic viscosity, m^2/s

ξ	strain, cm
ρ	density, g/cm^3; reflectivity
σ	Poisson's ratio
τ	time constant, s; shear stress, dyn/cm^2
τ_0	yield stress, dyn/cm^2
ϕ	angle
ψ	stream function, s^{-1}
Ω	angular velocity, s; natural frequency, s^{-1}
ω	circular frequency of pulse waves, s^{-1}; specific humidity

Superscripts

$-$	average value; molar average value
$\cdot$	rate
$'$	gradient
$''$	transfer flux
$'''$	volumetric rate of generation
$*$	per meter
0	under standard conditions

Subscripts

A	species, arterial blood or absorbed
a	additional; air
B	species
b	bulk; blood; black body; DuBois surface
c	convection; core; capillary
cr	critical value
cv	control volume
D	diffusion
e	effective value; environment
f	formation or fluid
H	high value
i	ith layer or species
in	flow in
k	conduction
L	low value
l	liquid phase
m	metabolic; mass transfer
o	initial value or state
out	flow out
R	reflection
γ	radiation
s	surface (skin surface)
sat	saturated state
sur	surroundings

sys	system
T	total quantity or transmission
t	thermal boundary layer or tissue
V	venous blood
v	vapor phase
λ	monochromatic wave
1	inlet, initial state; branch vessel
2	exit, final state; branch vessel
∞	ambient; free stream

MOMENTUM, HEAT, AND
MASS BALANCE EQUATIONS

The principle of conservation of property states that

$$\begin{Bmatrix} \text{Rate of property} \\ \text{flow across} \\ \text{control surface} \\ \text{into control} \\ \text{volume} \end{Bmatrix} - \begin{Bmatrix} \text{rate of property} \\ \text{flow across} \\ \text{control surface} \\ \text{out of control} \\ \text{volume} \end{Bmatrix} + \begin{Bmatrix} \text{rate of property} \\ \text{generation} \\ \text{within control} \\ \text{volume} \end{Bmatrix}$$

$$= \begin{Bmatrix} \text{rate of property} \\ \text{stored within} \\ \text{control volume} \end{Bmatrix} \qquad (B.1)$$

$$- \int_A \int b\rho \, \mathbf{v} \cdot \mathbf{n} \, dA + B'''V = \frac{d}{dt} \int \int_V \int \rho b \, dV \qquad (B.2)$$

where B is the property; $b = B/m$, specific volume; ρ, density; $\mathbf{v}$, flow velocity; $\mathbf{n}$, outward normal vector; A, control surface area; V, volume of control volume; and B''', volumetric rate of property generation within control volume. Equation (B.2) is the integral form of the balance equation. The differential forms of mass, momentum, and energy are summarized in Tables B.1 through B.4.

Table B.1 Equation of continuity

Coordinates	Equation	Eq. no.
Rectangular (x, y, z)	$\dfrac{\partial \rho}{\partial t} + \dfrac{\partial}{\partial x}(\rho v_x) + \dfrac{\partial}{\partial y}(\rho v_y) + \dfrac{\partial}{\partial z}(\rho v_z) = 0$	(B.3)
Cylindrical (r, θ, z)	$\dfrac{\partial \rho}{\partial t} + \dfrac{1}{r}\dfrac{\partial}{\partial r}(\rho r v_r) + \dfrac{1}{r}\dfrac{\partial}{\partial \theta}(\rho v_\theta) + \dfrac{\partial}{\partial z}(\rho v_z) = 0$	(B.4)
Spherical (r, θ, ϕ)	$\dfrac{\partial \rho}{\partial t} + \dfrac{1}{r^2}\dfrac{\partial}{\partial r}(\rho r^2 v_r) + \dfrac{1}{r \sin \theta}\dfrac{\partial}{\partial \theta}(\rho v_\theta \sin \theta) + \dfrac{1}{r \sin \theta}\dfrac{\partial}{\partial \phi}(\rho v_\phi) = 0$	(B.5)

Table B.2 Equation of motion

Coordinates	Equation	Eq. no.

In terms of shear stress τ

Rectangular (x, y, z) x component:

$$\rho\left(\frac{\partial v_x}{\partial t} + v_x\frac{\partial v_x}{\partial x} + v_y\frac{\partial v_x}{\partial y} + v_s\frac{\partial v_x}{\partial z}\right)$$
$$= -\frac{\partial P}{\partial x} - \left(\frac{\partial \tau_{xx}}{\partial x} + \frac{\partial \tau_{yx}}{\partial y} + \frac{\partial p_{zx}}{\partial z}\right) + \rho g_x \tag{B.6}$$

y component:

$$\rho\,b2\frac{\partial v_y}{\partial t} + v_x\frac{\partial v_y}{\partial x} + v_y\frac{\partial v_y}{\partial y} + v_z\frac{\partial v_y}{\partial z}$$
$$= -\frac{\partial P}{\partial y} - \left(\frac{\partial \tau_{xy}}{\partial x} + \frac{\partial \tau_{yy}}{\partial y} + \frac{\partial \tau_{zy}}{\partial z}\right) + \rho g_y \tag{B.7}$$

z component:

$$\rho\left(\frac{\partial v_z}{\partial t} + v_z\frac{\partial v_z}{\partial x} + v_z\frac{\partial v_z}{\partial z}\right)$$
$$= -\frac{\partial P}{\partial z} - \left(\frac{\partial \tau_{sz}}{\partial x} + \frac{\partial p_{xz}}{\partial y} + \frac{\partial p_{zz}}{\partial z}\right) + \rho g_z \tag{B.8}$$

Cylindrical (r, θ, z) r component:

$$\rho\left(\frac{\partial v_r}{\partial t} + v_r\frac{\partial v_r}{\partial r} + \frac{v_\theta}{r}\frac{\partial v_r}{\partial \theta} - \frac{v_\theta^2}{r} + v_z\frac{\partial v_r}{\partial z}\right)$$
$$= -\frac{\partial P}{\partial r} - \left(\frac{1}{r}\frac{\partial}{\partial r}(r\tau_{rr}) + \frac{1}{r}\frac{\partial \tau_{u\theta}}{\partial \theta} - \frac{\tau_{\theta\theta}}{r} + \frac{\partial \tau_{rz}}{\partial z}\right) + \rho g_r \tag{B.9}$$

θ component:

$$\rho\left(\frac{\partial v_\theta}{\partial t} + v_r\frac{\partial v_\theta}{\partial r} + \frac{v_\theta}{r}\frac{\partial v_\theta}{\partial \theta} + \frac{v_r v_\theta}{r} + v_z\frac{\partial v_\theta}{\partial z}\right)$$
$$= -\frac{1}{r}\frac{\partial P}{\partial \theta} - \left(\frac{1}{r^2}\frac{\partial}{\partial r}(r^2\tau_{r\theta}) + \frac{1}{r}\frac{\partial \tau_{\theta\theta}}{\partial \theta} + \frac{\partial \tau_{\theta z}}{\partial z}\right) + \rho g_\theta \tag{B.10}$$

z component:

$$\rho\left(\frac{\partial v_z}{\partial t} + v_r\frac{\partial v_z}{\partial r} + \frac{v_\theta}{r}\frac{\partial v_z}{\partial \theta} + v_z\frac{\partial v_z}{\partial z}\right)$$
$$= -\frac{\partial P}{\partial z} - \left(\frac{1}{r}\frac{\partial}{\partial r}(r\tau_{rz}) + \frac{1}{r}\frac{\partial p_{\theta z}}{\partial \theta} + \frac{\partial \tau_{zz}}{\partial z}\right) + \rho g_z \tag{B.11}$$

Table B.2 Equation of motion

Coordinates	Equation	Eq. no.

Spherical (r, θ, ϕ) r component:

$$\rho \left(\frac{\partial v_r}{\partial t} + v_r \frac{\partial v_r}{\partial r} + \frac{v_\theta}{r} \frac{\partial v_r}{\partial \theta} + \frac{v_\phi}{r \sin \theta} \frac{\partial v_r}{\partial \phi} - \frac{v_\theta^2 + v_\phi^2}{r} \right)$$
$$= -\frac{\partial P}{\partial r} - \left[\frac{1}{r^2} \frac{\partial}{\partial r} (r^2 \tau_{rr}) + \frac{1}{r \sin \theta} \frac{\partial}{\partial \theta} (\tau_{r\theta} \sin \theta) \right.$$
$$\left. + \frac{1}{r \sin \theta} \frac{\partial \tau_{r\phi}}{\partial \phi} - \frac{\tau_{\theta\theta} + \tau_{\phi\phi}}{r} \right] + \rho g_r$$

(B.12)

θ component:

$$\rho \left(\frac{\partial v_\theta}{\partial t} + v_r \frac{\partial v_\theta}{\partial r} + \frac{v_\theta}{r} \frac{\partial v_\theta}{\partial \theta} + \frac{v_\phi}{r \sin \theta} \frac{\partial v_\theta}{\partial \phi} + \frac{v_r v_\theta}{r} - \frac{v_\phi^2 \cot \theta}{r} \right)$$
$$= -\frac{1}{r} \frac{\partial P}{\partial \theta} - \left(\frac{1}{r^2} \frac{\partial}{\partial r} (r^2 \tau_{r\theta}) + \frac{1}{r \sin \theta} \frac{\partial}{\partial \theta} (\tau_{\theta\theta} \sin \theta) \right.$$
$$\left. + \frac{1}{r \sin \theta} \frac{\partial \tau_{\theta\phi}}{\partial \phi} + \frac{\tau_{r\theta}}{r} - \frac{\cot \theta}{r} \tau_{\phi\phi} \right) + \rho g_\theta$$

(B.13)

ϕ component:

$$\rho \left(\frac{\partial v_\phi}{\partial t} + v_r \frac{\partial v_\phi}{\partial r} + \frac{v_\theta}{r} \frac{\partial v_\phi}{\partial \theta} + \frac{v_\phi}{r \sin \theta} \frac{\partial v_\phi}{\partial \phi} + \frac{v_\phi v_r}{r} + \frac{v_\theta v_\phi}{r} \cot \theta \right)$$
$$= -\frac{1}{r \sin \theta} \frac{\partial P}{\partial \phi} - \left(\frac{1}{r^2} \frac{\partial}{\partial r} (r^2 \tau_{r\phi}) + \frac{1}{r} \frac{\partial \tau_{\theta\phi}}{\partial \theta} + \frac{1}{r \sin \theta} \frac{\partial \tau_{\theta\theta}}{\partial \phi} \right.$$
$$\left. + \frac{\tau_{r\phi}}{r} + \frac{2 \cot \theta}{r} \tau_{\theta\phi} \right) + \rho g_\phi$$

(B.14)

In terms of velocity gradients for Newtonian fluids with constant ρ and μ

Rectangular (x, y, z) x component:

$$\rho \left(\frac{\partial v_x}{\partial t} + v_x \frac{\partial v_x}{\partial x} + v_y \frac{\partial v_x}{\partial y} + v_z \frac{\partial v_x}{\partial z} \right)$$
$$= -\frac{\partial P}{\partial x} + \mu \left(\frac{\partial^2 v_x}{\partial x^2} + \frac{\partial^2 v_x}{\partial y^2} + \frac{\partial^2 v_x}{\partial z^2} \right) + \rho g_x$$

(B.15)

y component:

$$\rho \left(\frac{\partial v_y}{\partial t} + v_x \frac{\partial v_y}{\partial x} + v_y \frac{\partial v_y}{\partial y} + v_z \frac{\partial v_y}{\partial z} \right)$$

$$= -\frac{\partial P}{\partial y} + \mu\left(\frac{\partial^2 v_y}{\partial x^2} + \frac{\partial^2 v_y}{\partial y^2} + \frac{\partial^2 v_y}{\partial z^2}\right) + \rho g_y \tag{B.16}$$

z component:

$$\rho\left(\frac{\partial v_z}{\partial t} + v_x\frac{\partial v_z}{\partial x} + v_y\frac{\partial v_z}{\partial y} + v_z\frac{\partial v_z}{\partial z}\right)$$
$$= -\frac{\partial P}{\partial z} + \mu\left(\frac{\partial^2 v_z}{\partial x^2} + \frac{\partial^2 v_z}{\partial y^2} + \frac{\partial^2 v_z}{\partial z^2}\right) + \rho g_z \tag{B.17}$$

Cylindrical (r, θ, z) r component:

$$\rho\left(\frac{\partial v_r}{\partial t} + v_r\frac{\partial v_r}{\partial r} + \frac{v_\theta}{r}\frac{\partial v_r}{\partial \theta} - \frac{v_\theta^2}{r} + v_z\frac{\partial v_r}{\partial z}\right)$$
$$= -\frac{\partial P}{\partial r} + \mu\left[\frac{\partial}{\partial r}\left(\frac{1}{r}\frac{\partial}{\partial r}(rv_r)\right) + \frac{1}{r^2}\frac{\partial^2 v_r}{\partial \theta^2} - \frac{2}{r^2}\frac{\partial v_\theta}{\partial \theta} + \frac{\partial^2 v_r}{\partial z^2}\right] + \rho g_r \tag{B.18}$$

θ component:

$$\rho\left(\frac{\partial v_\theta}{\partial t} + v_r\frac{\partial v_\theta}{\partial r} + \frac{v_\theta}{r}\frac{\partial v_\theta}{\partial \theta} + \frac{v_r v_\theta}{r} + v_z\frac{\partial v_\theta}{\partial z}\right)$$
$$= -\frac{1}{r}\frac{\partial P}{\partial \theta} + \mu\left[\frac{\partial}{\partial r}\left(\frac{1}{r}\frac{\partial}{\partial r}(rv_\theta)\right) + \frac{1}{r^2}\frac{\partial^2 v_\theta}{\partial \theta^2} + \frac{2}{r^2}\frac{\partial v_r}{\partial \theta} + \frac{\partial^2 v_\theta}{\partial z^2}\right] + \rho g_\theta \tag{B.19}$$

z component:

$$\rho\left(\frac{\partial v_z}{\partial t} + v_r\frac{\partial v_z}{\partial r} + \frac{v_\theta}{r}\frac{\partial v_z}{\partial \theta} + v_z\frac{\partial v_z}{\partial z}\right)$$
$$= -\frac{\partial P}{\partial z} + \mu\left[\frac{1}{r}\frac{\partial}{\partial r}\left(r\frac{\partial v_z}{\partial r}\right) + \frac{1}{r^2}\frac{\partial^2 v_z}{\partial \theta^2} + \frac{\partial^2 v_z}{\partial z^2}\right] + \rho g_z \tag{B.20}$$

Spherical (r, θ, ϕ) r component:

$$\rho\left(\frac{\partial v_r}{\partial t} + v_r\frac{\partial v_r}{\partial r} + \frac{v_\theta}{r}\frac{\partial v_r}{\partial \theta} + \frac{v_\phi}{r\sin\theta}\frac{\partial v_r}{\partial \phi} - \frac{v_\theta^2 + v_\phi^2}{r}\right)$$
$$= -\frac{\partial P}{\partial r} + \mu\left(\nabla^2 v_r - \frac{2}{r^2}v_r - \frac{2}{r^2}\frac{\partial v_\theta}{\partial \theta} - \frac{2}{r^2}v_\theta\cot\theta - \frac{2}{r^2\sin\theta}\frac{\partial v_\phi}{\partial \phi}\right) + \rho g_r \tag{B.21}$$

θ component:

$$\rho\left(\frac{\partial v_\theta}{\partial t} + v_r\frac{\partial v_\theta}{\partial r} + \frac{v_\theta}{r}\frac{\partial v_\theta}{\partial \theta} + \frac{v_\phi}{r\sin\theta}\frac{\partial v_\theta}{\partial \phi} + \frac{v_r v_\theta}{r} - \frac{v_\phi^2\cot\theta}{r}\right)$$
$$= -\frac{1}{r}\frac{\partial P}{\partial \theta} + \mu\left(\nabla^2 v_\theta + \frac{2}{r^2}\frac{\partial v_r}{\partial \theta} - \frac{v_\theta}{r^2\sin^2\theta} - \frac{2\cos\theta}{r^2\sin^2\theta}\frac{\partial v_\phi}{\partial \phi}\right) + \rho g_\theta \tag{B.22}$$

Table B.2 Equation of motion

Coordinates	Equation	Eq. no.
ϕ component:	$\rho\left(\dfrac{\partial v_\phi}{\partial t} + v_r\,\dfrac{\partial v_\phi}{\partial r} + \dfrac{v_\theta}{r}\dfrac{\partial v_\phi}{\partial \theta} + \dfrac{v_\phi}{r\sin\theta}\dfrac{\partial v_\phi}{\partial \phi} + \dfrac{v_\phi v_r}{r} + \dfrac{v_\theta v_\phi}{r}\cot\theta\right)$ $= -\dfrac{1}{r\sin\theta}\dfrac{\partial P}{\partial \phi} + \mu\left(\nabla^2 v_\phi - \dfrac{v_\phi}{r^2\sin^2\theta} + \dfrac{2}{r^2\sin\theta}\dfrac{\partial v_r}{\partial \phi}\right.$ $\left.+ \dfrac{2\cos\theta}{r^2\sin^2\theta}\dfrac{\partial v_\theta}{\partial \phi}\right) + \rho g\phi$	(B.23)

Table B.3 Equation of energy

Coordinates	Equation	Eq. no.

In terms of energy and momentum fluxes

Rectangular

$$\rho C_v\left(\frac{\partial T}{\partial t} + v_x\frac{\partial T}{\partial x} + v_y\frac{\partial T}{\partial y} + v_z\frac{\partial T}{\partial z}\right) = -\left(\frac{\partial q_x}{\partial x} + \frac{\partial q_y}{\partial y} + \frac{\partial q_z}{\partial z}\right)$$

$$- T\left(\frac{\partial p}{\partial T}\right)_p\left(\frac{\partial v_x}{\partial x} + \frac{\partial v_y}{\partial y} + \frac{\partial v_z}{\partial z}\right) - \left(\tau_{xx}\frac{\partial v_x}{\partial x} + \tau_{rr}\frac{\partial v_y}{\partial y} + \tau_{zz}\frac{\partial v_z}{\partial z}\right)$$

$$- \left[\tau_{xz}\left(\frac{\partial v_x}{\partial y} + \frac{\partial v_y}{\partial x}\right) + \tau_{xz}\left(\frac{\partial v_x}{\partial z} + \frac{\partial v_z}{\partial x}\right) + \tau_{yz}\left(\frac{\partial v_y}{\partial z} + \frac{\partial v_z}{\partial y}\right)\right] + q'''$$

(B.24)

Cylindrical

$$\rho C_v\left(\frac{\partial T}{\partial t} + v_r\frac{\partial T}{\partial r} + \frac{v_\theta}{r}\frac{\partial T}{\partial \theta} + v_z\frac{\partial T}{\partial z}\right) = -\left(\frac{1}{r}\frac{\partial}{\partial r}(rq_r'') + \frac{1}{r}\frac{\partial q_\theta''}{\partial \theta} + \frac{\partial q_z''}{\partial z}\right)$$

$$- T\left(\frac{\partial p}{\partial T}\right)_p\left(\frac{1}{r}\frac{\partial}{\partial r}(rv_r) + \frac{1}{r}\frac{\partial v_\theta}{\partial \theta} + \frac{\partial v_z}{\partial z}\right) - \left[\tau_{rr}\frac{\partial v_r}{\partial r} + \tau_{\theta\theta}\frac{1}{r}\left(\frac{\partial v_\theta}{\partial \theta} + v_r\right)\right]$$

$$+ \tau_{zz}\frac{\partial v_z}{\partial z}\right] - \left\{\tau_{r\theta}\left[r\frac{\partial}{\partial r}\left(\frac{v_\theta}{r}\right) + \frac{1}{r}\frac{\partial v_r}{\partial \theta}\right] + \tau_{rz}\left(\frac{\partial v_z}{\partial r} + \frac{\partial v_r}{\partial z}\right)$$

$$+ \tau_{\theta z}\left(\frac{1}{r}\frac{\partial v_z}{\partial \theta} + \frac{\partial v_\theta}{\partial z}\right)\right\} + q'''$$

(B.25)

Spherical

$$\rho C_v\left(\frac{\partial T}{\partial t} + v_r\frac{\partial T}{\partial r} + \frac{v_\theta}{r}\frac{\partial T}{\partial \theta} + \frac{v_\phi}{r\sin\theta}\frac{\partial T}{\partial \phi}\right) = -\left[\frac{1}{r^2}\frac{\partial}{\partial r}(r^2 q_r'')\right.$$

$$+ \frac{1}{rr\sin\theta}\frac{\partial}{\partial \theta}(q_\theta''\sin\theta) + \frac{1}{r\sin\theta}\frac{\partial q_\phi''}{\partial \phi}\right] - T\left(\frac{\partial p}{\partial T}\right)_p\left[\frac{1}{r^2}\frac{\partial}{\partial r}(r^2 v_r)\right.$$

$$+ \frac{1}{r\sin\theta}\frac{\partial}{\partial \theta}(v_\theta\sin\theta) + \frac{1}{r\sin\theta}\frac{\partial v_\phi}{\partial \phi}\right] - \left[\tau_{rr}\frac{\partial v_r}{\partial r} + \tau_{\theta\theta}\left(\frac{1}{r}\frac{\partial v_\theta}{\partial \theta} + \frac{v_r}{r}\right)\right.$$

$$+ \tau_{\theta\theta}\left(\frac{1}{r\sin\theta}\frac{\partial v_\theta}{\partial \phi} + \frac{v_r}{r} + \frac{v_\theta\cot\theta}{r}\right)\right] - \left[\tau_{r\theta}\left(\frac{\partial v_\theta}{\partial r} + \frac{1}{r}\frac{\partial v_r}{\partial \theta} - \frac{v_\theta}{r}\right)\right.$$

$$+ \tau_{r\phi}\left(\frac{\partial v_\phi}{\partial r} + \frac{1}{r\sin\theta}\frac{\partial v_r}{\partial \phi} - \frac{v\phi}{r}\right) + \rho_{\theta\phi}\left(\frac{1}{r}\frac{\partial v_\phi}{\partial \theta} + \frac{1}{r\sin\theta}\frac{\partial v_\theta}{\partial \phi} - \frac{\cot\theta}{r}v_\phi\right)\right]$$

$$+ q'''$$

(B.26)

In terms of the transport properties for Newtonian fluids of constant ρ, μ, and k ($C_p = C_v$ for constant ρ)

Rectangular

$$\rho C_v\left(\frac{\partial T}{\partial t} + v_x\frac{\partial T}{\partial x} + v_y\frac{\partial T}{\partial y} + v_z\frac{\partial T}{\partial z}\right) = k\left(\frac{\partial^2 T}{\partial x^2} + \frac{\partial^2 T}{\partial y^2} + \frac{\partial^2 T}{\partial z^2}\right)$$

$$+ 2\mu\left[\left(\frac{\partial v_x}{\partial x}\right)^2 + \left(\frac{\partial v_y}{\partial y}\right)^2 + \left(\frac{\partial v_z}{\partial z}\right)^2\right] + \mu\left[\left(\frac{\partial v_x}{\partial y} + \frac{\partial v_y}{\partial x}\right)^2\right.$$

$$+ \left(\frac{\partial v_x}{\partial z} + \frac{\partial v_z}{\partial x}\right)^2 + \left(\frac{\partial v_y}{\partial z} + \frac{\partial v_z}{\partial y}\right)^2\right] + q'''$$

(B.27)

Cylindrical

$$\rho C_v\left(\frac{\partial T}{\partial t} + v_r\frac{\partial T}{\partial r} + \frac{v_\theta}{r}\frac{\partial T}{\partial \theta} + v_z\frac{\partial T}{\partial z}\right) = k\left[\frac{1}{r}\frac{\partial}{\partial r}\left(r\frac{\partial T}{\partial r}\right) + \frac{1}{r^2}\frac{\partial^2 T}{\partial \theta^2} + \frac{\partial^2 T}{\partial z^2}\right]$$

$$+ 2\mu\left\{\left(\frac{\partial v_r}{\partial r}\right)^2 + \left[\frac{1}{r}\left(\frac{\partial v_\theta}{\partial \theta} + v_r\right)\right]^2 + \left(\frac{\partial v_z}{\partial z}\right)^2\right\} + \mu\left\{\left(\frac{\partial v_\theta}{\partial z} + \frac{1}{r}\frac{\partial v_z}{\partial \theta}\right)^2\right.$$

$$+ \left(\frac{\partial v_z}{\partial r} + \frac{\partial v_r}{\partial z}\right)^2 + \left[\frac{1}{r}\frac{\partial v_r}{\partial \theta} + r\frac{\partial}{\partial r}\left(\frac{v_\theta}{r}\right)\right]^2\right\} + q'''$$

(B.28)

Table B.3 Equation of energy

Coordinates	Equation	Eq. no.
Spherical	$$\rho C_v \left(\frac{\partial T}{\partial t} + v_r \frac{\partial T}{\partial r} + \frac{v_\theta}{r} \frac{\partial T}{\partial \theta} + \frac{v_\phi}{r \sin \theta} \frac{\partial T}{\partial \phi} \right) = k \left[\frac{1}{r^2} \frac{\partial}{\partial r} \left(r^2 \frac{\partial T}{\partial r} \right) \right.$$ $$\left. + \frac{1}{r^2 \sin \theta} \frac{\partial}{\partial \theta} \left(\sin \theta \frac{\partial T}{\partial \theta} \right) + \frac{1}{r^2 \sin^2 \theta} \frac{\partial^2 T}{\partial \phi^2} \right] + 2\mu \left[\left(\frac{\partial v_r}{\partial r} \right)^2 \right.$$ $$+ \left(\frac{1}{r} \frac{\partial v_\theta}{\partial \theta} + \frac{v_r}{r} \right)^2 + \left(\frac{1}{r \sin \theta} \frac{\partial v_\phi}{\partial \phi} + \frac{v_r}{r} + \frac{v_\theta \cot \theta}{r} \right)^2 \right]$$ $$+ \mu \left\{ \left[r \frac{\partial}{\partial r} \left(\frac{v_\theta}{r} \right) + \frac{1}{r} \frac{\partial v_r}{\partial \theta} \right]^2 + \left[\frac{1}{r \sin \theta} \frac{\partial v_r}{\partial \phi} + r \frac{\partial}{\partial r} \left(\frac{v_\phi}{r} \right) \right]^2 \right.$$ $$+ \left. \left[\frac{\sin \theta}{r} \frac{\partial}{\partial \theta} \left(\frac{v_\phi}{\sin \theta} \right) + \frac{1}{r \sin \theta} \frac{\partial v_\theta}{\partial \phi} \right]^2 \right\} + q'''$$	(B.29)

Table B.4 Mass transfer equations

Coordinates	Equation	Eq. no.
In terms of mass fluxes		
Rectangular	$$\frac{\partial c_A}{\partial t} + \left(\frac{\partial N_{Ax}}{\partial x} + \frac{\partial N_{Ay}}{\partial y} + \frac{\partial N_{Az}}{\partial z} \right) = m_A'''$$	(B.30)
Cylindrical	$$\frac{\partial c_A}{\partial t} + \left(\frac{1}{r} \frac{\partial}{\partial r} (r N_{Ar}) + \frac{1}{r} \frac{\partial N_{A\theta}}{\partial \theta} + \frac{\partial N_{Az}}{\partial z} \right) = m_A'''$$	(B.31)
Spherical	$$\frac{\partial c_A}{\partial t} + \left[\frac{1}{r^2} \frac{\partial}{\partial r} (r^2 N_{Ar}) + \frac{1}{r \sin \theta} \frac{\partial}{\partial \theta} (N_{A\theta} \sin \theta) + \frac{1}{r \sin \theta} \frac{\partial N_{A\phi}}{\partial \phi} \right] = m_A'''$$	(B.32)
In terms of the transport properties for Newtonian fluids of constant $\mathcal{D}_{AB}$		
Rectangular	$$\frac{\partial c_A}{\partial t} + \left(v_x \frac{\partial c_A}{\partial x} + v_y \frac{\partial c_A}{\partial y} + v_z \frac{\partial c_A}{\partial z} \right)$$ $$= \mathcal{D}_{AB} \left(\frac{\partial^2 c_A}{\partial x^2} + \frac{\partial^2 c_A}{\partial y^2} + \frac{\partial^2 c_A}{\partial z^2} \right) + m_A'''$$	(B.33)
Cylindrical	$$\frac{\partial c_A}{\partial t} + \left(v_r \frac{\partial c_A}{\partial r} + v_\theta \frac{1}{r} \frac{\partial c_A}{\partial \theta} + v_z \frac{\partial c_A}{\partial z} \right)$$ $$= \mathcal{D}_{AB} \left(\left[\frac{1}{r} \frac{\partial}{\partial r} \left(r \frac{\partial c_A}{\partial r} \right) + \frac{1}{r^2} \frac{\partial^2 c_A}{\partial \theta^2} + \frac{\partial^2 c_A}{\partial z^2} \right] + m_A''' \right.$$	(B.34)
Spherical	$$\frac{\partial c_A}{\partial t} + \left(v_r \frac{\partial c_A}{\partial r} + v_\theta \frac{1}{r} \frac{\partial c_A}{\partial \theta} + v_\phi \frac{1}{r \sin \theta} \frac{\partial c_A}{\partial \phi} \right)$$ $$= \mathcal{D}_{AB} \left[\frac{1}{r^2} \frac{\partial}{\partial r} \left(r^2 \frac{\partial c_A}{\partial r} \right) + \frac{1}{r^2 \sin \theta} \frac{\partial}{\partial \theta} \left(\sin \theta \frac{\partial c_A}{\partial \theta} \right) \right.$$ $$+ \left. \frac{1}{r^2 \sin \theta} \frac{\partial^2 c_A}{\partial \phi^2} \right] + m_A'''$$	(B.35)

UNIT CONVERSION

The International System of Units (SI) has evolved from the mks (meter-kilogram-second) system, in which the meter is the unit of length, the kilogram is the unit of mass, and the second is the unit of time. In the cgs (centimeter-gram-second) system, the centimeter, gram, and second are the units of length, mass, and time, respectively. The different disciplines involved in biomedical engineering use different sets of units, with biology employing predominantly cgs, while engineering (in the United States) splitting between SI and English systems. However, the SI system is rapidly becoming the standard system of units throughout the world. To facilitate conversion from one system to another, conversion factors are incorporated in Table C.1.

Table C.1 Unit conversion

Quantity	Units and equivalences		
	mks(cgs*)	English	SI
Length	1 m	3.281 ft	1 m
	0.3048 m	1 ft	0.3048 m
Area	1 m²	10.76 ft²	1 m²
	0.09290 m²	1 ft²	0.09290 m²
Velocity	1 m/s	3.281 ft/s	1 m/s
	0.3048 m/s	1 ft/s	0.3048 m/s
Density	1 kg_m/m^3	0.06243 lb_m/ft^3	1 kg/m³
	16.02 kg_m/m^3	1 lb_m/ft^3	16.02 kg/m³
Mass velocity	1 kg_m/m^2 s	0.2048 lb_m/ft^2 s	1 kg/m² s
	4.882 kg_m/m^2 s	1 lb_m/ft^2 s	4.882 kg/m² s
Temperature	°C [= 5(°F − 32)/9]	°F [= 32 + 9°C/5]	°C [= 5(°F − 32)/9]
Heat	1 kcal	3.968 BTU	4.186 × 10³ J
	0.2520 kcal	1 BTU	1055.1 J
	2.389 × 10⁻⁴ kcal	3.412 BTU	1 J
Dynamic viscosity	1 kg_f s/m²	2.372 × 10⁴ lb_m/ft h	1.2706 × 10⁸ Pa s
	4.2159 × 10⁻⁵ kg_f s/m²	1 lb_m/ft h	5.3568 × 10³ Pa s
	7.870 × 10⁻⁹ kg_f s/m²	1.8668 × 10⁻⁴ lb_m/ft h	1 Pa s
Surface tension	1 kg_f/m	0.6720 lb_f/ft	9.807 N/m
	1.488 kg_f/m	1 lb_f/ft	14.594 N/m
	0.1020 kg_f/m	0.0685 lb_f/ft	1 N/m
Heat transfer coefficient	1 kcal/m² h °C	0.2048 BTU/ft² h °F	1.1630 W/m² K
	4.882 kcal/m² h °C	1 BTU/ft² h °F	5.6785 W/m² K
	0.8598 kcal/m² h °C	0.1761 BTU/ft² h °F	1 W/m² K

Quantity			
Thermal resistance	1 h °C/kcal	0.4538 h °F/BTU	0.8603 K/W
	2.2036 h °C/kcal	1 h °F/BTU	1.8958 K/W
	1.1624 h °C/kcal	0.5275 h °F/BTU	1 K/W
Heat flux	1 kcal/m² h	0.3687 BTU/h ft²	1.1623 W/m²
	2.712 kcal/m² h	1 BTU/h ft²	3.1525 W/m²
	0.8604 kcal/m² h	0.3172 BTU/h ft²	1 W/m²
Mass	1 kg_m	2.205 lb_m	1 kg
	0.4536 kg_m	1 lb_m	0.4536 kg
Specific heat capacity	1 kcal/kg_m °C	1 BTU/lb_m °F	4188 J/kg K
	2.388×10^{-4} kcal/kg_m °C	2.388×10^{-4} BTU/lb_m °F	1 J/kg K
Power	1 kcal/h	3.968 BTU/h	11.63 W
	0.2520 kcal/hr	1 BTU/h	0.2931 W
	860.0 kcal/h	3412 BTU/h	1 W
Thermal conductivity	1 kcal/m h °C	0.6720 BTU/h ft °F	1.731 W/m K
	1.488 kcal/m h °C	1 BTU/h ft °F	2.576 W/m K
	1.731 kcal/m h °C	0.3882 BTU/h ft °F	1 W/m K
Kinematic viscosity, thermal diffusivity, mass diffusivity	1 m²/h	10.76 ft²/s	2.778×10^{-4} m²/s
	3600 m²/h	3.874×10^4 ft²/s	1 m²/s
	0.09290 m²/h	1 ft²/s	2.581×10^{-5} m²/s
Force	1 kg_f	2.205 lb_f	9.808 N
	0.4536 kg_f	1 lb_f	4.448 N
	0.1020 kg_f	0.2248 lb_f	1 N
Pressure	1 kg_f/cm²	14.22 lb_f/in²(psi)	9.804×10^4 Pa
	0.07031 kg_f/cm²	1 psi	6894.8 Pa
	1.020×10^{-5} kg_f/cm²	1.450×10^{-4} psi	1 Pa

*For the cgs system, the units of meters (m) and kilograms (kg) must be converted into centimeters (cm) and grams (g), respectively.

Example

Volume

$$1 \text{ liter} = 10^3 \text{ cm}^3 = 10^{-3} \text{ m}^3$$

Viscosity

$$1 \text{ poise} = \frac{1(\text{dyn})(s)}{\text{cm}^2} = 1 \frac{g}{(\text{cm})(s)}$$

$$= 100 \text{ cp}$$

Force

$$1 \text{ dyne} = 1 \frac{(g)(\text{cm})}{s^2}$$

$$1 \text{ N} = 1 \frac{\text{kg m}}{s^2} = 10^5 \text{ gynes}$$

Energy

$$1 \text{ J} = 0.239 \text{ cal} = 10^7 \text{ ergs} = 1 \text{ N m}$$

$$1 \text{ cal} = 4.184 \text{ J} = 4.18 \times 10^7 \text{ ergs}$$

$$1 \text{ erg} = 1 \text{ dyn cm}$$

Heat rate

$$1 \text{ W} = \frac{J}{s} = 0.860 \frac{\text{kcal}}{h} = 0.239 \frac{\text{cal}}{s}$$

Pressure

$$1 \text{ Pa} = 1 \frac{N}{m^2} = 10 \frac{\text{dyn}}{\text{cm}^2}$$

$$1 \text{ atm} = 760 \text{ mmHg} = 1.013 \times 10^5 \text{ Pa}$$

$$1 \text{ psi} = 6.895 \times 10^4 \frac{\text{dyn}}{\text{cm}^2}$$

Temperature

$$K = {}^\circ C + 273.5$$

Thermal insulation

$$1 \text{ clo} = 0.155 \frac{m^2 K}{2} \text{ for clothing}$$

Notes: cp = centipoise; N = newton; J = joule; W = watt; Pa = pascal; K = Kelvin; C = Celsius.

PROPERTY TABLES

D.1 RHEOLOGICAL PROPERTIES

D.1.1 Hookean solids

$$\rho = \text{stress} = \frac{\text{force}}{\text{area}}$$

$$\epsilon = \text{strain} = \frac{\text{elongation}}{\text{initial length}}$$

$$\sigma = \text{Poisson's ratio} = -\frac{\text{secondary (lateral) strain}}{\text{primary (longitudinal) strain}}$$

$$\tau = \text{shear stress}; \gamma = \text{shear strain}$$

Moduli

$$E = \text{Young's modulus of elasticity} = \frac{\rho}{\epsilon} \quad \text{(see Table D.1)}$$

$$G = \text{shear modulus} = \frac{\tau}{\gamma}$$

$$L = \text{bulk modulus} = \frac{\text{volume change}}{\text{hydrostatic compressive force}}$$

Table D.1 Young's modulus of major blood-wall constituents

Material	Young's modulus,* dyn/cm²
Collagen	10^8-10^9
Elastin	3×10^6 to 6×10^6
Smooth muscle	10^5-10^7
Endothelial lining	Too soft to contribute to the elastic properties of the wall

* For comparison, $E = 3 \times 10^9$ dyne/cm² for polyethylene and $E = 15 - 50 \times 10^6$ dyn/cm² for soft vulcanized rubber.

For an isotropic material

$$K = \frac{E}{3(1 - 2\sigma)} \qquad G = \frac{E}{2(1 + \sigma)}$$

For blood walls

$$\sigma \cong 0.5 \qquad K = \infty \text{ (incompressible)} \qquad E = 3G$$

For most metals

$$\sigma = 0.3 = 0.4$$

See Table D.2 for size and shape of red blood cells.

For human red cells

Membrane thickness $= 75 \text{ Å}$

$$E = 7.2 \times 10^5 \frac{\text{dyn}}{\text{cm}^2} \left(\text{with a range of } 9.6 \times 10^4 \text{ to } 1.1 \times 10^7 \frac{\text{dyn}}{\text{cm}^2} \right)$$

Table D.2 Size and shape of red blood cells

Animal	Diameter, μm	Thickness, μm	Volume, μm³	Diameter/ thickness ratio
Human	7.8	1.84–2.06	88	3.8–4.2
Dog	7.2	1.70–1.95	69	3.7–4.2
Rabbit	6.6	1.84–2.15	63	3.1–3.6
Cat	5.6	1.75–2.10	43	2.7–3.2
Goat	4.0	1.95	25	2.1

D.1.2 Fluids

$\dot{\gamma}$ = rate of shear or velocity gradient

τ_0 = yield stress, the critical shear stress for transition from the Hookean solid to the Newtonian fluid

μ = coefficient of viscosity = $\dfrac{\tau}{\dot{\gamma}}$ for Newtonian fluids

μ_a = apparent viscosity = $\dfrac{\text{applied tangential stress}}{\text{rate of shear produced}}$ for nonNewtonian fluids

Plasma

ρ = 1.035 g/ml

μ = 0.11–0.16 poise (0.012 poise commonly used) with Newtonian behavior

Red blood cells

ρ = 1.08–1.10 $\dfrac{\text{g}}{\text{ml}}$

μ = 0.06 poise for hemoglobin solution which behaves as a Newtonian fluid

Human blood

Hemoglobin content = $\dfrac{34.0 \text{ g}}{100 \text{ mL RBC}}$ = $\dfrac{15.6 \text{ g}}{100 \text{ ml blood}}$

Hematocrit (volume fraction RBC) = 0.46

ρ = 1.05–1.06 $\dfrac{\text{g}}{\text{ml}}$

μ_a = apparent viscosity = 0.035–0.055 poise   and $\dot{\gamma}$ = 230 s^{-1}

τ_0 = 0.01–0.06 $\dfrac{\text{dyn}}{\text{cm}^2}$ $\left(0.05 \dfrac{\text{dyn}}{\text{cm}^2} \text{ commonly used}\right)$ for Bingham model

n = 0.68–0.80 (0.75 commonly used) for power-law model

τ_0 = 0.0594–0.109 $\dfrac{\text{dyn}}{\text{cm}^2}$ and η = 0.161–0.172 poise$^{1/2}$ for Casson model over

$\dot{\gamma}$ = 1–10^5 s^{-1} (small deviation from Newtonian behavior at high shear rates)

Surface tension = 47 $\dfrac{\text{dyn}}{\text{cm}}$.

specific heat = 0.92 $\dfrac{\text{cal}}{\text{g °C}}$

Table D.3 Rates of shear stress (sec^{-1}) in blood circulation with parabolic velocity profile

	Human being		Dog	
Structure	At the wall	Mean	At the wall	Mean
Aorta			400	270
Ascending	190	130		
Descending	120	80		
Large arteries	700	470	600	400
Capillaries	800	530	700	450
Large veins	200	130	50	35
Venae cavae	60	40	200	140

Table D.4 Physical and mechanical properties of tooth structure

	Value	
Property	Enamel	Dentin
Modulus of elasticity, dyn/cm^2	4.62×10^{11} to 8.27×10^{11}	1.172×10^{11}
Poisson's ratio	0.25	0.25
Density, g/cm^3	2.8	1.96
Coefficient of thermal expansion, K^{-1}	12.0×10^{-6}	7.5×10^{-6}
Thermal conductivity, cal/s cm °C	2.23×10^{-3}	1.36×10^{-3}
Specific heat, cal/g °C	0.17	0.38
Thermal diffusivity, cm^2/s	4.69×10^{-3}	1.83×10^{-3}

Table D.5 Light and sonic velocity*

Tissue	Sonic velocity, m/s
Muscle	1585
Liver	1590
Spleen	1555
Kidney	1560
Brain	1540
Fat	1440
Bone, skull	3360
Water	1500

*Data for comparison: speed of light in air = 10^{10} m/s; speed of ultrasound waves in water and high water-content tissues = 1500 m/s.

D.2 PHYSIOLOGICAL DATA

Table D.6 Oxygen consumption in some human organs

Organ	O_2 consumption, ml/min 100 g of organ	Comments
In vivo		
Left ventrile	7.8–10.3	Resting subject
Brain	3.4–3.8	
Kidney*	~10	
Muscle	0.15–0.20	Resting subject
	~11	Exercise of 180 W
Heart, entire	~5	Resting subject
In vitro		
Smooth muscle	~1	
Salivary glands	~6	
Liver	~6	
Stomach	3–4	
Colon	3–4	
Rectum	3–4	
Lung†	~2	

*Renal blood flow through each kidney, 500 ml/min.

†Dead space, 150 ml; Breathing frequency, 12/min; Pulmonary blood flow rate, 5 L/min; Pulmonary capillary blood volume, 75 ml.

Systemic circulation	O_2	CO_2
Arterial blood concentration, ml/ml blood	0.195	0.480
Venous blood concentration, ml/ml blood	0.145	0.520

Pulmonary circulation	O_2	CO_2
Metabolic rate, ml/min	250	200
Arterial blood concentration, ml/ml blood	0.195	0.48
Venous blood concentration, ml/ml blood	0.145	0.52
Alveola gas concentration, ml/ml dry gas	0.153	0.050
Partial pressure in alveoli, mmHg	100	40
Partial pressure in venous blood, mmHg	40	46
Partial pressure in arterial blood, mmHg	100	40

Metabolism and oxygen consumption	Metabolism, kcal/h	O_2 consumption, L/min: RQ, 0.82
Basal	72	0.25
Resting	90	0.31
Walking (4.48 km/h)	228	0.79
Walking (5.6 km/h on 2.5% incline)	306	1.18
Record heavy work (3.2-km race for world record)	1550	5.35

Table D.7 Miscellaneous physiological data

Parameter	Measurement
Brain Tissue	
Temperature	37°C
Blood flow rate	0.0067–0.011 g/s cm³ of tissue volume
Thermal conductivity	0.0015 cal/s cm °C
Metabolic heat production	0.006 cal/s cm³ of tissue volume
Human body	
Density	1.08 g/cm³
Specific heat	0.83 cal/g C
Thermal conductivity	1.4×10^{-3} W/cm °C
Water content	80%
Skin (male adult of 70 kg and 170 cm)	
Weight	4 kg
Surface area	1.8 m²
Volume	3.6 L
Water content	70–75%
Thickness	0.5–5 mm
Metabolic heat production	10 kcal/h
Thermal conductivity	$(1.5 \pm 0.3) \times 10^{-3}$ cal/cm s °C
Thermal diffusivity	7×10^{-4} cm²/s (surface layer 0.26 mm thick)
$k\rho C_p$ (thermal inertia)	90×10^{-5} to 400×10^{-5} cal²/cm⁴ s °C²
Specific heat	0.83 cal/g °C
Emissivity (infrared)	0.99
Blood perfusion rate of segmental body areas [as percentage of total area (Du Bois)]	
Head	7
Arms	14
Hands	5
Trunk	35
legs	32
Feet	7

D.3 THERMAL PROPERTIES

Table D.8 Latent heats

Phase change	Temperature, °C	Latent heat, cal/g	Heat of reaction, cal/mol
Vaporization of water	37	575	9.85
Melting of ice	0	80	1.44

Table D.9 Thermal conductivity *k*

Substance	$k \times 10^3$, ca/cm °C s	Comments
Blood (with 40 volume percent RBC)	1.32	At 37.8°C
	1.14	At 4.4°C
Biological tissues	1.4	At 37°C
Water	1.47	For comparison at 27°C
Clay	3.11	For comparison at 27°C
Stainless steel (AISI 304)	34.9	
Ice	5.60	0°C
Tooth enamel	2.23	
Dentin	1.36	

Table D.10 *kρc* of some human tissues (Houdas and Ring, 1982)

Tissue	kρc $\times 10^4$, J²/cm⁴ °C² s	Comments
Skin	157	In vitro
	96–131	Tourniquet-occluded in vivo
Fat	38–46	
Bone	77–118	
Muscle	98–197	
Forearm	253	In vivo (segment)
Heel	122	In vivo
Clay	16.7	
Water	256	For comparison at 27°C
Stainless Steel (AISI 304)	5500	

Table D.11 Thermal conductivities of various substances

Substance	Thermal conductivity $\times 10^3$, W/(m K)
Still air dry	25
Arctic mammals	36–106
Various wild mammals	38–51
Merino sheep	37–48
Newborn merinos, down, Cheviot and Scottish blackface sheep	65–107
Cattle	76–147
Rabbit	38–100
Kangaroo	43–64
Grouse feathers	29–58
Penguin	31–46
Gosling	36–46
Artificial fur	40–67
Woven fabric	40

Source: Cena, K. and Clark, J. A. (1978) Thermal insulation of animal coats and human clothing, *Phy. in Med. & Biol.* 23; 565–591.

Table D.12 $(\rho C_p k)$ values and thermal sensation

Material	$\rho C_p k$, J²/cm⁴ °C² s, at 27°C	Threshhold of pain (cold)	Temperature, °C	
			Range of comfort	Threshhold of pain (warmth)
Carbon steel	1.57	14	29–32	45
Concrete	2.83×10^{-1}	4	27–34	54
Rubber (soft)	2.65×10^{-4}	−12	24–35	67
Wood				
Oak	1.45×10^{-3}	−20	22–35	74
Pine	8.45×10^{-4}	−53	17–39	84
Cork	0.84×10^{-4}	−140	5–42	150
Water	2.56×10^{-2}			

Table D.13 Metabolic heat production in various body parts at resting condition

Body part	Metabolic heat production	
	Rate, kcal/h	Percentage
Respiratory pathways and circulatory systems	13.5	15
Brain tissues	18	20
Musclar tissues	18	20
Viscera (in body cavities)	40.5	45
Total body	90	100

Table D.14 Heat dissipation from body surface at resting condition

Mechanism	Percentage of heat dissipation*	
	$T_a = 20\text{–}25°C$	$T_a > 28°C$
Radiation, conduction, convection	75	0
Evaporation	25	100

*T_a = ambient temperature.

D.4 MASS TRANSFER PROPERTIES

Table D.15 Mass diffusivity

System	Temperature, °C	D_{AB}, cm²/s
Water vapor–air	25	0.258
CO_2–H_2O	20	1.77×10^{-5}

Table D.16 Permeability of membranes, K in kidney

Structure	K
Glomerulus	6 ml/min mmHg (filtration)
Proximal tubule	4×10^{-4} ml/min mosm
Descending loop of Henle	1×10^{-4}
Ascending loop of Henle	40 ml/min-mosm (active transport)
Distal tubule	1×10^{-4}
Collecting duct	1×10^{-4}

SUBJECT INDEX

AUTHOR INDEX